George Alexander Otis

A Report on Excisions of the Head of the Femur for Gunshot Injury

George Alexander Otis

A Report on Excisions of the Head of the Femur for Gunshot Injury

ISBN/EAN: 9783337015497

Printed in Europe, USA, Canada, Australia, Japan

Cover: Foto ©berggeist007 / pixelio.de

More available books at **www.hansebooks.com**

CIRCULAR No. 2.

A REPORT

ON

EXCISIONS OF THE HEAD OF THE FEMUR

FOR

GUNSHOT INJURY.

WASHINGTON:
GOVERNMENT PRINTING OFFICE.
1869.

CIRCULAR No. 2.

WAR DEPARTMENT,
SURGEON GENERAL'S OFFICE,
Washington, January 6, 1869.

The following Report, relating the experience acquired in the War of the Rebellion in regard to Excisions of the Head of the Femur for Gunshot Injury, is published for the information of the Medical Officers of the Army.

JOSEPH K. BARNES,
Surgeon General,
United States Army.

A REPORT

ON

EXCISIONS OF THE HEAD OF THE FEMUR

FOR

GUNSHOT INJURY.

BY GEORGE A. OTIS,
ASSISTANT SURGEON AND BREVET LIEUTENANT COLONEL, U. S. ARMY.

SURGEON GENERAL'S OFFICE,
WASHINGTON, D. C.,
January 2, 1869.

BREVET MAJOR GENERAL J. K. BARNES,
SURGEON GENERAL, U. S. ARMY.

GENERAL: In compliance with your instructions, I have prepared a report on excisions at the hip-joint for gunshot injury. The report contains an account of all such operations, performed during the war of the rebellion, of which it has been possible to obtain reliable descriptions—a comparison of the results of those operations with those of amputating at the hip, or of abstaining from operative interference—and a review of the excisions at the hip in military surgery in other countries.

The instances afforded by the war of the removal of the head, or of the head, neck, and trochanters of the femur, on account of the immediate or remote effects of gunshot fractures of the upper extremity of the bone are numerous; the operations were performed under very varied circumstances; and their results must have great weight in the determination of one of the most important questions of modern military surgery. Indeed, when it is considered that most of the abstracts of this great array of cases furnish all essential particulars respecting the nature of the injury, the date and mode of operation, the after treatment, progress, and termination, that the condition of each of the survivors of the operation has been traced for years, and that the aggregate results are compared with those of the only remaining alternatives in the treatment of such injuries, viz: temporization and amputation, it must be admitted that this report comprizes the elements for solving the grave problem of the appropriate treatment of gunshot injuries involving the hip-joint.

I describe the operations in three categories: primary, intermediate, and secondary excisions. This classification, adopted by this Office in discussing all the major amputations and excisions, has been criticised by students of the closet and by surgeons in civil life, but by no military surgeons of practical experience. If a study of the histories of over twenty thousand major amputations, and of more than four thousand excisions of the larger joints, performed during the late war, may permit me to speak authoritatively on this point, I would say that no doctrine in military surgery is supported by more ample evidence than that which teaches that in operations for traumatic causes, there is a wide difference in the results of those performed immediately after the reception of the injury, those performed during the existence of inflammatory action, and those done after the symptomatic fever and inflammatory symptoms have abated. Boucher was the first to formally define these periods, in a memoir addressed to the French Academy of Surgery in 1752. His division was adopted by Guthrie, and commands the approval of the vast majority of modern military surgeons. It has been well said by Rutherford Alcock, that the mode in which authors define the periods and circumstances which constitute primary and secondary operations is one of the main causes of erroneous estimates of the comparative value of primary and secondary operations, and of the discrepancy of opinion on the subject. It is notorious that some authors limit the designation of primary operations to those performed within ten or twelve hours from the reception of the injury, while others extend the period to three or four days. M. Legouest has well observed that more exact definitions on this subject are among the *desiderata* of modern surgical science. That the operations done in the intermediate period must be separated, in estimating results, from those performed in the primary and secondary stages, and must be looked upon as compulsory operations, may henceforward be assumed. Critics may cavil at the "scientific accuracy"[*] of such classification, but when the facts are at hand to demonstrate its utility, their strictures are of little value.[†]

---

[*] See The British and Foreign Medico-Chirurgical Review, No. LXXIII, p. 173, July, 1868.

[†] Those not weary of a thread-bare argument may consult: GUTHRIE, *Treatise on Gunshot Wounds*, 3d ed. London, 1827, p. 230; ALCOCK, *Notes on the Medical History and Statistics of the British Legion in Spain*, London, 1838, p. 66; SÉDILLOT, *Traité de Médecine Opératoire*, T. 1, p. 316; LEGOUEST, *Dict. Encycloped. des. Sci. Med.* Paris, 1865, Art. Amputations; BERNHARD BECK, *Kriegs-Chirurgische Erfahrungen während des Feldzuges 1866 in Süddeutschland*, Freiburg, 1867, p. 266. [Dr. Beck agrees with M. Legouest, that instead of three divisions only, the intermediary period should be sub-divided.] Dr. S. FENWICK, of Newcastle-on-Tyne, *Monthly Journal of Medical Science, and Archives Générales de Médecine*, 4ᵉ Série, T. XVI, p. 302; NÉLATON, *Élémens de Pathologie Chirurgicale*, T. 1, p. 224; PITHA and BILLROTH, *Allgemeinen und Speciellen Chirurgie*, Erlangen, 1865, Erster Band, Zweiter Abt., 5, 369; NEUDÖRFER, *Handbuch der Kriegschirurgie*, Erste Hälfte, Allgemeiner Theil, Seite 345; BALLINGALL, *Outlines of Military Surgery*, 5th ed., p. 424; W. FERGUSSON, F. R. S. *A System of Practical Surgery*, London, 1857, p. 197; BÉRARD, DENONVILLIERS, and GOSSELIN, *Compendium de Chirurgie Pratique*, T. II, p. 504. The latter authors, and Sir William Fergusson and M. Nélaton, are among the prominent surgeons in civil life who recognize the distinction sought to be established by the military surgeons.

Convinced of the importance of distinguishing the intermediate operations, and of adopting precise definitions, I have used for purposes of statistical comparison those cases only in which the notes permit the period of the operation to be determined or approximated with reasonable accuracy. I describe as primary excisions those performed in the interval between the reception of the injury and the commencement of the inflammatory symptoms, a period rarely exceeding in duration twenty hours, although in a few exceptional instances it is prolonged to thirty-six or forty-eight hours. I classify as intermediate excisions those performed during the persistence of the inflammatory stage, a more variable period, extending over one, two, or three months, according to the extent of the injury, the constitutional powers of the patient, and the conditions under which he is treated. I place in the category of secondary excisions those performed after inflammation has subsided, the traumatic phenomena have abated, and the local lesions have become analogous to those resulting from chronic diseases.

Although excision of the upper extremity of the femur for gunshot injury was first practised only forty years ago, the bibliography of the subject is inaccessible to many of the medical officers, and it is thought best to review concisely what has been written upon it. The operations performed during the war will then be described. Some information additional to that contained in *Circular* No. 7, S. G. O., 1867, in regard to amputation at the hip-joint will be recorded. Numerous abstracts of cases of gunshot fractures involving the hip-joint, and treated without operative interference, will be given; and, in the concluding observations, the comparative value of excision, amputation, and temporization, will be discussed.

HISTORICAL REVIEW.

On February 9th, 1769, thirty years after Morand had proposed the operation of amputation at the hip-joint, Charles White, F. R. S., one of the Corporation of Surgeons in London, and Surgeon to the Manchester Infirmary, read before the Royal Society of London, an account of a successful excision of the head of the humerus, performed by him on April 14th, 1768, and, at the conclusion of his paper, made this statement: "I have likewise, in a dead subject, made an incision on the external side of the hip-joint, and continued it down below the great trochanter, when, cutting through the bursal ligament, and bringing the knee inwards, the upper head of the os femoris hath been forced out of its socket, and easily sawn off;· and I have no doubt but this operation might be performed upon a living subject with great prospect of success." This seems to have been the first formal proposition to excise the upper extremity of the femur.*

Prior to this, however, in 1730, J. D. Schlichting had extracted the remnants of a carious head of a femur by dilating a sinus on the hip of a strumous girl of fourteen, long a victim of coxalgia.[1] Other instances in which the diseased head of the femur was spontaneously eliminated or was extracted without a formal operation were reported by Vogel,[2] in 1771, Kirkland,[3] in 1780, and Hoffman,[4] in 1782. In 1783, Mr. Joseph Brandish[5] described a similar case, which is remarkable because the necrosis of the epiphysis was the result of a gunshot wound. The patient recovered.

In 1786, Vermandois[6] made experiments in excising joints of the lower animals, and repeatedly removed the head of the femur successfully from dogs. Chaussier[7] and Köler[8] undertook similar experiments, and found that decapitation of the femur in dogs was not more dangerous than the removal of the head of the humerus.

Petit-Radel,[9] Wachter,[10] and Rossi,[11] about this time advocated the operation, and described methods for its performance on the human subject, the two former recommending a longitudinal incision over the great trochanter, as suggested by Charles White, while the latter advised that the articulation should be laid open by raising a triangular flap.

In 1806, Dr. James Jeffray,[12] of Glasgow, published a description of the chain saw

---

* The paper is recorded in the *Philosophical Transactions for* 1769, Vol. LIX, p. 45, and is republished the following year in CHARLES WHITE's *Cases in Surgery*, London, 1770, p. 66.
[1] *Philosophical Transactions*, Vol. XL, 11, p. 270.
[2] VOGEL, *Observationes quaedam chirurgicas defendit*. Kiliæ, 1771, and *Bibl. Chir. du Nord*. pp. 391, 393.
[3] KIRKLAND, *Thoughts on Amputations*, etc., London, 1780.
[4] GÜNTHER, *Lehre von den blutigen Operationen*. Abth. II. Leipzig, p. 204.
[5] *London Medical Journal*, Vol. VII, p. 138.
[6] *Ancien Journal de Chirurgie Médecine et Pharmacie*, T. LXVI, pp. 70, 200, Janvier, 1786.
[7] CHAUSSIER, *Mém. de la Société Méd. d'Emulation*, T. III, p. 399, and *Magazin Encyclopédique*, T. VII, p. 248.
[8] KÖLER, *Experimenta circa Regenerationem Ossium*. Gottingen, 1786, pp. 84-94.
[9] PETIT-RADEL, *Encyclopédie Méthodique*, T. I, Art. Chirurgie.
[10] WACHTER, G. H. *Dissertatio Chirurgica de Articulis Extirpandis*, Groningen, 1810. pp. 91-94.
[11] ROSSI, F. *Éléments de Médecine Opératoire*, Turin, 1806, T. II, p. 224.
[12] PARK and MOREAU, *Cases of the Excision of Carious Joints*, Glasgow, 1806, p. 175.

invented by him in 1790, which subsequently came to be regarded as an indispensable implement in many excisions.

In 1816, Dr. H. T. Schmalz,[1] of Pirna, in Saxony, repeated Schlichting's operation of extracting a carious detached head of a femur, by dilating a sinus on the hip of a scrofulous boy. The child was three years in getting well. It is related also that at this period Heine[2] and Klüge[3] extracted the coxal extremity of the femur partly destroyed by caries and necrosis; but these cases are of doubtful authenticity.

About this time also Roux,[4] Montfalcon,[5] Percy,[6] and Champion,[7] discussed the methods by which excision of the head of the femur might be most safely accomplished, and M. Briot,[8] who imagined that the suggestion of the operation had originated with himself, describes a spontaneous luxation in a case of coxalgia in a child of fourteen, in which he proposed to operate, but was dissuaded by his colleagues. The Moreaus, of Bar-sur-Ornain, to whom surgery is so much indebted for the introduction of excisions of the joints, repeatedly demonstrated on the dead subject the facility with which the femur might be decapitated, the joint being exposed by forming a large quadrilateral flap.[9]

Yet, with all these experiments and discussions, the first real excision of the head of the femur upon the living human subject appears to have been performed by Mr. Anthony White,[10] of Westminster Hospital, London, who, in April, 1822, with the sanction of Mr. Travers, and in opposition to the advice of Sir E. Home, removed the head and neck of the femur with a portion just below the trochanter minor, from the dorsum of the ilium. The patient was a boy, who, four years and a quarter before the operation was performed, "being at that time nine years old, was thrown down. The injury was followed by disease of the hip, which was treated with leeches, blisters, rest, and other usual means. Large abscesses formed, and burst around the joint with extreme pain and copious discharge of pus; and the head of the femur was dislocated far on the dorsum ilii. The patient was

---

[1] HEDENUS, A. G. *Commentatio Chirurgica de Femore in Cavitate Cotyloidea Amputando*, Lipsiæ, 1823, p. 63.
[2] LEPOLD, *Ueber die Resection des Hüftgelenkes*, Würzburg, 1834.
[3] WAGNER, *Art. Decopitatio*, in BUSCH'S *Encyclopædisches Worterbuch*. Vol. IX, p. 188.
[4] ROUX, *De la Résection ou du Retranchement des Portions d'Os Malades*, Paris, 1812, p. 49.
[5] MONTFALCON, *Mémoire sur l'État Actuel de Chirurgie*, Paris, 1816, p. 103.
[6] PERCY, *Dictionnaire des Sciences Médicales*, T. LXXII, p. 554.
[7] CHAMPION, *Traité de la Résection*, Paris, 1817.
[8] BRIOT, *Histoire de l'État et des Progrès de la Chirurgie Militaire en France pendant les Guerres de la Révolution*, Besançon, 1817, p. 177.
[9] This fact is recorded by Champion (*loc. cit.*) For many years following 1786, the Moreaus had urged upon the French Academy of Surgery the value of excisions of the large joints; but their reports were discredited, their propositions violently condemned, and their papers lost amid the unpublished records of that learned body. In 1803, Moreau (fils) printed his *Observations particulières relatives à la Résection des Articulations affectées de Carie*, and, in 1816, his *Essai sur l'Emploi de la Résection des Os*. But notwithstanding arguments and demonstrations and successful operations, military surgeons refused to attempt their methods of saving limbs throughout the protracted and bloody wars which convulsed Europe for years after their announcement. Well might the younger Moreau quote from Condillac : *Il est rare que l'on arrive tout-à-coup à l'évidence ; dans toutes les sciences et dans tous les arts, on a commencé par une espèce de tâtonnement !*
[10] This case is often confounded by the French and German writers with the suggestion of the operation by Charles White. The history of the case was not published for many years. The first notice of it in print appears to be in a letter from Lionel J. Beale, inserted in the *London Medical Gazette*, Vol. IX, p. 853, March, 1832. A further account is given in "*A System of Surgery*," by J. M. Chelius, translated by J. F. South, Vol. II, p. 979, London, 1847. The history in the text is quoted from the "*Descriptive Catalogue of the Pathological Specimens in the Museum of the Royal College of Surgeons of England*," Vol. II, p. 230, London, 1847, and was furnished by Anthony White himself, with the specimen. The latter is numbered 941, in the Museum of the College. A long and minute history is published in *The Lancet* for 1849, Vol. I, p. 360, and translated by M. LÉON LEFORT, *De la Résection de la Hanche*, p. 459. Paris, 1861.

reduced to a very debilitated state; and during the two years and a half in which the discharge continued, became exceedingly emaciated; but for some months before the operation no fresh abscesses formed, and the progress of the local disease appeared to be checked. The operation was effected by dividing and separating the integuments from a little above the point of lodgment down to that opposite the site of the acetabulum. At this point the bone was divided with a small straight saw about two inches below the top of the great trochanter, raised with a spatula, and then carefully detached from the ilium. The knee, which had long been immoveably imbedded in the opposite thigh, was now with facility brought into a straight line, and the whole limb was secured with a long splint, and treated as a compound fracture. The wound quickly healed, the various sinuses soon ceased to discharge, and the health of the patient rapidly improved. Within twelve months a most useful compensation for the loss of the original joint was obtained. Perfect flexion, and extension, and every other motion except the power of turning the knee outwards, were restored; but the femur did not grow after the operation." The patient lived five years after the operation, and died consumptive.

In 1828, the operation was repeated by Mr. Hewson,[1] of Dublin, in a case of caries of the head of the femur. A lunated flap was raised, and the bone was sawn a little above the trochanter minor. The patient died three months afterward, with large purulent collections extending into the pelvis through an opening in the cotyloid cavity.

The third authenticated instance of excision of the head of the femur, and the first example of the performance of the operation for gunshot injury, occurred in 1829. It was a primary operation, performed by Dr. Oppenheim,[2] of Hamburg, then in the Russian military service. The history of the case, which was to a certain extent encouraging, obtained but little publicity until printed, eight years after the operation, in Dieffenbach's Journal. As the case marks the introduction of this operation into military surgery, forty years ago, all its details are of interest:

CASE.—A Russian Chasseur was wounded at the battle of Eski-arna-utlar, fought by the Russians and Turks on May 5, 1829. He was struck over the left hip by a musket ball. He was conveyed to the ambulance station of Dr. Oppenheim immediately after the reception of the injury. The entrance wound was small, and there was no counter opening. Exploration with the finger could not detect the missile. The foot was everted, there was excessive pain about the hip, and everything led to the belief that the ball had broken the neck of the femur and lodged in it. Dr. Oppenheim enlarged the entrance wound better to examine the injury of bone, and then found such extensive fracture that he extended his incision several inches in order to extract the large bone fragments. He found that the projectile had struck the femur over the trochanter in oblique direction, corresponding with the axis of the neck. There was little injury to the soft parts, and the large nerves and vessels were unharmed. The broken extremity of the shaft was protruded through the wound and sawn just above the lesser trochanter. The upper fragment could then be carefully examined. Besides numerous splinters, the upper extremity was shattered into three principal portions. In the lower one the ball was imbedded. The surgeon then separated all the muscular attachments from the great trochanter; then, not without difficulty, he reached the cotyloid cavity and divided the round ligament, and removed the shattered head; the other two principal fragments and the ball were then removed without difficulty. The upper margin of the acetabulum had been carried away, and this portion of its edge was rough to the touch. The remainder of the cavity appeared to be uninjured. The wound having been cleansed, was closed by four points of suture, the lower angle being left open for the escape of pus. Three days subsequently Dr. Oppenheim saw this patient at the hospital at Warna, whither he had been moved immediately after the operation. His condition was unsatisfactory. On removing the dressings, the wound presented some superficial gangrenous patches; the limb was hot, swollen, and painful; the pulse small and frequent; the face was pallid, the tongue dry, the lips contracted and at times tremulous. There had been no stool since the operation, and a saline draught was ordered, to be followed by alternate doses of calomel and nitrate of potassa with opium. The opiates were to be given in large doses. The sutures were clipped and the

---

[1] HARGRAVE, *A System of Operative Surgery*, Dublin, 1831, pp. 267–514.
[2] LEPOLD, *Ueber die Resection des Hüftgelenkes*, Würzburg, 1834, appears to be the first writer who mentions the case. In 1835, it is reported in the *Gazette Médicale de Paris*, p. 165. In 1836, it is published in the *Zeitschrift für die gesammte Medicin*, von J. T. DIEFFENBACH, and in KLEINERT's *Allgemeines Repertorium*, X. Jahrg. VI, Heft, S. 109. The account in the text is taken from BONINO, *De la Résection de la Tête du Fémur*, in *Annales de la Chirurgie Française et Étrangère*, T. X, p. 387.

thigh was enveloped in compresses saturated with an aromatic lotion. The next day there was some improvement; the patient had several stools and slept tranquilly for hours, and then took food with relish. On the sixth day, the swelling of the limb had subsided, an healthy suppuration was established, the wound had acquired an healthy aspect, and the patient's strength was satisfactory. He was supplied with liberal diet; and took mineral acids and the wine of cinchona. Up to the seventeenth day all went well, when a patient in the ward was attacked by the plague. The wounded man became immediately much worse; his fever kindled up with intensity, his thigh swelled, the suppuration became sanious, he was delirious at night, and died the following day, May 23, 1829.

In the hospital at Warna, Oppenheim saw another patient with a comminution of the head of the femur caused by a musket ball, and proposed to repeat the operation of excision, but the chief surgeon of the hospital, Dr. Janoffsky, Councillor of State, positively forbade the undertaking.[1]

Mr. Guthrie has claimed[2] that he was the first to recommend excision as a substitute for amputation at the hip-joint. Unquestionably, his great authority was influential in establishing the operation as a legitimate resource in military surgery; but I cannot find a recommendation of the operation in any of his writings prior to 1831. In that year Mr. James Syme[3] used the following language:

"When the head of the thigh-bone has been broken into pieces by a musket bullet, without any injury of the great blood-vessels or nerves, or extensive laceration of the muscles, it would certainly be better to extract the fragments than to perform amputation at the joint, as the patient would thus not only retain a limb that might possibly be of use to him, but also avoid the shock necessarily attending the removal of so large a portion of the body."

In 1832, M. Seutin, chief of the Belgian military surgeons, excised unsuccessfully the head, neck, and trochanters, and several inches of the upper part of the shaft of the femur, on account of an extensive and complicated gunshot fracture.[4] This case gave rise to warm and even acrimonious discussions. The whole subject of excisions had been but little studied by the majority of surgeons, and its application to so important a joint as the hip was pronounced an act of almost culpable temerity. The principal details of M. Seutin's case were as follows:

CASE.—Private Lisieux, 25th Infantry, was wounded at the siege of Antwerp, in December, 1832, by a ball from a wall-piece, which struck the outer part of the left thigh and made its exit at the perinæum, having shattered the upper extremity of the femur. The French surgeons wished to amputate at the hip, but M. Seutin preferred to undertake an operation which he regarded as very much less dangerous. Thirty-six hours after the reception of the wound, Seutin made an incision from the iliac crest to a point three inches below the great trochanter, and removed the upper third of the femur, which was found broken into sixteen fragments. Difficulty was experienced in disarticulation. Little blood was lost. The limb was placed on a double inclined plane. The patient survived nine days, and died from gangrene of the limb.

A period of fifteen years now elapsed before another excision of the head of the femur was performed on account of gunshot injury. At the expiration of this period, in 1847, Kajetan Textor,[5] of Würzburg, operated on a man forty years of age, with caries following a gunshot wound received two years previously. The head and neck of the bone were

---

[1] S. OPPENHEIM, *Ueber die Resection des Hüftgelenkes*, Würzburg, 1840, p. 26.
[2] GUTHRIE, G. J. *Commentaries on the Surgery of the War in Portugal, Spain, France, and the Netherlands.* London, 1855, p. 645.
[3] SYME, *Treatise on the Excision of Diseased Joints*, Edinburgh, 1831, p. 125.
[4] PAILLARD, *Relation Chirurgicale du Siége de la Citadelle d'Anvers*, Paris, 1833, p. 105, and H. LARREY, *Histoire Chirurgicale du Siége de la Citadelle d'Anvers*, in T. XXXIV. *Rec. de Mém. de Méd. de Chir., et de Phar. Mil.* Paris, 1833, p. 105, and SEUTIN, *Bulletin de Thérapeutique*, Paris, 1833. See also Froriep's Notizen, Band XXXV, No. 15.
[5] KARL TEXTOR, *Der Zweite Fall von Aussagung des Schenkelkopfes mit Volkommen Erfolg.* Würzburg, 1856, and OSCAR HEYFELDER, *Lehrbuch der Resectionen*, Wien, 1863, S. 82. FOCK, C. *Bemerkungen und Erfahrungen über die Resection im Hüftgelenk*; in B. LANGENBECK's *Archiv für Klinische Chirurgie*, Berlin, 1861. B. I, p. 189.

removed. No further particulars are recorded, except that death ensued in ten days after the operation.

During this period, however, the subject of excisions for disease had received much attention, especially from the British and German surgeons. Michael Jaeger,[1] Kajetan Textor,[2] and Bernard Heine,[3] contributed greatly by their demonstrations and experiments to a favorable appreciation of coxo-femoral resection as a resource in morbus coxarius. Textor operated four times, and once with complete success.[4]

In Great Britain, Sir Benjamin Brodie performed excision for hip disease, in 1836, and in 1845, this practice received a great impulse from the teachings and example of Mr. Fergusson.[5] His first operation was a brilliant success, and he had soon many imitators. Meanwhile three great British authorities in military surgery, Mr. Guthrie, Sir Philip Crampton, and Sir George Ballingall,[6] continued earnestly to advise excisions of the head of the femur in such cases of gunshot injury as had proved invariably fatal up to that time, unless life had been preserved by amputation at the hip joint. But it was not until a later period that an opportunity was afforded to the British surgeons of submitting their views to a practical test.

In France, excisions at the hip continued to meet with disfavour. The operation had been emphatically condemned by the older surgeons, especially by Boyer,[7] whose authority was almost unquestioned, and it was not until 1847, that Roux[8] ventured to disregard the precepts of the masters. It may be noticed, however, that Gerdy,[9] in 1839, while pronouncing decapitation of the femur for caries or spontaneous luxation utterly inadmissible, believes that "it is not impossible that this resection, practised on account of gunshot wounds, may sometimes be useful, and perhaps, in certain cases, preferable to amputation at the joint."

---

[1] RUST, *Handbuch der Chirurgie*, Bd. V, s. 559.
[2] TEXTOR, *Ueber die Wiedererzeugung der Knochen nach Resectionen*, Würzburg, 1842.
[3] V. GRAEFE'S und V. WALTHER'S *Journal*, Bd. XXIV, Hft. 4.
[4] FOCK, C., *loco citato*.
[5] FERGUSSON, W. *A System of Practical Surgery*, 4th Ed. London, 1867, p. 461. *The Transactions of the Medico-Chirurgical Society for* 1845, Vol. XXVIII, p. 571. *London Medical Times* April 7, 1849.
[6] Medical officers are familiar with Mr. Guthrie's views on this subject as expressed in his *Commentaries* and in his *Lectures*. Sir George Ballingall's opinion is as follows: "The hazardous character of wounds involving the hip-joint is well known to every experienced surgeon, and the removal of the thigh at the hip-joint, recommended for these wounds, is an operation which one can never undertake but with reluctance. The experience which we have of the excision of the head of the femur, in cases of caries, is now considerable, and appears to me to be encouraging; and since I have become familiar with the excision of other joints, I have frequently reflected upon the possibility of substituting the operation of excision for that of amputation at the hip-joint, in some of those cases of gunshot wounds where the latter has been recommended. I am now encouraged to speak with more confidence on this point, from finding the operation advocated by Mr. Guthrie, one of the first authorities in military surgery. That the operation may be performed with facility, I make no doubt. Without being aware of my friend's sentiments on this subject, I was in the habit of showing in my class for a number of years, that by a perpendicular incision along the bone on the outside of the joint, and another crossing it at right angles, the head of the femur may be excised, even when the head and neck of the bone are not previously broken or comminuted. When this is the case, the operation would of course be greatly facilitated, and would probably be most advantageously performed by raising a semilunar flap of the soft parts as proposed by Mr. Guthrie. I should scarcely expect by such an operation to save a very useful limb, but it should never be forgotten that in cases requiring amputation at the hip-joint, it is not only the patient's limb, but his life which is deeply involved."
—*Outlines of Military Surgery*, 5th ed., p. 397.
[7] BOYER, *Traité des Maladies Chirurgicales*, T. IV, p. 541.
[8] ROUX, *Gazette des Hôpitaux*, 1847, No. 28. The operation by Maisonneuve. (*Gazette des Hôpitaux*, 1847, p. 99, et 1849, p. 54,) and (*Arch. Gén. de Méd.* 4ᵐᵉ Série, T. XXV, p. 539,) often cited as an excision, was what is sometimes called "Barton's operation," a section through the trochanters to establish a pseudarthrosis.
[9] GERDY, *De la Résection des Extrémités Articulaires*, Paris, 1839, p. 158. Gerdy states that he has knowledge of an excision for coxalgia done in Paris at about this period, "with the result that might have been anticipated." The unfortunate operator apparently did not dare to publish his case.

In 1849, Stromeyer,[1] though an opponent of excision of the head of the femur in gunshot fractures, became so discouraged by the results of amputation at the hip, in the Schleswig Holstein war, that he permitted, according to Esmarch's expression, an attempt at resection. The following are the details of the case as given by Esmarch and Schwartz:

CASE.—A Danish private, O., was shot in the left hip, at Kolding, on April 23d, 1849; the great trochanter was comminuted, and the bone fractured obliquely through the two trochanters. Profuse suppuration having ensued and the discharge being confined, it was determined to enlarge the wound and to extract loose fragments. On March 15th, Dr. Harald Schwartz made a longitudinal incision, four inches long, over the trochanter, and finding that fissures extended into the joint, he seized the upper fragment with forceps and exarticulated, and then thrust the lower fragment through the wound, and sawed the shaft two inches below the lesser trochanter. The patient improved at first, but was seized with a rigor on the third day, and died May 20, 1849. It was found that the ischium had been injured by the bullet, and that its spongy substance was infiltrated with sanies. The right shoulder and ankle joints were full of pus.

In 1850, Dr. Ross removed the head and neck of the femur unsuccessfully for caries consequent upon a gunshot wound received more than a year previously in the attack on Fredericia in Holstein. The patient was much exhausted by suppuration when the operation was undertaken. It was believed by the medical attendants that had the operation been performed earlier, there would have been a fair prospect of success. The following details of the case are given by Dr. Fock,[2] of Magdeburg:

" One of the sharpshooters, aged 23d years, received a gunshot wound of the left hip on the 8th of May, 1848, before Fredericia. The ball had penetrated behind the great trochanter, and was extracted soon after, together with several splinters of bone. During the following months a number of long episodes came away with a profuse discharge of pus. Temporary attacks of retention of urine followed, and in the summer of 1849, ascites supervened, which, however, was soon controlled. In June, 1850, the patient was conveyed to Altona. Dr. Ross found the left lower extremity to be shortened about three inches and rotated inwards, whether the limb was flexed or extended. Behind the great trochanter there were three flatulous openings, and two sinuses in front, through which a probe detected rough surfaces of bone. On June 10, 1850, Dr. Ross made a free incision on the outside of the thigh over the trochanter major, and exposed a large suppurating cavity in which lay the carious head of the femur and a large portion of the neck. The diseased epiphysis was twisted off by strong forceps. On the evening of June 11th, considerable hæmorrhage took place. Bleeding recurred on the 12th and though again speedily checked, the patient, already much exhausted, died in a state of extreme collapse at two in the morning of June 13, 1850. At the autopsy it was found that the patient had a fatty liver, numerous calculi in the bladder, a calculus the size of a hazel nut in the left ureter, fatty degeneration of the left kidney, and atrophy of the right. The acetabulum was flattened, and the trochanter minor, studded with osteophytes, lay against it. The great trochanter was eburnated, and lay in a sort of socket formed by the ligaments and indurated muscles."

The sixth example of excision of the head of the femur for gunshot injury that has been published, occurred in 1854, and was a strictly primary operation. The following are the facts as recorded by Lohmeyer:[3]

" In the case of a line officer who was wounded upon the parade ground in Nordheim, Professor Baum undertook the operation. The missile penetrated the greater trochanter from before, comminuted the neck of the femur, and afterwards, to all appearances, injured the joint. Under such circumstances, it was questionable whether resection or exarticulation of the femur should be performed. In view of the unfavorable results which until the present time have been observed to follow exarticulation of the femur for gunshot wounds, Prof. Baum decided upon resection, and it was performed about seven hours after the reception of the injury. A vertical incision about six inches long was carried through the orifice of entrance, a second extended perpendicularly from the orifice anteriorly to the track of the crural nerve. The soft parts were separated from the great trochanter and the bone was sawn at a point sufficiently below to include the fissures of bone. The capsule was cut through anteriorly, the round

---

[1] STROMEYER, *Maximen der Kriegsheilkunst*, Hannover. 1861, p. 503. It is remarkable that Stromeyer appears to claim the operation for his own: "Der einzige Fall, in welchem ich die Resection des Hüftgelenks in dem Jahre 1849 vornehmen liess," u. s. w.

HARALD SCHWARTZ, *Beiträge zur Lehre von den Schusswunden*, Schleswig, 1854. S. 142, oder ESMARCH: *Ueber Resectionen nach Schusswunden.* Kiel, 1851, S. 125.

[2] LANGENBECK, *Archiv für Klinische Chirurgie*, Berlin, 1861, S. 214. Dr. Fock quotes from *Deutsche Klinik*, 1850, S. 451. I have been unable to procure the original report. Esmarch and Reid cite the case, and it appears to be accepted as authentic by all the modern German writers. Dr. Demme, of Bern, (*Studien*, 1861, S. 256,) ascribes this operation to Reid, who merely reports it. In his second edition, (*Specielle Chirurgie*, 1864, Zweite Abth. S. 355,) Demme corrects his error. Dr. S. W. Gross, (*Am. Jour. Med. Sci.*, Vol. LIV, p. 445,) fails to observe this correction, and duplicates the case.

[3] LOHMEYER, *Die Schusswunden*. Zweite Ausgabe, S. 199. Göttingen. 1859.

ligament was divided, and the upper fragment exarticulated with the aid of strong forceps. The patient was much depressed after the operation. The following night he was very restless, and death supervened the next morning, about twenty-two hours after the reception of the injury, and without any signs of reaction."

In this year (1854) began the Crimean war, during which the British surgeons repeatedly performed excision of the head of the femur. It is noticeable, however, that they did not undertake the operation until the siege of Sevastopol commenced, and hospitals were established, and until they had witnessed numerous unsuccessful hip-joint amputations after the battles of the Alma and of Inkermann.

The French, Russian, and Piedmontese surgeons in the Crimean war, abstained from excisions at the hip, if we may judge from the silence on the subject of Chenu,[1] Pirogoff,[2] and Gherini.[3]

The first excision of the hip undertaken in the Crimea, and the seventh recorded example of the performance of the operation on account of gunshot injury, was performed by Dr. George H. B. Macleod,[4] of Glasgow. The following account of the case is abridged from his description and that by Staff Surgeon Matthew:[5]

CASE.—Private Couch, of the rifle brigade, a delicate looking young lad, was struck on June 18, 1855, in the assault on the Redan, by a ball which fractured the left ulna, and then struck the great trochanter of the femur of the same side. On July 5th, he came under the care of Dr. G. H. B. Macleod, who found a ragged wound over the trochanter, discharging pus profusely. At the bottom of this wound was a circular cavity in the bone, which was black and split in all directions. The limb was not shortened or distorted. The ball had either been removed prior to admission to hospital, or had fallen out spontaneously. The pain in the hip was so great that the patient earnestly requested some operation to be performed. He was consequently put under chloroform, and a thorough exploration of the wound made. The great trochanter was found to be badly broken, and the fracture appeared to extend into the joint. It was therefore decided in consultation to remove the head of the bone. This was done with little difficulty through a straight incision carried directly downwards over the great trochanter, of about 8 or 9 inches long, met at the centre by another running backwards for about 2¼ inches, at right angles to the first. But little blood was lost, and he went on remarkably well, (the hectic symptoms having much diminished,) till the 9th July, when he was suddenly seized with choleraic symptoms, and died the following day.

Mr. George E. Blenkins, Surgeon of the Grenadier Guards, operated on the next case. The patient lived five weeks and was doing well when suppuration of the knee-joint supervened. In some very judicious observations on the subject in the additions to the recent eighth edition of Cooper's Surgical Dictionary,[6] Mr. Blenkins expresses himself as strongly in favor of the operation, and pronounces as suitable cases for excision "all severe compound fractures of the head, neck, and trochanters, or of the head and neck of the thigh bone, or of the neck and upper part of the shaft without fracture of the head, which latter is a much more common injury, the head of this bone not being often implicated in the fracture." He observes that the objection that a useless limb is left is invalid, since the operation must be viewed, "as far as results at present inform us, as the only one which affords a chance of saving life." He adds, "it may be even questioned whether absolute inutility of the limb is always the inevitable result of such operations. It is to be hoped that after a sufficient interval of time to allow of consolidation of the parts, and the limb has been duly exercised, certain amount of motion, strength, and usefulness of the new joint may be regained." This prediction, since completely verified, is a remarkable exam-

---

[1] CHENU, *Rapport au Conseil de Santé des Armées*, Paris, 1865.
[2] PIROGOFF, *Grundzüge der Allgemeinen Kriegschirurgie*, Leipzig, 1864.
[3] GHERINI, *Vade Mecum per le Ferite d'Arma da Fuoco*. Milano, 1866.
[4] MACLEOD, *Notes on the Surgery of the War in the Crimea*, London, 1858, p. 333.
[5] MATTHEW, *Medical and Surgical History of the British Army which served in Turkey and the Crimea, in the years* 1854, 1855–'56. London, 1858, Vol. II, p. 379.
[6] *Cooper's Dictionary of Practical Surgery and Encyclopædia of Surgical Science*, New Edition, brought down to the present Time. By SAMUEL A. LANE, assisted by various eminent Surgeons. London, 1861. Vol. I, p. 838.

ple of the foresight of a good pathologist and surgeon of great practical experience. The following notes of the case are condensed from those furnished by Dr. Macleod,[1] Mr. Guthrie,[2] and Mr. Matthew in the British Surgical History:[3]

"Private Charles Monsterey, 3d battalion Grenadier Guards, aged 24 years, was struck in June, 1855, in the trenches before Sevastopol, by a fragment of shell, which extensively comminuted the trochanters and neck of the right femur, and greatly lacerated the surrounding muscles. It was at once recognized as an appropriate case for excision, and the operation was performed half an hour after his arrival in camp. The wound was extended in a longitudinal direction, to the extent nearly of five inches, and the shaft of the femur sawn through at the junction of the upper fifth with the rest of the shaft. The muscles were next detached from the trochanter, and the capsule lastly divided. Very little blood was lost during the operation. Examination afterwards of the excised bone showed it to be fractured in fourteen pieces. The case continued to do well for the first three weeks; healthy granulations sprang up, both from the end of the divided shaft, and the surrounding cavity and acetabulum. At this period pain and swelling of the knee-joint of the same limb supervened, the capsule of that joint became filled with purulent matter, the cartilages eroded, and he sank gradually, worn out with hectic symptoms, at the end of the fifth week, in spite of every effort to support him. The case was doing remarkably well, and there was every hope of his recovery, until empyema came on."

The next case was operated on by Staff Surgeon J. Crerar. In his minute report of the case to Mr. Guthrie,[4] Dr. Crerar implies that his patient was badly nourished, remarking that "a better and more liberal supply of animal and vegetable food is required, if England expects her soldiers to survive severe operations." The following is an abridgement of the abstract of the case:

"Private William Smith, 1st Royals, was wounded in the Greenhill trenches, before Sevastopol, August 6, 1855, by a piece of a grenade, which struck the left great trochanter. Crepitus was distinct on moving the limb. It was impracticable to determine the extent of the fracture upwards or downwards. After consultation with Deputy Inspector General Taylor and Surgeon Payntor, Surgeon J. Crerar, 62th Regiment, excised the head, neck, and trochanters of the femur, through a vertical incision nine inches in length, crossed at the trochanter by another incision of two and a half inches. The head being exarticulated and fragments removed, the shaft was protruded through the wound and cleanly sawn off. There was little hæmorrhage and little shock. The wound was brought together by adhesive strips. The patient lived until August 21, 1855, a little over a fortnight. Wine, arrowroot, and chicken broth, constituted the nourishment during the after treatment. There was no attempt at reparation at the seat of operation. The muscles of the thigh were infiltrated with pus. The excised bones are in the Museum of the Royal College of Surgeons of London."

The fourth excision at the hip in the Crimean war was the well-known successful operation of Mr. Thomas C. O'Leary, Surgeon of the 68th British Light Infantry. In 1863, Dr. George Williamson[5] reports the patient's limb to have been "about two and a half inches shorter than the other, and capable of bearing some considerable portion of the weight of the body. He could swing it and advance it, but the knee could not be bent. Rotation was admitted to a very limited extent, but performed with considerable pain. The wound was soundly healed, and the man was discharged from service." It was supposed that the fracture extended into the joint; but this supposition was disproved at the operation. The excised parts were sent to the Museum of the Royal College of Surgeons, and are represented in the wood-cuts inserted below, which are copied from those in Mr. O'Leary's description of the case in the *Lancet*. The following abstract of the case is quoted from the British Surgical History:[6]

---

[1] MACLEOD, *Notes*, etc., p. 341.
[2] GUTHRIE, *Commentaries*, p. 620.
[3] *Med. and Surg. Hist.*, etc., p. 378, Vol. II.
[4] GUTHRIE, *Commentaries*, 5th English Ed., p. 622. The case is also reported by MACLEOD, *Notes*, etc., p. 342.
[5] WILLIAMSON, *Military Surgery*, London, 1863, p. 231.
[6] *Med. and Surg. Hist.*, etc, (already cited,) Vol. II, p. 378. The case is more minutely reported in the London *Lancet* for July 12, 1856, p. 46, and is also recorded by GUTHRIE, (*Commentaries*, p. 621,) and Dr. MACLEOD, (*Notes, &c.*, p. 343,) and in many periodicals.

EXCISIONS AT THE HIP.

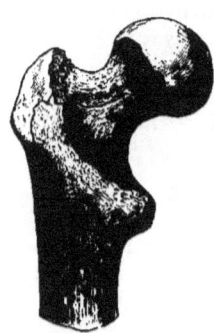

Fig. 1. Upper extremity of left femur fractured by a fragment of shell, and excised by Mr. O'Leary. (From a drawing in the Lancet.)

Fig. 2. Another view of the specimen from Mr. O'Leary's case.

CASE.—"Private Thomas McKevena, 68th Regiment, age 25, was admitted into the regimental hospital on the 19th of August, 1855, from the trenches, where he had been struck by a fragment of a shell over the great trochanter of the left femur. The portion of shell, which weighed about three-quarters of a pound, had remained in the wound, which admitted the introduction of the forefinger, and extended down to the bone. At the bottom of it some fragments of bone could be felt lying loose. From the examination it was evident that the neck of the bone was fractured, and on the following day it was decided to excise the head of it. The man was placed under the influence of chloroform, and an incision carried downwards along the shaft of the bone, which was separated from its attachments; no difficulty was experienced in removing the head of the bone from the acetabulum, and the shaft was then sawn through about half an inch below the lesser trochanter. A considerable quantity of blood was lost, but no vessel required a ligature. When the operation was completed, the edges of the wound were brought together by the interrupted suture, and a bandage applied. The after-treatment was conducted with the greatest care. The leg was placed in a sling made of strong canvas, suspended from a beam over his cot, the heel being considerably elevated, and the injured limb slightly abducted. Approximation of the upper end of the shaft of the bone to the acetabulum was thus encouraged, and by uniform pressure on all sides of the limb, accumulation of matter among the tissues was prevented. When the injured parts were removed, it was found that the fracture extended obliquely downwards between the trochanters and upwards, within half an inch of the cartilage covering the head of the bone. Although the man's pulse did not for several weeks fall below 100, the functions were healthfully performed; he slept well, ate with appetite, and was cheerful throughout; his diet was generous and varied, and a liberal allowance of wine was given. The employment of an air-bed prevented the formation of bed-sores. At the end of the twelfth week he was able to leave his bed, and move about on crutches, and on the 16th of January he was transferred to a ship about to proceed to England with invalids. When he left the camp the wound was firmly united, two small sinuses only existing, which discharged a very small quantity of thin purulent matter. He was gradually regaining power over the limb, and was able, to a limited extent, to flex the leg upon the thigh, and the thigh upon the pelvis. Shortening to about four inches, and very slight inversion, were the chief deformity consequent upon the operation, and his general health was almost entirely re-established. On his arrival at Chatham, the limb is reported to have been about two and a half inches shorter than the other, and capable of bearing some considerable portion of the weight of the body. He could swing it and advance it, but the knee could not be bent. Rotation was admitted to a very limited extent, but performed with considerable pain. The wound was soundly healed, and the man was discharged from the service."

Staff Surgeon George Hyde, M. D., operated on the fifth of the Crimean cases. The following abstract of the case is given by Mr. Guthrie, and by Mr. T. P. Matthew in the British Surgical History:

"Corporal Benjamin Sheehan, 41st Regiment, was wounded in the attack on the Redan on the 8th September, 1855, while in the act of retreating from that work to the English trenches, and lay on the field till the following day, when he was brought to the hospital of the Royal Sappers and Miners. On examination it was found that a grape shot had entered over the great trochanter, and passing inwards and a little forwards, had passed out at the groin of the same side, about an inch below Poupart's ligament, externally to the course of the femoral vessels. The femur was fractured—its lower extremity protruded through the wound of entrance—and, on introducing the finger, the neck of the bone was found to be in a comminuted state. Excision of the head and neck of the femur was decided on, and performed about one p. m. on the 9th. An incision about four inches in length, commencing a little above the trochanter, was carried downwards along the outer side of the femur. The lower fragment was cleared of its attachments, pushed through this incision, and smoothed with the saw, about an inch of the shaft requiring removal. The head of the bone was next dissected from its socket. This part of the operation was considerably facilitated by an assistant catching a firm hold of the neck with a pair of tooth forceps, then rotating the head, and using slight force to dislodge it from the cavity, the operator dividing the capsular and round ligaments. The upper part of the trochanter and comminuted fragments were next dissected out, and the edges of the incision then brought together with sutures, and a bandage applied. It was not found necessary to tie any vessel, and there was very little hæmorrhage. The man bore the operation well, and was replaced in his bed in good spirits, and with a good pulse. On the 10th, he is reported to have passed a good night, and to be in good spirits; the pulse 106, soft, and the skin cool. On 11th, had slept some hours; pulse 106, soft; bowels open; tongue furred, but moist. The wound was dressed and looked well—some healthy discharge. On 13th was apparently going on well—pulse still 106—countenance good. In the evening complained of an increase of pain in the hip, but otherwise said he felt much as usual—pulse small and rapid—wine and arrowroot ordered. He died at 6 a. m. on the 14th. The autopsy showed a considerable cavity, filled with sanies, in the situation of the operation. All the fractured bone had been removed. The articular surface of the acetabulum was coated with a fœtid, pasty substance."

The sixth and last of the excisions at the hip in the Crimean war was performed on account of a gunshot fracture of the neck of the femur, not implicating the head, by Dr. Coombe, of the Royal Artillery. The particulars of the case are not given in the British Surgical History. Dr. Macleod states that the "operation was not a primary one; but the patient survived a fortnight, and died of exhaustion; the most marked feature in the case being, that the pulse remained very high—never below 120—during the period he lived, while his aspect was calm, and such as might have led one to expect a more subdued state of the circulation."

The foregoing twelve cases of excision of the head of the femur for gunshot injury are all that were recorded in print prior to 1861. Seven of the operations, those, namely, of Oppenheim, Scutin, Baum, Blenkins, Crerar, O'Leary, and Hyde, were primary; three, by Schwartz, Macleod, and Coombe, were intermediary; and two, by Textor and Ross, were secondary. The average duration of life after the operation, in the eleven unsuccessful cases, was nine days, a contrast to the speedily fatal results of the majority of hip-joint amputations. In two of the cases it was believed that death was due to causes not connected with the operation: to the plague and to cholera.

A single success in twelve operations was not a very encouraging record; but when surgeons reflected upon the excessive mortality of hip-joint amputations and of temporization in fractures of the upper extremity of the femur, they were much inclined to accept excision as a just illustration of the aphorism of Celsus: "*Anceps remedium melius quam nullum.*" Mr. Matthew, the compiler of the British Surgical History, declared that the operation must be regarded as one occasionally required in field practice, and as a substitute for amputation in many cases. Mr. Guthrie[1] insisted that when the head or neck of the femur were fractured by a musket ball, without injury to the great vessels or nerves, excision should be preferred to temporization or amputation. Mr. Blenkins[2] considered excision the only chance of preserving life in severe compound fractures of the head or neck of the femur. Dr. Macleod[3] regarded the "chance of saving life" as manifestly on the side of excision. Dr. Williamson[4] declared the operation to be the only alternative of the surgeon to save his patient from almost certain death, or the last very doubtful alternative of amputation at the joint. The report to the French Academy of Medicine on the subject by Larrey, Jobert, Velpeau, and Gosselin,[5] was very guarded; but in the discussion which followed, Baron Larrey[6] advised that excision should have farther trial. Legouest[7] declared that the efforts of the English and German surgeons to substitute excision for amputation in suitable cases deserved the highest praise, and that until farther

---

[1] GUTHRIE, *Commentaries, etc.*, already cited, p. 77. A monograph on excisions of the head of the femur would be incomplete if a quotation of Mr. Guthrie's graphic description was omitted: "Picture to yourselves a man lying with a small hole either before or behind in the thigh, no bleeding, no pain, nothing but an inability to move the limb, to stand upon it, and think that he must inevitably die in a few weeks, worn out by the continued pain and suffering attendant on the repeated formation of matter burrowing in every direction, unless his thigh be amputated at the hip-joint, or he be relieved by the operation which, I insist upon it, ought first to be performed."

[2] BLENKINS, *Cooper's Dictionary*, already cited, Vol. I, p. 833.

[3] MACLEOD, *Notes, etc.*, already cited, p. 346.

[4] WILLIAMSON, *Military Surgery*, p. 231.

[5] *Bulletin de l'Académie Impériale de Médecine.* T. XXVII, p. 53.

[6] *The same volume of the Bulletin*, p. 142. "Je n'hésite pas à reconnaître que l'on n'a pas encore assez expérimenté la résection de la hanche." * * * "Mieux vaudrait peut-être, en dernière analyse, tenter la résection, jusqu'à ce que l'expérience de l'avenir ait eu raison de l'expérience du passé."

[7] LEGOUEST, *Traité de Chirurgie d'Armée*, p. 754.

experience was acquired it was not only admissible but obligatory to resort to the new practice. Heyfelder, Beck, Paul, and Esmarch,[1] promulgated similar views, and Stromeyer[2] modified his earlier opinion and sanctioned the operation.

The information acquired prior to the late American war, in regard to excision of the head of the femur for traumatic cause has been carefully summed up by various authors. In 1857, an important article on the subject appeared in the British and Foreign Medico-Chirurgical Review.[3] In 1860, Dr. L. A. Sayre,[4] of New York, published a very complete tabular statement of these operations, and M. Léon Lefort[5] communicated to the French Academy of Medicine an elaborate memoir in advocacy of excisions at the hip. In 1861, Dr. Charles K. Winne,[6] of Buffalo, (now Assistant Surgeon and Brevet Lieutenant Colonel U. S. Army,) printed an excellent paper in which the histories of the excisions at the hip for gunshot wounds were concisely recorded. In the same year the exhaustive dissertation by Professor Fock, of Magdeburg, appeared in the first number of Langenbeck's Archives, and Dr. R. M. Hodges,[7] in his admirable monograph on Excisions, devoted a chapter to a comprehensive inquiry respecting those at the hip. Dr. Oscar Heyfelder,[8] of St. Petersburg, whose extended work was translated and edited with additions by M. Bœckel,[9] and Dr. Rocco Gritti,[10] are among the other authors who have very fully discussed the subject. Latterly the operation and its results are described by all systematic writers on surgery.

The revival of the operation of excision of the head of the femur by Mr. Fergusson in 1845, excited much attention from surgeons in all parts of the world; and, as Mr. Barwell[11] justly remarks, there was for a time greater danger that patients might be unnecessarily subjected to the treatment than that the operation should be too much dreaded.

It was first performed in this country by Dr. Henry J. Bigelow,[12] in 1852, and afterwards by Dr. Sayre, Dr. Markoe, Dr. Church, Dr. A. B. Mott, Dr. Krackowizer, Dr. Holston, Dr. Kinloch, and others; so that in 1861, Dr. Hodges enumerated forty-four operations by American surgeons. It would be foreign to the purpose of this report to dwell upon the excisions at the hip for disease or deformity, further than to indicate that, at the outbreak of the late war, this operation was generally admitted to be a legitimate surgical resource. The papers which have appeared during the war or since its termination in relation to excisions at the hip for traumatic cause will be noticed further on; and the more important publications respecting the operation in its general relations will be referred to in the bibliographical list at the conclusion of this report.

---

[1] J. F. HEYFELDER, *Ueber Resectionen*, Breslau, 1854, S. 137; BECK, *Die Schusswunden*, Heidelberg, 1850, S. 336; PAUL, *Die Conservative Chirurgie der Gleider*, Breslau, 1859, S. 193; ESMARCH, *loc. cit.*
[2] L. STROMEYER, *Maximen der Kriegsheilkunst*, 1861, S. 504: "Diese Resection müsse in Zukunft doch wohl gemacht werden, da die Exarticulation so constant tödtliche Folgen gehabt habe, dass die Meisten das Vertrauen dazu verloren hätten."
[3] *The British and Foreign Medico-Chirurgical Review*, Am. ed. Vol. XX, p. 231.
[4] *The Transactions of the American Medical Association*, Vol. XIII, p. 558.
[5] LEFORT, *De la Résection de la Hanche dans les Cas de Coxalgie et de Plaies par Armes à Feu. Mém. de l'Acad. de Méd.* Paris, 1861, p. 556.
[6] WINNE, *Statistical Inquiry as to the Expediency of Excision of the Head of the Femur. Am. Jour. of Med. Sciences*, Vol. XLII, p. 26.
[7] HODGES, *The Excision of Joints*, Boston, 1861, p. 92.
[8] O. HEYFELDER, *Lehrbuch der Resectionen*, Wien, 1863, S. 73.
[9] BŒCKEL, *Traité des Resections*, Strasbourg, et Paris, 1863, p. 59.
[10] GRITTI, *Delle Fratture del Femore per Arma da Fuoco, Studiate sotto il Punta di Vista della Chirurgica Militare*, Milano, 1866, p. 83.
[11] BARWELL, *A Treatise on Diseases of the Joints*, Am. ed., 1861, p. 436.
[12] HODGES, *Op. cit.*, p. 92.

EXCISIONS AT THE HIP IN THE WAR OF THE REBELLION.

A few months after the termination of the War, a tabular statement of the operations of excision of the head of the femur for gunshot injury that had been reported to the Surgeon General's Office, was printed in the report on the nature and extent of the materials available for a surgical history of the war, in *Circular No. 6, S. G. O.*, 1865. This table comprised particulars of thirty-two operations. While the report was in press, a communication from Dr. Warren Webster, U. S. A., proved conclusively that the case numbered XIX in the table, that of Private Joslyn, was not a decapitation of the femur, but a successful excision of the upper part of the shaft. It was impracticable, from typographical considerations, to withdraw the entry from the table; but a note was inserted to call attention to the mistake in diagnosis which had led to the erroneous classification. After the publication of the report, surgical critics agreed, and probably with justice, that *Case* XXX should be excluded; since, immediately after excision, amputation was performed on account of the extended lesions of the shaft. Quite recently it has been ascertained by conversations with the surgeons engaged, that *Case* XX, of an unknown Confederate, and *Case* XXIV, that of Private Beard, 12th Mississippi Regiment, were identical. In regard to the other cases recorded in the table, further research has only accumulated proofs of the accuracy of the original reports, and supplied additional particulars. Eliminating the three erroneous entries from the tabular statement, there remain twenty-nine well-attested excisions of the head of the femur for gunshot injury, with three successes. It is now proposed to give as full abstracts of these cases as practicable, together with the histories of thirty-four other excisions at the hip for gunshot injury that were performed during the war, and have since been learned from various sources.

These sixty-three excisions of the head, or of the head, neck, and trochanters of the femur for the immediate or remote effects of gunshot injuries at or near the hip, are all that were performed during the war of which sufficient details were preserved to make them available for statistical purposes. It has been thought prudent to exclude all reports of cases in which a possibility of duplication existed, and hence some reports from very reliable observers have been set aside, the absence of names and dates rendering it possible, though improbable, that they were describing cases already recorded. After detailing the histories of the sixty-three authenticated operations, the rejected cases will be briefly noticed.

Of the sixty-three operations, forty-eight were performed by Union, and fifteen by Confederate surgeons. It is probable that the history of the operation in the U. S. Armies approaches completeness; and that it is incomplete in reference to the Confederate Armies. In collecting accounts of the operations in the latter, I have to acknowledge gratefully very efficient and cordial assistance from numerous surgeons in the Confederate service, and especially from Dr Hunter McGuire, Medical Director of General Jackson's army;

Dr. John D. Jackson, Inspector of Hospitals; Dr. J. T. Gilmore, of General Longstreet's corps; Dr. Claude H. Mastin, Medical Inspector of General A. S. Johnston's army; Dr. Julian J. Chisolm, Professor of Military Surgery; and Dr. J. J. Dement, Inspector of Hospitals.

Of the sixty-three excisions, thirty-two were primary, twenty-two were intermediate, and nine were secondary operations. In each category the cases are related in chronological order.

In twenty-three cases the operation was performed on the right side, in thirty-six cases on the left side, and in four cases the particular site of the operation is not recorded.

In forty-three cases the injuries for which excisions were performed were inflicted by conoidal musket balls; in twelve cases by musket balls the form of which is not designated; in one case the nature of the missile was unknown; in one it was a fragment of a mortar bomb, and in six the femur was shattered by fragments of shell.

PRIMARY EXCISIONS.

There were thirty-two primary excisions, the operations being performed in every case within twenty-four hours of the reception of the injury, and in most instances within two or three hours. Two of the operations succeeded. This gives the high mortality rate of 93.75 for this particular series of primary cases. Too much importance must not be attached, however, to this naked percentage, as will be shown hereafter. In the thirty unsuccessful cases, the mean duration of life after the operation was a little over seven days. This average is increased by the protracted struggle for life of one patient, who survived sixty days. The majority died in two or three days from the combined shock of the injury and of the operation.

The account of the first case was communicated to the compiler by Professor G. C. Blackman, in 1868. It was not reported from Camp Dennison, or registered at St. John's Hospital. A brief notice of it, written at the time of the operation by Dr. T. McMillin, now Assistant Surgeon, U. S. A., has alone preserved a record of it. The case is interesting as having been the first instance in which excision at the hip for gunshot injury was resorted to in this country. The operation was not undertaken under very hopeful circumstances, copious hæmorrhage having taken place, and the patient being greatly prostrated:

CASE I.—Private John McCulloch, aged 35 years, a recruit at the depot for volunteers at Camp Dennison, Ohio, was wounded on August 30, 1861, by the accidental discharge of a musket. The ball, taking effect at the distance of a few yards only, severely shattered the upper part of the femur, and lacerated the soft parts extensively. The patient was conveyed to St. John's Hospital, in Cincinnati, and on arriving was greatly depressed by loss of blood. Professor George C. Blackman determined that removal of the shattered bone offered the best resource for the preservation of life, and, the patient having been rendered insensible by chloroform, excision of the head, neck, and trochanters was practised without delay, through a vertical incision on the exterior of the thigh. The patient died August 30, 1861, four hours after the completion of the operation.

In the next case, also, the operation was performed on account of an accidental gunshot injury. The abstract is compiled from a report by Surgeon James T. Calhoun, 74th New York Volunteers,* and a letter from Dr. J. W. S. Gouley. The case seems to have been well selected. The subject was healthy, and had to be removed but a short distance

---

* An account of the case was published by Dr. Calhoun, in 1862, in the *New Jersey Medical Reporter*, Vol. VIII, p. 76, and it is recorded also in *Circular* No. 6, S. G. O., 1865, p. 62.

to a permanent hospital; the local lesions were limited; the operation was rapidly accomplished; and the fatal issue must be traced to the complications through which many of the best planned surgical manœuvres are frustrated, the supervention of pyæmia:

CASE II.—Private Timothy Greely, Co. C, 5th Excelsior, (74th New York Volunteers,) aged 20 years, was wounded October 5th, 1861, by a round musket ball from a smooth-bore musket, which entered near the fold of the left nates, struck the left femur at the digital fossa, splintered the neck into the articulation, and made its exit outside the vessels anteriorly. He was conveyed to the E street Infirmary, Washington, on the same day. A stream of blood and another of clear and pellucid synovia issued from the wound of exit. There was but little constitutional irritation, the pulse was but slightly depressed, and the patient congratulated himself on having escaped with what he regarded as a slight injury. A consultation of the surgical staff of the hospital having decided that an excision of the head of the femur was expedient, on the morning of the following day, October 6th, Assistant Surgeon John W. S. Gouley, U. S. A., assisted by Surgeon C. H. Laub, U. S. A., Assistant Surgeon C. B. White, U. S. A., Surgeon T. Sim, and Surgeon H. C. Brown, of the Excelsior Brigade, proceeded to operate. Insensibility having been induced by chloroform, Dr. Gouley made an incision seven inches long, commencing above and behind the trochanter major and continued downwards in the axis of the limb. The neck of the bone was found to be badly shattered, but the fracture did not extend to the shaft. A section through the great trochanter and base of the neck was made with the chain saw. The head of the bone was then disarticulated and removed, and the fragments of the neck were extracted. There was very little loss of blood. The wound having been approximated and dressed simply, the patient was put to bed, and the limb was kept in position by pads and cushions. Surgical fever set in soon after the operation; pyæmia was developed, and the patient gradually sank, and died on October 12, 1861. His friends would not permit an autopsy. The pathological specimen and Dr. Gouley's notes of the case were destroyed in the conflagration which shortly afterwards consumed the Infirmary.

In the next case it is obvious that no operation could avail, and undoubtedly none would have been undertaken had the grave injury of the pelvic cavity involved by the wound been apprehended. Hence this case, like a number of others to be hereafter described, is not entitled to weight in estimating the utility of excisions in properly selected cases of injuries at the hip. The abstract is compiled from letters from the operator:*

CASE III.—Private ———— ————, was wounded on August 28th, 1862, in the engagement between General Rufus King's Division of the First Army Corps and the advance of General Jackson's column on the Washington turnpike, near Gainesville. A conoidal musket ball had splintered the neck and trochanters of the left femur, and was supposed to have lodged about the acetabulum, though the operator discovered in the sequel that it had penetrated into the cavity of the pelvis. The symptoms of shock were very grave and the prognosis very unfavorable; but the chief medical officer of the Division, Surgeon Peter Pineo, U. S. Volunteers, determined to remove the upper extremity of the femur. The upper fourth of the femur was excised a few hours after the reception of the injury. The excision was done under chloroform, with little apparent loss of blood, through a vertical incision on the outside of the limb. The femur was sawn about two inches below the lesser trochanter. It was now discovered that the ball had passed through the innominatum, and that internal hæmorrhage was going on. During the night of August 28th, General King's Division was driven back to Manassas, and this patient with other wounded fell into the hands of the enemy. It is probable that he survived but a very short time. Dr. Pineo secured the specimen, and it is preserved in the Surgical Section of the Army Medical Museum as No. 71. It is figured at p. 233 of the Catalogue of the Surgical Section, and another view is given in the accompanying wood cut, (FIG. 3.) The trochanter major is separated into five fragments, and a long oblique fissure produces a complete solution of continuity in the shaft of the femur.

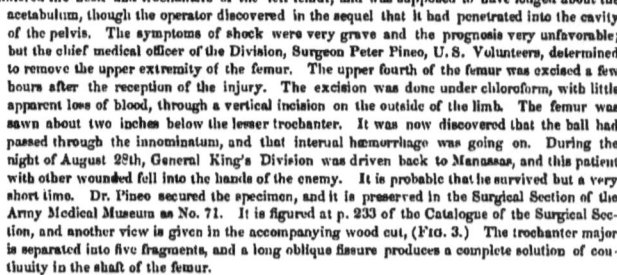

FIG. 3.—Perforation of the trochanteric region of the left femur by a musket-ball, with fissuring of the shaft. *Spec.* 71, Sect. I, A. M. M.

The account of the next case so closely resembles that of the preceding, that, but for the unquestionable authority of the operators, it would be inferred that they were identical. The descriptions of the injuries are very similar; but, in Dr. McNulty's operation, the section of the femur appears to have been made higher up than in the specimen forwarded by Dr. Pineo, and there is a difference of two days in the dates assigned to the operations. This, and the five succeeding cases, it is believed have not been heretofore published.

---

* The case is No. 5, of the tabular statement of excisions of the head of the femur, in *Circular No. 6*, S. G. O., 1865, and is recorded also in Vol. 1, p. 13, of the *Photographs of the Army Medical Museum*.

CASE IV.—Private ———— ————, of the 1st Army Corps, was wounded at the second battle of Bull Run, August 30, 1862, by a conoidal musket ball, which entered the left hip below and in front of the trochanters, and fractured the femur at the junction of the head and neck. He was conveyed to the hospital of General King's Division of the 1st Army Corps in the brick house which had been occupied as the rebel headquarters at the first battle of Bull Run. The prostration from shock was great, yet it was thought that exarticulation of the femur was the only resource, and that excision would be less hazardous than amputation. Accordingly, fifteen hours after the reception of the injury, Surgeon John McNulty, U. S. Volunteers, Medical Director 19th Army Corps, proceeded to excise the head and neck of the left femur through a vertical incision on the exterior of the limb about six inches in length, the patient being rendered insensible by chloroform. On dividing the round ligament to enucleate the head of the bone, it was discovered that the ball had penetrated the pelvic cavity through the lower portion of the acetabulum. The femur was sawn through at the junction of the shaft and neck by a narrow-bladed saw. After the removal of the shattered fragments of bone, the patient suffered much less pain. There were no symptoms of peritonæal inflammation at the date of the operation, but they were developed subsequently. The patient died August 31st, 1862, eighteen hours after the operation, from the shock of the injury and of the operation.

In the following case, also, the experienced and accomplished surgeon who performed the operation, and his colleagues who recommended it as the only resource to preserve life, were deceived as to the extent of the injury, and the case offers another illustration of the great difficulties of diagnosis in some gunshot lesions about the hip.

CASE V.—It has not been possible to learn the name and military description of the subject of this operation. He was a private soldier of the 1st Army Corps, and a Frenchman, for the operator recalls the broken English in which he begged for the operation, and expressed his relief and thanks after it was performed. He had a terrible comminution of the upper extremity of the left femur, inflicted by a fragment of shell at the battle of Antietam, September 17, 1862. Surgeon John McNulty, U. S. Vols., Medical Director 12th Army Corps, excised the head and neck of the femur, five hours after the reception of the injury, through a vertical incision six inches long. Chloroform was used. As in the two other operations performed by Dr. McNulty, it was found that the lesions extended into the pelvis. Consequently there could be little or no hope of a successful result. This patient survived the operation only ten hours.

The next case is a discouraging one, certainly, for the operation was performed under favorable circumstances, on a healthy subject, who, as an officer, received most careful after-treatment, and yet the result was rapidly fatal:

CASE VI.—Captain Frederick M. Barber, Co. H, 16th Connecticut Volunteers, aged 32 years, was wounded at the battle of Antietam, September 17, 1862, by a musket ball, which entered a little behind the right trochanter major, and shattered the trochanters and neck of the femur. He was conveyed to the field hospital of the Third Division of the Ninth Corps. His general health was good, and there was but little shock. There was no swelling of the soft parts. The fracture was very accessible to exploration, and appeared to be limited to the epiphysis. The case was one in which excision seemed peculiarly applicable, and, after a consultation of several surgeons of the Division, that operation was decided upon. On the morning of September 18th, the patient being anæsthetized by chloroform, Surgeon Melancthon Storrs, 8th Connecticut Volunteers, proceeded to make a straight incision four inches long passing through the wound of entrance. The comminuted fragments of the neck and trochanter were extracted, the round ligament was divided, the head of the femur was removed, and the fractured upper extremity of the shaft was sawn off by the chain saw. The edges of the wound were then approximated by adhesive straps and simple dressings were applied. But little blood was lost, and the patient rallied promptly from the operation, and appeared quite comfortable during the day. Surgical fever soon set in, however, the patient sank rapidly under the constitutional irritation, and died on September 20, 1862.

The next case is reported by the operator, Professor Hunter McGuire, of the Medical College of Virginia, who has communicated the following observations on excisions of the head of the femur for gunshot injury in connection with this case and of another, (CASE XV,) to be related further on:[*] "The results in these two cases has not lessened my conviction of the propriety of this operation in certain cases of gunshot wounds of the hip-joint. When the blood vessels or nerves are injured, or the soft parts badly torn, it would be best to let the patient alone. It would be useless and cruel under such circumstances to

---

[*] It is much to be regretted that the original notes of these cases, which were minute and complete, were captured with other professional papers of Dr. McGuire, in one of the engagements in the Shenandoah Valley, and were probably destroyed. Diligent but unavailing inquiries have been made for them and for the surgical papers of Dr. Gilmore, lost under similar circumstances.

add to his sufferings by any operation. But when the injury is principally confined to the bone, and the question arises whether to let the patient alone, to amputate at the hip, or to excise the joint, I should not hesitate in preferring the last resource. The first is *inevitably fatal*, the second scarcely less so, and the last gives the best and almost only chance of life. The question of the usefulness of the limb after excision is of secondary importance. The issue here is of life and death. During the war, I did not perform this operation more frequently, partly because my conveniences for the after-treatment were often poor and uncertain, and I felt a natural reluctance to operate under such circumstances. Sometimes it would have been necessary to move the patient after the operation, perhaps a short distance, a mile or two only. But I have no doubt that such removal might have made all the difference between life and death. Sometimes a tendency to pyæmia, erysipelas, or gangrene, among the wounded rendered it advisable to avoid grave mutilations as much as possible; and, still oftener, a general scorbutic condition of the men, materially lessened the chances for success, and suggested the same forbearance in operative interference. It is chiefly to these causes: want of facilities for proper after-treatment, frequent necessity for the removal of patients, tendency to hospital gangrene, pyæmia, etc., and the frequently bad sanitary condition of soldiers in the field, that the greater success of excision of the coxo-femoral joint for disease, than when performed for gunshot injuries, is to be attributed."

CASE VII.—A Confederate private soldier was wounded at the battle of Fredericksburg, Virginia, December 13th, 1862, by a fragment of shell, which struck the trochanter of the right femur and fractured it and the neck of the bone, and severely lacerated the soft parts, but without injuring any of the important vessels or nerves. He was conveyed to a Field Infirmary, where, a few hours after the reception of the wound, he was placed under the influence of chloroform, and Surgeon Hunter McGuire, Medical Director of Jackson's Corps, having ascertained the extent of the injury, decided that although the lesions of the soft parts rendered the case a very unpromising one, yet that excision was the only resource that offered any hope, and accordingly proceeded to excise the head, neck, and trochanters of the femur, dividing the shaft just below the trochanter minor with a chain saw. The wound was left open; the limb placed in a comfortable position by means of pillows and cushions, without the use of splints, and the patient was treated at the temporary hospital at which the operation was performed. Notwithstanding the most careful attention to the after-treatment, he succumbed to the effects of the shock two or three days after the operation.

The facts in regard to the following case are gathered from the Casualty Lists of the engagement at Arkansas Post; communications from Surgeon Milton T. Carey, 48th Ohio Volunteers; Surgeon W. W. Slaughter, 60th Indiana Volunteers; and from the records of the hospital transport D. A. January. The necessity of removing the patient after the operation, is greatly to be deplored:

CASE VIII.—Private John Coon, Co. C, 60th Indiana Volunteers, aged 20 years, a robust man, was wounded at the battle of Arkansas Post, January 11, 1863, by a conoidal musket ball which entered the right buttock and passed forwards and outwards, striking the femur on the inter-trochanteric line and comminuting the neck and upper part of the shaft of the bone. A few hours after the reception of the injury the patient was conveyed to an hospital steamer and a consultation was held, at which it was determined to excise the injured portions of bone. The loss of blood which had taken place, and the patient's exposure to inclement weather, were regarded as very unfavorable circumstances, but it was considered that, on the whole, an excision was the best thing to be done. An ounce of brandy and other restoratives were administered, and half an hour subsequently the wounded man was placed under the influence of chloroform, and Surgeon Milton T. Carey, 48th Ohio Volunteers, proceeded with the operation. A semi-circular incision. beginning two inches above the prominence of the great trochanter, was made to extend downwards in the direction of the shaft of the femur. The muscular attachments at the trochanters were then divided, and the capsular ligament freely incised. Some difficulty was then experienced in dividing the ligamentum teres; but this was finally accomplished, and the head of the femur was turned out of the acetabulum and removed. A careful dissection was then made to ascertain how much of the shaft was injured. The extent of splintering having been determined, the shaft was sawn below the trochanter minor by means of a chain saw. The edges of the wound were then brought together, and a retentive apparatus was applied. After the operation the patient seemed much prostrated, but he rallied after a few hours, and was conveyed on the hospital transport D. A. January to Memphis, Tennessee, and placed in the military general hospital at that place, where he died ten days subsequently, January 21, 1863.

The following case was reported by Dr. J. T. Gilmore, of Mobile, Alabama, formerly Surgeon C. S. A., and chief surgeon of the Division of General McLaws, in General Longstreet's Corps. Dr. Gilmore's experience of field surgery in the late war was great, and this Office is under many obligations to him for careful and conscientious reports of his observations. Dr. Gilmore had full notes of this and two other cases of resection of the head of the femur, of which brief descriptions will be inserted farther on. Unfortunately these notes, with all of Dr. Gilmore's professional papers, were captured on the occasion of General Early's defeat at Fisher's Hill, and it is to be feared were destroyed, since the most diligent inquiries by this Office have failed to recover them:*

CASE IX.—A Confederate soldier of Kershaw's South Carolina Brigade, was wounded at the battle of Chancellorsville, Virginia, May 3d, 1863, by a musket ball which shattered the neck of the femur. It having been decided, after an examination of the wound under chloroform, that the case was a favorable one for the operation of excision, the important nerves and vessels being intact, and the injury limited mainly to the neck of the bone, the operation was performed by Surgeon James, 16th South Carolina Regiment, on the day after the reception of the wound. The patient sank from the shock of the injury and operation and died, May 6th, 1863.

The particulars of the next case were communicated by the operator, Colonel Henry S. Hewit, senior surgeon of the volunteer staff, and Medical Director of the Army of the Ohio, who remarks of it that, notwithstanding its fatal issue, it was one of the cases that came under his observation which had much weight in convincing him of the superiority of excision over coxo-femoral amputation in severe gunshot injuries of the upper extremity of the femur, uncomplicated by grave lesions of the vessels or other soft parts. "At any rate," Colonel Hewit continues, "the less degree of severity of the operation and the greater average success are arguments in its favor. The probable prolongation of life in unsuccessful cases is also a point worthy of consideration." Dr. Hewit adds that he believed at the time, and still adheres to the opinion, that under more favorable circumstances this patient would have recovered. He "considered the death due to vital and constitutional exhaustion, with the additional burthen of wounds and operation." He adds that the pathological specimen was placed in his saddle-bags, and unfortunately lost with his horse "in attempting to send the latter through the enemy's lines," for Dr. Hewit voluntarily remained a prisoner with his wounded:†

CASE X.—Sergeant James M. Tolman, Co. H, 18th Wisconsin Volunteers, aged 30 years, was wounded on the 14th of May, 1863, at the engagement near Jackson, Mississippi, by a conoidal musket ball, which comminuted the head and neck of the left femur, lodging in the latter, and producing fissures which extended about two inches below the lesser trochanter. The important nerves and vessels of the region and the walls of the pelvis had escaped injury. The patient was a somewhat cachetic subject, debilitated by malarial disorders. It was deemed that the gravity of the injuries of the upper extremity of the femur rendered operative interference imperative. About twenty-four hours after the reception of the injury, the patient was placed under the influence of chloroform, and Surgeon Henry S. Hewit, U. S. Volunteers, proceeded to exarticulate the head of the femur. The incision commenced a little above and anterior to the trochanter major and extended downwards in a curved direction with the convexity backwards, and passed through the wound of entrance. The splintered fragments of the head and neck and the ball were removed, and then the fissured upper extremity of the shaft was sawn two and a half inches below the lesser trochanter. The operation was well borne, and the patient was removed the same day to an hospital in the city of Jackson, where he was supplied with every comfort and provided with the most careful attendance. He did apparently well until the third day, when he began to sink, the wound from this time forward exhaling a faint cadaveric odor. He died four days after the operation, May 19, 1863.‡

---

* This case is referred to in "*A Contribution to the History of Hip-joint Operations performed during the late Civil War,*" by Paul F. Eve, M. D., p. 8. Dr. Gilmore is incorrectly reported as the operator.
† This case is recorded in *Circular No. 6, S. G. O.*, 1865, p. 66.
‡ The printed report of the Adjutant General of Wisconsin, for 1865, p. 300, gives the date of death as May 16, 1863; but Dr. Hewit states that this is incorrect, an assertion sustained by the entry in the hospital register.

PRIMARY OPERATIONS.   25

The account of the next case is compiled from a memorandum furnished by Dr. McNulty, and from particulars collected by Surgeon General Dale, of Massachusetts. The case has not been heretofore published:

CASE XI.—Captain Thomas R. Robeson, 2d Massachusetts Volunteers, aged twenty-four years, an athletic man, was wounded on the morning of July 3d, 1863, at the battle of Gettysburg, his regiment having become warmly engaged under a musketry fire at short range. A rifled musket ball struck him over the right trochanter major, shattering the neck and head of the femur, and, as was subsequently ascertained, fractured the pelvis and penetrated its cavity. He was carried a short distance to the rear, where the stretcher-bearers became exhausted and laid him down. Sergeant Francis O'Doherty, of his regiment, coming shortly afterwards wounded to the rear, impressed some stragglers and had the wounded man conveyed to a field station of medical officers of the 12th Corps. In the afternoon he was brought into the 12th Corps Hospital, and was examined by Surgeon John McNulty, U. S. Vols., the Medical Director of the Corps. The sufferings of the patient were intense, and he urgently demanded some operative interference for his relief. Although the prospect was very discouraging, it was decided to comply with his request. An exploration of the wound indicated that there was some injury of the pelvic wall. The patient was placed under the influence of chloroform very soon after his admission to the hospital, and a few hours subsequent to the reception of the injury, Dr. McNulty made an incision over the trochanter major six inches long, passing through the wound of entrance and continued downwards in the axis of the limb, turned out the shattered superior extremity of the femur, and sawed the bone just below the trochanters. The fragments of the head and neck were then removed. There was more bleeding in this than in Dr. McNulty's other operations of excision of the head of the femur, yet the hæmorrhage could not be called profuse. The patient survived the operation fourteen hours. During this interval he appeared to be unconscious.

The facts in the next case are derived from communications from Dr. J. T. Gilmore, of Mobile, Alabama, Dr. J. J. Knott, of Atlanta, Georgia, and from the records of the Knoxville Hospitals. The method of operating is unprecedented, and, it is to be hoped, unlikely to be imitated:*

CASE XII.—At the assault on Knoxville, Tennessee, on November 16th, 1863, a soldier of a Michigan cavalry regiment was wounded and made a prisoner. He was a man about eighteen years of age, five feet eight inches in height, with light hair and blue eyes, and was in robust health when he received the injury. A Minié ball entering about the centre of the nates, passed forwards shattering the head and upper part of the neck of the femur, but did not injure the acetabulum. No hæmorrhage of importance followed the wound. It was considered that the case demanded excision of the head of the femur, and the operation was performed on the day of the reception of the injury by Surgeon J. S. D. Cullen, P. A. C. S. "The operator made his incision posteriorly, directly through the thickest part of the gluteal muscles, on a line parallel with the os femoris, instead of laterally. In making his incision, which, at the least calculation, was ten inches in length, he cut the gluteal artery, near its point of exit from the pelvis." The artery was ligated finally, though not until there had been much loss of blood. The head and neck of the femur were then excised. "When the siege was abandoned," another report states, "General Longstreet retired to Russelville, and this patient was left behind. I am positive that he could not have recovered, for the suppuration that followed the operation was immense, and he was suffering from hectic fever when I last saw him, some six days after the operation." There can be but little doubt that the patient referred to in this account, was Private Isaac Melcer, Co. A, 8th Michigan Cavalry, aged 18 years, who was found abandoned on the retirement of the Confederate army from Knoxville, and was taken to Hospital No. 2, in that city, and entered as a case of "gunshot fracture of left hip." No other Michigan cavalry soldier is reported at the period referred to with this or any similar wound. This man died on December 2d, 1863. The register of the Knoxville hospital gives no particulars of the case.†

The particulars of the next case have been obtained from the casualty list of the 1st Division, 4th Army Corps, for the engagement at Buzzard's Roost Gap, Georgia, February 25th, 1864; a letter from Brevet Brigadier General Samuel Breck, Assistant Adjutant General, U. S. A., a report of Surgeon S. G. Menzies, 1st Kentucky Infantry, and a letter from the operator, enclosing a brief printed history:‡

CASE XIII.—Private Bartholomew Dempsey, Battery I, 4th U. S. Artillery, was wounded on February 25th, 1864, at Buzzard's Roost, Georgia, by a piece of shell, which passed through the upper portion of the right thigh, crushing the trochanter

---

* This case is referred to in "*A Contribution to the History of the Hip-Joint Operations Performed during the late Civil War*, by Paul F. Eve, M. D.," p. 9.

† The reader can judge whether the following extract from a letter to the reporter from a Confederate surgeon refers to this case: "But one case of excision of the head of the femur fell under my observation during the war: that of a Federal soldier at Knoxville, Tennessee, in November, 1863. I will not give you the name of the operator, as the operation was performed in such a bungling manner as to reflect but little credit on him. I have no doubt that the patient died."

‡ *Chicago Medical Examiner*, October, 1864, p. 612.

and neck of the femur, and producing fissures which extended to the head of the bone. The wounded man was taken to a private house in the neighborhood, at a place called "Big Spring," or in another report "Burke's Spring," a place ten miles northwest of Dalton, where, shortly after the reception of the injury, it was decided, on the recommendation of the surgeon-in-chief of General Cruft's Division, Surgeon S. G. Menzies, 1st Kentucky Volunteers, that excision of the injured extremity of the femur should be performed. Chloroform having been administered, Surgeon Nathan W. Abbott, 66th Illinois Volunteers, made a longitudinal incision, five inches in length, commencing two inches above the trochanter major. After dissecting aside the muscular attachments, and removing many fragments of the neck and trochanteric portions of the femur, the shaft of the bone was smoothly divided by the chain saw, at a point an inch or a little more below the lesser trochanter. Then, with a straight bistoury, the capsular and round ligaments were divided, and the fractured head of the femur was exarticulated. The wound was then approximated by sutures and adhesive strips. The patient rallied satisfactorily from the shock of the injury and operation, and his condition was encouraging on the following morning, when the Union forces retired, sending all the wounded who could be moved to the hospitals at Chattanooga. Private Dempsey alone was left at Big Spring. On the evening of February 26th, Surgeon Menzies sent Assistant Surgeon P. F. Ravenot, 75th Illinois Volunteers, with a cavalry escort, from General Cruft's camp to Big Spring, a distance of five miles, to learn of Private Dempsey's condition, and, if possible, to bring him off. The escort was dispersed and Dr. Ravenot was captured. The fate of Dempsey could not be definitely ascertained. He is dropped from the rolls of his company as "missing in action at Buzzard Roost Gap." Dr. Abbott afterwards heard, indirectly, that Dempsey survived the operation four or five weeks; but was not satisfied that this information was reliable. That the case had a fatal termination there can be no doubt. The excised portions of bone were preserved by Dr. Barnes, of Contralia, Illinois, who was present at the operation. Dr. Barnes proposes, Dr. Abbott writes, to send the specimen to the Army Medical Museum. As this report is passing through the press, a statement has been received from Bvt. Major F. B. Atwood, 16th U. S. Infantry, that he had learned from parties who attended Private Dempsey after he was wounded, that he died on February 28th, 1864, at the house of a Mr. Rogers, ten miles northwest of Dalton, Georgia.

The particulars of the following case have been communicated by Dr. Claude H. Mastin, of Mobile, formerly Inspector of Hospitals in the Confederate service, by the operator, Surgeon J. J. Dement, P. A. C. S., now a leading practitioner in Huntsville, Alabama, and by Dr. J. B. Duggan, of Toombsborough, Georgia, who attended the patient in a fatal illness long subsequent to the operation. The case is of peculiar interest as the second instance hitherto reported of a successful primary excision of the upper extremity of the femur for gunshot injury, and one of the two examples of the successful result of such operations during the late war. The history of the case has not been published heretofore:

CASE XIV.—Private Cannon, Co. A, 49th Georgia Regiment, Thomas's Brigade, aged about 24 years, was wounded at the battle of the Wilderness, Virginia, on the morning of May 5th, 1864, by a conoidal musket ball, which entered the left hip one inch below the trochanter major, extensively comminuting the femur and lodging in the muscles of the inside of the thigh. Just after he was wounded the Confederate line was forced back a short distance, and the wounded man laid upon the disputed ground, swept by the fires of the contending armies, until the following morning, when the Confederates recovered their position and carried off their wounded. Cannon was taken to a Field Infirmary, and, after a consultation by Surgeons J. J. Dement, Holt, J. J. Winn, and F. B. Henderson, it was determined to enlarge the wound and to remove the detached fragments of bone. Accordingly, the patient having been chloroformed, Surgeon Dement made an incision two inches upward from the entrance wound, and extending from the wound downwards four inches. On ascertaining the condition of the injured parts, it was decided to exarticulate the head of the femur. This was readily accomplished, and then the neck and upper extremity of the shaft were removed. The fragments of the upper extremity of the femur, when put together, measured four and a half inches. The hæmorrhage during the operation was trivial. All the medical gentlemen present remarked upon the slight degree of shock that was induced by the operation. The depression was less than commonly follows an ordinary amputation low down. The limb and body were confined by a roller baudage to a straight splint extending from the axilla to the foot. A full dose of sulphate of morphia was then administered. In a few hours the patient was placed in an ambulance wagon, and was conveyed to Orange Court House, twenty-five miles distant, and thence by railway to Staunton, Virginia, about seventy miles farther. The after-treatment was conducted at the general hospital at that place. Little can be learned of the after-treatment save that the patient was supplied with rich diet, with a liberal allowance of wine, and that no untoward complication occurred except the formation of abscesses attendant on an exfoliation of a ring of bone from the upper end of the shaft. When this was eliminated, the wound rapidly healed. At the end of nine months the cicatrix was firm. The limb was shortened three inches, and was useless for purposes of locomotion. The patient was in fine health, and moved about on crutches. He went to his home, in Toombsborough, Georgia, in February, 1865, and earned a livelihood by his trade of shoemaking. He enjoyed good health until November 12th, 1865, when he had an attack of diphtheria, which terminated fatally on November 23d, 1865.

The facts concerning the next case were furnished by Dr. Hunter McGuire, Professor of Surgery in the Medical College of Virginia, and have not been previously published:

CASE XV.—A Confederate private soldier of Ewell's Corps was wounded at the battle of the Wilderness, May 5th, 1864, by a conoidal musket ball, which broke the neck of the left femur into several fragments, and lodged in the bone at the junction of the head and neck. A few hours after receiving his wound, he was placed under the influence of chloroform at a field hospital, and was examined by Surgeon Hunter McGuire, the Medical Director of the Corps, who decided that the case was well adapted for the operation of excision of the head of the femur, and accordingly proceeded to remove, through a longitudinal incision, the head and shattered fragments of the neck, and to smooth off with a saw the jagged upper extremity of the shaft. The operation was accomplished with but trifling hæmorrhage. It is Dr. McGuire's impression, but, owing to the loss of his notes he cannot state positively, that in the subsequent rapid movements of the army, it was necessary to send the patient to the rear, and besides the disadvantages of removal, be failed to receive such nourishment and careful treatment as his case demanded. He died of pyæmia, May 22d, 1864.*

The facts in the next case were communicated by Dr. J. T. Gilmore, of Mobile, Alabama.† This is the only one of the three excisions of the head of the femur observed and reported by Dr. Gilmore in which he operated himself:

CASE XVI.—Private O'Rourke, 18th Mississippi Regiment, aged 24 years, healthy and of fine constitution, was wounded at the battle of the Wilderness, May 6th, 1864, by a musket ball, which entered the right thigh a little behind the trochanter, shattered the neck of the femur, and lodged. There was little injury to the soft parts, and the important vessels and nerves were unharmed. He was taken to a field hospital, and his injury was examined under chloroform, by Surgeon J. T. Gilmore, Chief Surgeon, 1st Division, Longstreet's Corps. The limb was everted and shortened, the fracture appeared to be confined to the epiphysis, and there was no bleeding. Believing that removal of the injured bone offered the best chance of preserving life, Surgeon Gilmore proceeded to excise the head and neck of the femur. A curvilinear incision four or five inches long, with its convexity backwards, was carried downwards from a point a little above and behind the trochanter, and was made to pass through the entrance wound. The muscles inserted in the trochanter were then divided, the head was readily disarticulated, and the femur was then smoothly divided through the trochanters by a chain saw. The operation was accomplished with the loss of but little blood. Yet the patient did not react, but gradually sank, and died, May 9th, 1864, three days after the operation.

A memorandum of the next case was found in a Confederate case-book. The distinguished operator died at the close of the war, but one of his assistants, Dr. W. F. Richardson, of Richmond, has communicated the principal facts connected with the operation. This and the four succeeding cases have not been heretofore recorded in print:

FIG. 4.—Longitudinal and oblique fissuring of the upper extremity of the right femur by a conoidal musket-ball. Spec. 5409. Sect. I, A. M. M.

CASE XVII.—Private J. J. Phillips, Co. G, 61st Virginia Regiment, was wounded on the second day of the battle of the Wilderness, May 6th, 1864, by a conoidal musket ball, which entered at the posterior upper portion of the left thigh, fractured the femur, and lodged. He was immediately conveyed to Richmond by rail, and was admitted to the Receiving and Wayside Hospital on May 7th. The wound was at once thoroughly explored under chloroform, and excision of the shattered bone was decided on. Surgeon Charles Bell Gibson, C. S. A., performed the operation. A long vertical incision over the trochanter major exposed the injured bone. It was found that the ball had produced extensive longitudinal splintering, and had itself split, a small fragment lodging in the medullary canal, while a larger portion had buried itself in the gluteal muscles, about two inches from the point of impact upon the bone. The muscles inserted into the trochanter having been divided, the head of the femur was exarticulated, and the upper extremity of the shaft was smoothed off with a saw. The operation was accomplished without much hæmorrhage, and the patient rallied promptly from the shock. He had an anodyne, and passed a good night, and, on the following day, May 8th, he appeared to be doing well. On the 9th, however, there was much constitutional irritation, and on the moring of the 10th, it was apparent that the man was sinking. He died at 4 A. M. May 11th, 1864. The pathological preparation was contributed by Dr. Richardson to the Army Medical Museum, and is a fine illustration of the characteristic longitudinal fissuring produced in the femur by conoidal balls. It is represented in the adjoining wood-cut, (FIG. 4.)

Assistant Surgeon John S. Billings, U. S. A., while Acting Medical Inspector of the Army of the Potomac, reported the fol-

---

* The loss of Dr. McGuire's notes of these operations has already been referred to in connection with CASE VII. Dr. McGuire regarded the reports of the cases here given as too imperfect to be recorded; and was reluctant that any observation of which he had not a distinct remembrance should be published. But as there can be little question as to the main facts, he has not insisted upon the withdrawal of his cases from this report.

† This case is cited in "*A Contribution to the History of the Hip-Joint Operations Performed during the late Civil War,*" p. 9, where *two* operations are erroneously ascribed to Dr. Gilmore. A writer in the *British and Foreign Medico-Chirurgical Review*, July, 1868, p. 174, accredits to Dr. Gilmore *three* operations with one success. Dr. Gilmore reported only his own operation and those by Drs. James and Cullen, (*Cases* IX, XII, and XVI.) All three were fatal.

lowing case to the Medical Director of that Army, Surgeon T. A. McParlin, U. S. A., who was requested to learn the particulars of it. His inquiries were unavailing:*

CASE XVIII.—An unknown private soldier of the 5th Corps, Army of the Potomac, was wounded in the engagement at Laurel Hill, near Spottsylvania Court-house, Virginia, on May 10th, 1864, by a musket ball, which fractured the trochanteric portion of the left femur. He was conveyed to the field hospital of the 5th Corps, at Cussin's, on the Block House road. He was placed under the influence of chloroform, and the head, neck, and trochanters of the left femur were excised. Assistant Surgeon J. S. Billings, U. S. A., saw him on the following morning, when he appeared to be in a comfortable condition. Dr. Billings recollects that he was a young and healthy looking man. The attendants mentioned the character of the operation and the name of the operator; but Dr. Billings cannot recall these particulars. On revisiting the hospital, three days subsequently, Dr. Billings learned that the patient had died on that morning, May 13th, 1864.

The facts in the next case are derived from a memorandum in a Confederate casebook, and a communication to this office by Dr. W. F. Richardson, one of the surgeons of the hospital in which the patient was treated. Dr. Richardson writes: "I refused to have anything to do with this operation, though the patient was in my ward. The man himself objected to resection. Dr. Gibson was my superior officer; yet I was willing to risk a court martial rather than to assist in an operation my judgment condemned:"

FIG. 5.—Comminution of the neck and trochanters of the left femur by a conoidal musket ball.—*Spec.* 5498, Sect. I, A. M. M.

CASE XIX.—Private G. W. Mayo. Co. D, 25th Battalion Virginia Reserves, was wounded at the affair between Yellow Tavern and the outer defences of Richmond, Virginia, May 12th, 1864, by a conoidal musket ball, at short range, which entered the right buttock and passed forwards and outwards through the thigh, striking the femur between the trochanters, and producing very extensive splintering of the neck and shaft. He was admitted to the Receiving and Wayside Hospital, at Richmond, early the next day, and his wound being examined under chloroform, Surgeon Charles Bell Gibson, C. S. A., determined to proceed at once with the operation of excision of the head and upper extremity of the femur. The injured bone being exposed by a long straight incision, the muscular and ligamentous attachments were divided, and the head of the femur was disarticulated. Numerous detached fragments were then removed, and the shaft of the femur was sawn at a point five or six inches below the trochanter minor. The operation was rapidly accomplished, but the shock, added to the depression already existing from the injury, was such that the patient did not react. He died at 9 A. M. on May 15th, 1864, about forty-five hours after the operation. The pathological specimen was preserved, and has lately been contributed to the Army Medical Museum by Dr. W. F. Richardson. It is represented in the annexed wood-cut, (FIG. 5.)

The abstract of the next case is compiled from the casualty list of the 9th Corps for Spottsylvania, and from a memorandum of operations sent from the field hospital of the 3d Division, 9th Corps:†

CASE XX.—Lieutenant John A. McGuire, Co. I, 148th Pennsylvania Volunteers, was wounded on May 12th, 1864, at the battle of Spottsylvania, by a musket ball, which smashed the trochanters and neck of the right femur. He was carried to the hospital of the 3d Division of the 9th Corps, where, after an exploration of the wound under chloroform, and a consultation of the senior surgeons of the Division, it was determined to excise the injured bone. The head, neck, and trochanters were accordingly removed through a longitudinal incision by Surgeon George W. Snow, 35th Massachusetts Volunteers. The patient survived the operation three days. His death is reported on May 15th, 1864.

---

* The "classified return of wounds and injuries received in action" of the 3d Division, 5th Corps, for May 10, 1864, has an entry of one case in the column for "excisions" on account of a fracture of the upper extremity of the left femur by a musket ball. The accompanying nominal list of casualties contains the names of three men with gunshot fracture of the upper third of the left femur, all of whom died. But there is nothing to indicate in which of the three operative interference was undertaken. Interrogatories have been addressed, without result, to the principal operators of the Division, Surgeon L. W. Read, U. S. Vols.; Surgeon J. J. Comfort, 1st Pennsylvania Rifles; Surgeon Charles Bower, 35th Pennsylvania Vols.; and Surgeon Benjamin Rohrer, 10th Pennsylvania Reserves.

† A letter from the operator, Dr. G. W. Snow, dated Newburyport, Massachusetts, November 4, 1868, stated that he was unable to furnish any further particulars of this operation, as he had kept no private record of his surgical observations in the army.

PRIMARY OPERATIONS.   29

The next case is reported by Assistant Surgeon J. S. Billings, U. S. A., who saw the patient in his inspection of the 18th Army Corps Hospital:*

CASE XXI.—An unknown soldier of the 18th Army Corps was wounded in the assault on the enemy's intrenched lines at Cold Harbor, June 3d, 1864, by a fragment of shell, which completely comminuted the trochanter and neck of the right femur. Shortly after the reception of the injury, he was conveyed to the field hospital of the 18th Corps, and immediately anæsthetized and examined. Excision of the head, neck, and trochanters of the right femur was then practised. Assistant Surgeon Billings, U. S. A., saw the patient soon after the operation, and observed that he had rallied encouragingly, and was in a comparatively comfortable condition. On June 7th, the wounded of the 18th Corps were placed in wagons and sent to the rear. Dr. Billings visited the hospital with a view of preventing the removal of this patient; but was informed by the director of transportation that the man had died in the previous night, June 6th, 1864, three days after the operation.

The next case is reported in the eighteenth volume of the Transactions of the American Medical Association, p. 263, in Dr. Eve's paper, by Dr. Charles James O'Hagan, of Greenville,† North Carolina. Dr. John D. Jackson, of Danville, Kentucky, has communicated some additional facts:

CASE XXII.—Private ——— ———, 56th North Carolina Regiment, Ransom's Brigade, was wounded on the night of the 17th of June, 1864, in front of Petersburg, by a conoidal ball probably, in the right thigh. The ball entered on the inner aspect of the limb, and passed obliquely upwards and outwards, producing a comminuted fracture of the neck of the femur, and driving the fragments of bone into the surrounding tissues. The shaft of the bone was not shattered, and, as the man was very much worn and exhausted, as most of the Confederate troops were at that time, it was considered advisable to perform resection of the head and neck of the femur, as offering a better chance of recovery than amputation. The operation was performed on June 18th, twelve hours after the reception of the wound, and the bone was sawn through the trochanters. He bore the operation well, and, although weak, was hopeful. He was sent to the Fair Grounds Hospital, from which he was removed in a few days, placed in a tent, and attended by Surgeon Ladd, 56th North Carolina Regiment, and Dr. C. J. O'Hagan. He survived the operation two months, and succumbed at last to suppuration, caused by the want of proper food and stimulants, and the general prevalence of pyæmic infection, which at that period intervened in nearly all the surgical cases in the neighborhood of that hospital. To this account, Dr. J. D. Jackson adds: "I recollect very distinctly of being present at the operation of Dr. Ladd, being then of the same division with him, but not of the same brigade. It was on the 18th or 19th of June, 1864, that it was done, the place being an unfinished brick church in the centre of Petersburg, which we were then occupying as an hospital. There were also present some four or five other surgeons, among whom I recollect Surgeon C. J. O'Hagan, Dr. Wilson, of Virginia, then the senior Surgeon of Ransom's Brigade, Dr. Luckie, of the same Brigade; and, if I mistake not, Dr. R. L. Brodie, then Medical Director of General Beauregard's army, was among the number. The man operated upon was of Dr. Ladd's own regiment; his age, and any other personal peculiarities, I have forgotten, though I think he was young and comparatively robust. The wound had apparently been done by a musket ball, and the range of the wound was, I think, from the inner and upper aspect of the thigh, and nearly transversely through, ranging slightly upwards, the aperture of exit being over the trochanter major. If I recollect aright, the trochanter was torn off, and most of the neck of the femur shattered to fragments, the shaft of the femur being entirely separated from the head. Chloroform was given, and Dr. Ladd operated by making a slightly curvilinear incision over the acetabulum and trochanter—the aperture of the wound being in its line—cut down upon the head of the femur, exarticulated and removed it, and cut off a sharp fragment of the remaining end of the femur. The difficulty of performing the operation seemed to be small. The hæmorrhage was trifling. I do not recollect that I saw the patient again, he being sent off to the General Hospital at what was then known as the "Fair Grounds Hospital," situated in the suburbs of Petersburg. But I further remember distinctly of hearing Dr. Ladd, Dr. O'Hagan, and probably others of Ransom's Brigade speaking of his death, which was on the sixtieth day after the operation, and which all agreed at the time in ascribing to want of good food in proper quantity. Owing to the scarcity of our supplies, and the immense number of wounded men then crowding the city in consequence of the battles fought in front of Petersburg on the 17th, 18th, and 19th of June, food really proper for wounded men was not to be obtained, and anything like delicacies were out of the question."

The facts in the next case were communicated by the operator, now a practitioner of Pulaski, Tennessee:‡

---

* It is hardly necessary to state that every effort has been made to identify this case. Besides searching the records of the 18th Corps, letters of inquiry have been addressed to the Medical Director of the Corps, Dr. George Suckley, Surgeon J. M. Rice, 25th Massachusetts Volunteers, Surgeon S. A. Richardson, 13th New Hampshire Volunteers, and Surgeon A. W. Woodhull, 9th New Jersey Volunteers, senior medical officers of the Corps. The 18th Corps, belonging to the Army of the James, arrived by a forced march from White House in season to participate in the sanguinary battle of Cold Harbor, and five days after the battle was hurried back to Petersburg. The hospital arrangements were necessarily greatly confused, a large number of the wounded being treated in the open air.

† Not "Glenville," as printed by Dr. Eve.

‡ Letter of Dr. J. F. Grant, dated February 29th, 1868. The case is published also in the Transactions of the American Medical Association, Vol. XVIII, p. 261.

CASE XXIII.—Private T. J. Hobson, Co. II, 32d Tennessee Regiment, aged 23 years, was wounded at the battle of Kenesaw Mountain, June 24, 1864, by a conoidal musket ball, which struck the femur and extensively comminuted the neck and trochanters. The fracture extended within the capsular ligament. The shock was very great. The patient was seen by Surgeon J. F. Grant, P. A. C. S., who found that amputation was not practicable except at the hip-joint, and deemed it expedient to undertake the operation of excision, as giving, in his judgment, the best chance for recovery. The army was then retreating, and if the patient was removed to the rear it was doubtful if surgical relief could be had. Accordingly, about twelve hours after the reception of the injury, the patient being placed under the influence of chloroform, Dr. Grant proceeded to operate, by making a linear incision ten inches long on the outside of the thigh over the trochanters. The articulation was exposed, the capsular ligament divided, and the head of the bone enucleated. The shattered fragments were then removed, and the shaft of the femur was divided by a straight saw just below the trochanter. The loss of blood was slight. Immediately after the completion of the operation, the patient was placed upon a box-car, and transported forty miles over a very rough road to the rear. Reaction was never complete, though the patient lingered three days, and died on June 27th, 1864.

The following case is briefly reported on the casualty list of the 9th Army Corps, and particulars concerning it have been communicated by Dr. James Oliver, formerly Surgeon 21st Massachusetts Volunteers. This and the following case have not been published:

CASE XXIV.—Sergeant Edwin T. Brown, Co. C, 21st Massachusetts Volunteers, aged about 30 years, was wounded in front of Petersburg, at 1 o'clock p. m. on July 23d, 1864, by a ragged fragment of a mortar bomb, which struck the left thigh over the trochanter major and comminuted the upper extremity of the femur. The wounded man was immediately conveyed to the hospital of the 1st Division of the 9th Army Corps. Surgeon Whitman V. White, 57th Massachusetts Volunteers, and Surgeon James Oliver, 21st Massachusetts Volunteers, saw the patient a short time after his admission to the hospital. He was a strong, healthy man, five feet ten inches in height, weighing about one hundred and sixty pounds, with a constitution of iron, and was in perfect health when injured. The soft parts about the seat of injury were lacerated and torn, and the upper extremity of the femur, to an extent of five inches, was crushed to fragments. No important arteries or nerves were wounded. Excision of the fractured bone was decided upon. On the afternoon of the day on which the injury was received, chloroform was administered, and Dr. White made a longitudinal incision and removed the shattered fragments. The ligamentum teres was divided and the head of the bone turned out. The broken extremity of the shaft of the femur was evened off with a chain saw. The patient reacted promptly from the shock of the operation. The limb was placed in proper position and stimulants were freely used. Dr. Oliver states that he saw the man several times on the following day, and he was in excellent spirits; that he talked and laughed, and did not complain of any pain. His appearance was remarkable after so severe a wound and operation. On the 26th his appetite failed, and he began to sink. He died at 9 A. M. on the 27th July, 1864, four days after the operation.

The account of the following case is taken from the register of the Washington Street Confederate Hospital, at Petersburg, and from a letter from Dr. W. L. Baylor, of Petersburg:

CASE XXV.—Private J. T. Goode, Co. K, 6th Virginia Infantry, aged 21 years, was wounded in the intrenched lines before Petersburg, July 31st, 1864, by a conoidal musket ball, which fractured the upper extremity of the left femur. A few hours after the reception of the injury, he was anæsthetized by a mixture of chloroform and ether, and the wound having been explored, resection was decided upon. Surgeon G. S. West, C. S. A., proceeded to perform the operation, assisted by Dr. W. L. Baylor and others. Upon making a linear incision in the axis of the limb, and exposing the fracture, it was found that it extended longitudinally much lower on the shaft than was anticipated. Dr. Baylor reports that one of the gentlemen present thinks that fully one-third of the femur was excised. The patient never fairly rallied from the shock of the operation; but he lingered until August 2d, 1864, when he died. Dr. Baylor adds that the circumstances were very unfavorable, the patient being poorly nourished, and nosocomial gangrene at the time pervading the surgical wards.

It is greatly to be deplored that military exigencies necessitated the removal of the subject of the operation described in the next abstract. The case appears to have been a typical one for the operation of primary excision. The patient was hale and vigorous, the injury to the soft parts was insignificant, and the injury to the bone was confined to the head and upper part of the neck. The man's condition for several days was most encouraging. Even after his transfer to the base hospital at City Point he did well, until it was thought necessary to move him again, hundreds of miles away to Philadelphia :*

CASE XXVI.—Private Edward A. McDonald, Co. F, 149th Pennsylvania Volunteers, aged 31 years, a robust, athletic man, was wounded on August 20th, 1864, in the fighting for the possession of the Weldon Railroad. A conoidal musket ball

---

* See *Circular* 6, Surgeon General's Office, 1865, p. 70, Case 23.

PRIMARY OPERATIONS.

entered the upper anterior part of the right thigh, and lodged in the head of the femur, after splintering its neck. He was carried to the hospital of the First Division of the Fifth Corps, and the wound was examined by the principal surgeons there, who decided that the injury was limited to the head of the femur, and that decapitation of the injured bone offered the best chance of preserving life. The patient was placed under the influence of chloroform a few hours after the reception of the injury, and Surgeon F. C. Reamer, 143d Pennsylvania Volunteers, assisted by Surgeon Thomas, 119th Pennsylvania, and others, proceeded with the operation. A V-shaped incision, arranged to traverse the entrance wound, exposed the muscular attachments of the neck and trochanter. These being divided, with the capsular and round ligaments, the head of the femur was exarticulated. Fragments of the neck were extracted, and then the femur was sawn through the trochanteric ridge by the chain saw. The wound was then partly closed by sutures and adhesive plasters, a pledget of lint being inserted at the lower end, and the limb was bandaged and suspended by a Smith's anterior splint. Little loss of blood had been incurred, and the patient reacted, and his condition appeared hopeful. Two days afterwards it was deemed necessary to remove the severely wounded from the advanced position of the 5th Corps, and McDonald was sent in an ambulance wagon several miles, over a rough road, to the railroad communicating with the base hospital at City Point. He remained three days in the base hospital, and was then placed on an hospital transport steamer and sent to Philadelphia. He was admitted to the Broad and Cherry Streets Hospital, Philadelphia, on the 27th of August. The injured limb was extended in a straight position by means of weights and pulleys. Concentrated nourishment and stimulants, with quinine and opium, were administered. On August 31st, a chill and other evidences of the invasion of pyæmia were observed, and this fatal complication made rapid progress. Death took place on September 4th, 1864. At the post mortem examination, extensive metastatic foci were found throughout both lungs.

The following case corroborates the evidence furnished by CASES III, IV, V, and XI, that when gun-shot injury involves the acetabulum as well as the head of the femur, excision of the latter will be almost certainly unavailing. The facts of the case have been reported by the operator, Surgeon A. A. White, 8th Maryland Volunteers, and by Assistant Surgeon J. S. Billings, U. S. A., and the late Assistant Surgeon George M. McGill, U. S. A. Additional information regarding the case has been furnished by Assistant Surgeon Charles K. Winne, U. S. A., and the pathological specimen was forwarded to the Army Medical Museum by the late Assistant Surgeon Samuel Adams, U. S. A. Owing to discrepancies in dates and descriptions in the hasty field reports, the case was duplicated in the tabular statement of excisions of the head of the femur published in Circular No. 6, S. G. O., 1865, pp. 68–70, and there figures both as CASE XX and Case XXIV. The pathological specimen also was erroneously referred in the Catalogue of the Surgical Section of the Museum. Assistant Surgeon A. A. Woodhull, U. S. A., in describing the specimen remarks[*] on its fitness for the operation of excision; but the evidence at that time filed in the Surgeon General's Office was insufficient for its identification. It is believed that the facts in the case are correctly recorded in the following abstract:[†]

CASE XXVII.—Private Charles Beard, 12th Mississippi (Confederate) Regiment, was wounded and made a prisoner in the engagement on the Weldon Railroad, August 23d, 1864. With nearly two hundred other wounded Confederates, he was received at the field hospital of the First Division of the Fifth Army Corps at Reams' Station, where it was found that a conoidal musket ball had entered the front of the right thigh, a little to the outside of the course of the great vessels, and had comminuted the neck of the femur and fractured the head, and lodged in the acetabulum, of which the lower portion of the rim was broken off. A few hours after the reception of the wound the patient was placed under the influence of chloroform, and, after a thorough examination, it was deemed expedient to excise the upper extremity of the femur. The operation was performed by Surgeon A. A. White, 8th Maryland Volunteers. An incision, commencing a little below the anterior superior spine of the ilium, was carried downwards below and behind the prominence of the trochanter major. From the lower extremity of the first, another incision, Dr. McGill states, was carried backwards. The muscular attachments were then dissected aside, and the chain saw was passed around the bone, which was divided just above the lesser trochanter. The head of the femur was then readily exarticulated, and the ball and splintered fragments were removed. The wound was then approximated by sutures and adhesive strips, and the limb

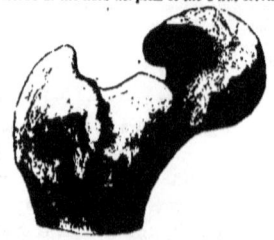

FIG. 6.—Fracture of the head of the right femur by a conoidal musket ball.—*Spec.* 1410, Sect. I, A. M. M.

[*] *Catalogue of the Surgical Section of the Army Medical Museum*, p. 243. "Specimen 1430, XII, A. B. b., 55. * * * An apparently suitable case for primary excision."

[†] See *Circular* 6, S. G. O., 1865, p. 70, *Case* 24.

was suspended by Smith's anterior splint. The patient reacted favorably; but very soon after the operation there was a marked rigor, and, on the following day, there was extreme irritability of stomach and retention of urine. The case terminated fatally on August 25th, 1864, two days after the reception of the injury. At the autopsy, it was found that the fracture of the acetabulum did not communicate with the interior of the pelvis; but the articular surface was intensely injected. Its cavity was filled with offensive sanious pus. The sawn extremity of the femur was black. One report states that the patient's appearance was of one who had undergone great privations, and was not in a favorable condition to undergo any severe operation. The excised portions of bone, represented in the accompanying wood-cut, (FIG. 6,) were sent to the Army Medical Museum without a memorandum; but were ultimately identified, and numbered 1410 in the Surgical Section.

The abstract of the following case is compiled from a special report from the operator, and from the register of Sheridan Field Hospital, at Winchester. This and the remaining primary cases have not been previously reported:

CASE XXVIII.—Sergeant Samuel Grimshaw, Co. H, 6th New York Cavalry, aged 31 years, was wounded at the battle of Cedar Creek, Virginia, October 19, 1864, by a fragment of shell, which, after lacerating the scrotum, entered at the upper inner part of the left thigh near the femoral artery, making a wound one and a half inches in length, and passing upwards and backwards, shattered the head and neck of the femur, and produced fissures extending four and half inches in the shaft, and lodged in the acetabulum. The shock to the nervous system was great. The patient was desponding, and he complained of severe pain. He was conveyed to a field hospital, and at eight o'clock, P. M., two hours after the reception of the injury, he was placed under chloroform, and Surgeon A. P. Clark, 6th New York Cavalry, made a straight incision seven inches in length over the trochanter major, which was carried down to the bone, and excised the head and four and a half inches of the shaft of the femur. The wound was then dressed, and the limb was supported by pasteboard splints. On the following day no bad symptoms were observed. Beef essence and stimulants were freely given, and afterwards sulphate of morphia was administered. He was removed to the Sheridan Field Hospital at Winchester on October 20th, and there died on November 5th, 1864, sixteen days after the reception of the injury.

The following case is recorded in the hospital register of the 10th Army Corps, and is farther described in a letter from the operator:

CASE XXIX.—Private Thomas G. Pease, Co. B, 117th New York Volunteers, was wounded in action on October 28th, 1864, in the engagement of the 10th Corps on the north bank of the James, near Fair Oaks Station. The trochanters and neck of the right femur were shattered by a musket ball, which lodged against the head in the cotyloid cavity. The soft parts were not injured badly, and it was determined by the surgeons on duty at the "flying hospital" of the 10th Corps that excision of the upper extremity of the femur was expedient. The operation was performed, a few hours after the reception of the injury, by Surgeon N. Y. Leet, 76th Pennsylvania Volunteers. The patient survived two days. His death is reported on October 29th, 1864.

The next case is a very gratifying one, the subject of it having already survived the operation four and a half years, in the enjoyment of tolerable health, the mutilated limb being of a certain amount of utility, enabling him to dispense with crutches and to walk with the assistance of a cane. The subjoined particulars are taken from the registers of the Tenth Corps Field Hospital, and of Chesapeake Hospital, and from letters from the operator and from the patient, Lieutenant Dwight Beebe, formerly Adjutant 3d New York Volunteers, now a resident of Havana, Schuyler County, New York. It was Mr. Beebe's impression that the neck and trochanters only were excised, and that the head of the femur was left in the cotyloid cavity; but Surgeon Leet states very positively and precisely that the head was exarticulated, and that the fragments excised included the upper part of the shaft, the neck, and head, and measured altogether about four inches in length. This account is corroborated by the entries on the Tenth Corps register, and the returns from the hospital signed by Surgeon A. C. Barlow, 62d Ohio Volunteers, and Surgeon M. S. Kittinger, 100th New York Volunteers, of the Army of the James. In such a matter of anatomical detail the statements of the surgeons must be accepted rather than the impressions of the patient. Mr Beebe also states that the operator was Surgeon Clark, of an Illinois regiment; but on this point the assertions of Dr. Leet and the record in the official reports are equally positive and precise:

CASE XXX.—Lieutenant Dwight Beebe, Adjutant 3d New York Veteran Volunteers, was wounded on October 27th, 1864, while in command of a skirmish line on the Darbytown road, near Richmond. A conoidal musket ball fractured the neck and trochanter of the right femur. He was immediately removed to the Flying Hospital of the Tenth Army Corps, near Chapin's Farm, when, after an exploration of the wound under chloroform, it was determined to excise the shattered extremity of the

RESULT OF A SUCCESSFUL PRIMARY EXCISION AT THE HIP FOR GUNSHOT INJURY
Case of Lieutenant Dwight Beebe

PRIMARY OPERATIONS.   33

femur. The operation was performed, only a few hours after the reception of the injury, by Surgeon N. Y. Leet, 76th Pennsylvania Volunteers. A straight incision was made on the outside of the limb, the muscles were divided, the head exarticulated, and the shaft was sawn by a chain saw about the level of the trochanter minor. Altogether, about four inches of the upper extremity of the femur was excised. The incision was then drawn together by stitches and adhesive plaster, cold water dressings were applied to the wound, and the limb was suspended by means of Smith's anterior splint. On the following day the patient was conveyed in an ambulance to James River, and thence transported on an hospital steamer to Fort Monroe. He was admitted to Chesapeake Hospital on the evening of November 1st, 1864. He was placed upon a fracture bed, and, on account of the interest as well as the gravity of the case, it received the most assiduous and careful attention. There was a very profuse discharge, and the patient soon became much exhausted. The upper extremity of the lower fragment necrosed, and a ring of bone seven-eighths of an inch in length exfoliated. This occurred in the middle of January, 1865. After the dead bone was eliminated, Lieutenant Beebe rallied and improved uninterruptedly. On May 17th, 1865, he was able to be moved comfortably, and obtained a leave of absence and went to his home in New York. He returned to the hospital July 9th, and in September, 1865, again went to the North with his regiment to be mustered out at the expiration of its term of service. He continued to improve in general health, and was able to hobble about on a cane. In August, 1867, the wound broke out afresh, and there was a profuse purulent discharge for about three weeks. The surgeons who were consulted ascribed this formation of abscess to irritation from using the limb too much. No elimination of dead bone was observed. The wound soon healed again firmly, and from that period until November, 1868, when Mr. Beebe was last heard from, there had been no inconvenience from the wound. "With the exception of the limb being quite weak," Lieutenant Beebe writes, "it only troubles me in damp weather, when it has a dull, heavy ache. The wound has firmly healed. The flesh on the outside of the thigh is quite numb. My knee is stiff. My general health is not good, and my physician cautions me to be very careful of myself. The limb measures three and one-eighth inches shorter than its fellow."

The abstract of the next case is taken from the registers of a field hospital of the 10th Army Corps, and of the Point of Rocks Hospital:

CASE XXXI.—Private Robert Cole, Co. B., 29th Connecticut Regiment, (colored troops), was wounded in an engagement on the north bank of the James, near Fair Oaks, on October 27th, 1864, by a musket ball, which shattered the upper extremity of the right femur, without injury to any important vessels or nerves. He was conveyed to the "flying hospital" of the Tenth Corps, where the wound was explored, and it was decided to excise the head, neck, and trochanters of the femur. The operation was performed by Surgeon Clark, a few hours after the reception of the injury, by a longitudinal incision over the trochanter major, and division of the superior portion of the shaft by a chain saw. Dressings to secure the immobility of the limb were applied, and the patient was removed to the base hospital of the Army of the James, at Point of Rocks, where he was received on October 28th, and died on October 29th, 1864, two days after the operation.

A minute of the last case was found in a pencil memorandum of the operations performed by surgeons of the 3d Division, 5th Corps, on March 29th, 1865, and the particulars have been supplied by the operator, Dr. William Fuller, of New Minden, Illinois, late Surgeon of the 1st Michigan Volunteers:

CASE XXXII.—Private Charles Morrison, Co. C, 185th New York Volunteers, was wounded in the engagement of the Fifth Corps, on the Quaker Road, south of Petersburg, Virginia, on the evening of March 29th, 1865. A conoidal musket ball struck the outside of the left thigh, fractured the trochanter, and separated the neck from the shaft. In less than two hours after the reception of the injury, he was placed on the operating table at the field hospital of the 1st Division of the 5th Corps, and his wound was examined while he was under the influence of chloroform. He was a robust man, in the best health. In the judgment of the operating staff, the case was a very favorable one for the operation of excision. Surgeon William Fuller, 1st Michigan Volunteers, was requested to perform the operation, and proceeded with it without delay. He entered his knife an inch above the great trochanter and made a longitudinal incision three and a half inches in length, divided the muscular attachments, and readily exarticulated the head of the femur. A fissure was found to extend downwards half an inch below the trochanter minor. The shaft was divided by a chain saw at this point. The ball could not be found, but, from the direction of its track, it was the opinion of the operator and his colleagues that it had entered the pelvis, through the obturator foramen. There was scarcely any hæmorrhage during the operation, no artery requiring ligation or torsion. A tent was introduced into the wound, which was then approximated by two sutures, and covered by a compress dipped in cold water. A full dose of morphia was then administered, and the patient was made as comfortable as possible in a bed in a hospital tent. In the middle of the night Surgeon Fuller returned to the hospital to visit his patient; but found that he had been removed to City Point, in compliance with orders from a superior authority. Dr. Fuller was subsequently informed by Surgeon Joseph Thomas, 118th Pennsylvania Volunteers, that the man died on the way to the base hospital, about twelve hours after the operation. There was some hæmorrhage a few hours after the operation, but it was not considerable. The report of the patient's death was premature. The records of the City Point Hospital show that he was received there, and survived until April 26, 1865, when he succumbed to the irritation and profuse suppuration following the operation.

INTERMEDIATE EXCISIONS.

This category comprises twenty-two excisions of the head of the femur, two of which resulted successfully, a mortality rate of 90.9. The shortest interval between the reception

5

of the injury and the date of operation was two days, the longest was twenty-eight days, the average interval was thirteen and a half days. Five of the patients were youths of eighteen or nineteen years, one was a veteran of sixty years, the majority were in the prime of life, between twenty-five and forty years of age. In the twenty unsuccessful cases, the average duration of life after the operation was twelve and a half days. One patient survived seventy-five days, and apparently succumbed to climatic influences, and not to the effects of the operation. In another case, in which the patient lived twenty days, and died with colliquative diarrhœa, the fatal event was ascribed to malarial disorders rather than to the effects of the wound or operation. Three of the cases were complicated by fractures of the pelvis. One patient died of profuse venous secondary hæmorrhage, two from frequently recurring capillary bleeding, one from peritonitis, one from diarrhœa, six from pyæmia, and nine from exhaustion. In all of the cases but one, the wounds were inflicted by conoidal musket balls.

The details of the first intermediate excision were communicated to the reporter by Surgeon M. Goldsmith, U. S. V. The operation is not recorded on the register of the hospital at which it was performed; but the facts communicated by Dr. Goldsmith rendered it possible to identify the case, and to obtain the name and military description of the patient from the hospital register:

CASE XXXIII.—A soldier of General Buell's army was wounded in a picket skirmish, about seven miles from Nashville, Tennessee, in March, 1862, by a conoidal musket ball, which shattered the neck and trochanters of the femur. He was immediately conveyed to Nashville, and placed in the College Hospital, under the care of Surgeon A. H. Thurston, U. S. Volunteers. Surgeon M. Goldsmith, U. S. Volunteers, saw him two days after the reception of the injury, and deemed the case peculiarly well adapted for the operation of excision. The surgeon in charge concurring in this opinion, the patient was anæsthetized, and Dr. Thurston proceeded to excise the head and splintered upper extremity of the femur, through a long straight incision. The operation was accomplished with but little hæmorrhage, and although the patient was much prostrated by the shock of the injury and of the operation, he reacted and was in a comfortable condition for several days. But surgical fever and suppuration soon set in, and he gradually sank, and died one week after the operation. There can be little doubt that the subject of this operation was Corporal Henry F. Smith, Co. B, 1st Wisconsin Volunteers, who, according to the records of the Nashville Hospital, was admitted for a gunshot wound of the hip, and was the only patient who died from wounds in Dr. Thurston's wards, at the period referred to.* Corporal Smith died on March 15, 1862. The operation was probably done on March 10th.†

The particulars of the next case were communicated by the operator, who was serving at the time as a Brigade Surgeon of Volunteers. He remarked that at the period of the operation, "pyæmia was destroying almost every [wounded?] soldier in the hospital, those who were subjected to operation and those who were not. An amputation of the great toe was not less fatal than one at the hip:"

CASE XXXIV.—Private D. M. Noe, Co. C, 46th Ohio Volunteers, aged 22 years, was wounded at the battle of Shiloh, Tennessee, April 6th, 1862, by a conoidal musket ball, which shattered the neck of the left femur. The patient was placed on board the hospital transport steamer Lancaster, under the charge of Surgeon George C. Blackman, U. S. Volunteers. On April 16th, 1862, chloroform having been administered, Dr. Blackman made a longitudinal incision four inches in length over the trochanter, and excised the head, neck, and trochanters, together with three inches of the shaft of the femur, the diaphysis being divided by a common amputating saw. The patient reacted well after the operation, and for five days the symptoms progressed favorably. Pyæmia was subsequently developed, and death ensued on April 24th, 1862, eight days after the operation.

A special report from the operator, Assistant Surgeon John S. Billings, U. S. A., furnishes the materials for the abstract of the next case. Dr. Billings remarked that ampu-

---

* The above account has been submitted to Dr. M. Goldsmith, who states that he thinks that the identification of the case is complete. Surgeon Thurston died during the war.

† The report of the Adjutant General of Wisconsin for 1865, p. 33, states that on March 8th, 1862, five companies of the 1st Wisconsin Volunteers were sent out beyond Nashville on picket duty. They were attacked by a cavalry force, and Private Willett Greenly was killed—"the first Union soldier killed in Tennessee"—while Corporal H. F. Smith and one other were wounded, and were sent to Nashville.

tation at the hip-joint would have been unjustifiable in this case, and that, had it been attempted, the patient, in his opinion, would have died upon the table. Dr. Billings also regarded the statistics of Mr. Guthrie and Mr. Macleod as utterly discountenancing coxo-femoral amputation except as a primary operation:[*]

CASE XXXV.—Private T. C. Christopher, Co. D, 18th South Carolina Regiment, aged 21 years, a robust muscular man, was wounded at the battle of Williamsburg, Virginia, May 5th, 1862, by a conoidal musket ball, which entered about two inches below and behind the left trochanter major, and passed forwards, upwards, and inwards. He was stooping at the time he received the injury. He was made a prisoner, and was sent with other wounded to the York River Landing, and thence on a hospital steamer to Washington, and was placed in Cliffburne Hospital on May 17th. He had received little attention prior to his admission, and was very despondent. His pulse was at 100 and was weak, and he complained of severe pain in the hip and knee. The tissues about the hip were much swollen; the limb was everted, and shortened one and a half inches. The opening made by the bullet was very small, and discharged a thin sanious pus. There was no orifice of exit. The patient was etherized and a careful exploration of the wound revealed a fracture of the inner portion of the neck and probably of the head of the femur. It was decided that excision should be performed, and the patient was placed upon a soothing and supporting regimen preparatory to the operation. Three days subsequently, May 20th, Assistant Surgeon J. S. Billings, U. S. A., assisted by several members of the hospital staff, proceeded to remove the shattered portions of the bone. A curvilinear incision, four inches in length, one inch behind the great trochanter, and nearly in the axis of the limb, clearly revealed the condition of the parts. Fragments of the inner extremity of the neck were removed piecemeal. The head was then removed from the cotyloid cavity, except a small fragment which was extracted from an inter-muscular space. The ball was now discovered lying in the obturator externus muscle, and was extracted. Very little blood was lost during the operation, and reaction took place fairly. Water dressing was applied to the wound, and a grain of sulphate of morphia was administered. Eversion of the limb was corrected by fastening the limb by straps of adhesive plaster to an upright piece of wood screwed to the foot of the bedstead, and, the foot of the bedstead being raised, adequate extension and counter-extension were secured. On May 21st the patient reported a comfortable night, but now had a very irritable stomach, with frequent vomiting. His skin was cool and clammy; his pulse small and feeble at 115. He was ordered aromatic spirits of ammonia in small doses, brandy, egg-nog, and beef essence, with sinapisms to the epigastrium. On May 23d, the irritability of the stomach had subsided. The patient was weaker, but complained of no pain. Stimulants and concentrated nourishment continued. On May 24th, the patient grew weaker rapidly. Capillary hæmorrhage took place from the surfaces of the incision, but was readily checked by the application of a solution of persulphate of iron. Enemata of beef essence and brandy were administered, and these articles were also given by the mouth. But the patient continued to sink rapidly, and died at seven in the evening of May 24th, five days after the operation. At the autopsy, made twelve hours after death, the soft parts surrounding the seat of injury were found dark in color and softened. The acetabulum was eroded. A clot of blood weighing about three ounces was found between the peritonæum and iliacus externus muscle. The innominatum and superior portion of the femur were removed, and, together with the excised fragments, were forwarded to the Army Medical Museum, and numbered 10 of the Surgical Section. A view of the specimen is given in the accompanying wood-cut, (FIG. 7.)

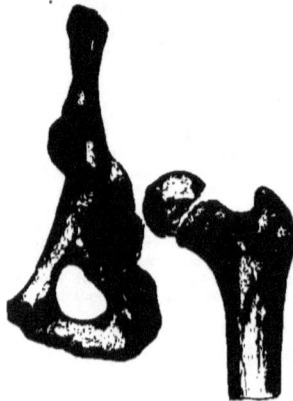

FIG. 7.—Transverse gunshot fracture of the anatomical neck of the left femur, with caries.—Spec. 10, Sect. 1, A. M. M.

The notes of the next case are also taken from a special report by Assistant Surgeon J. S. Billings, U. S. A. The operation is not referred to in the hospital register, and there were so many wounded with fractures of the upper extremity of the femur in the hospital at the period referred to that it has been impracticable to identify this case. A list of three names of patients with fractures in the trochanteric region of the left femur who were under treatment at Cliffburne Hospital, after the second battle of Bull Run, was submitted to Dr. Billings, who replied: "I cannot identify my patient with either of the names given. Should I select one, it would be 'Best,' for my patient was a German. I am very sure the dates in *Circular* No. 6 are correct. I had nothing to do with the records of the hospital at the time, Surgeon A. Bryant, U. S. V., being in charge. The man had no

---

[*] *Catalogue of the Surgical Section of the Army Medical Museum*, p. 245. *Circular* No. 6, S. G. O., 1865, p. 62, CASE 3.

symptoms of pyæmia, and everything about the wound and incision was normal, the latter having nearly healed. The discharge, to within twenty-four hours of his death, was what is called 'laudable' pus. I was very sure he would recover; and still believe that he would have done so but for some improper diet which was given to him:"*

CASE XXXVI.—A private soldier of General Pope's Army of Virginia, was admitted on September 2d, 1862, to the Cliffburne Hospital at Washington, D. C., with a gunshot fracture of the neck of the left femur, received at the second Bull Run battle, on August 29th, 1862. A conoidal musket ball had entered the left hip directly over the trochanter major and embedded itself in the neck of the femur. The trochanter major and the neck of the femur were split and comminuted, but the head was uninjured. The patient had suffered greatly from the journey from the battle-field to the hospital, and was prostrated by diarrhœa and malarial complications. The tissues about the hip-joint were but slightly swollen, and the wound discharged healthy pus. It was decided that excision was the most hopeful resource, and on September 4th, 1862, Assistant Surgeon John S. Billings, U. S. Army, proceeded to operate. Chloroform was administered, and a straight incision was made over the trochanter major, and the head and fragments of the neck were removed. The shaft of the femur was then divided by a chain saw at the level of the trochanter minor. The patient reacted well from the operation. He was placed on a fracture bed, and extension by means of a weight was made on the injured limb. The diarrhœa increased in severity despite all treatment, and the patient succumbed, exhausted, on September 24th, 1862, twenty days after the operation.

The two following cases are compiled from special reports furnished at the time by the operator, Assistant Surgeon B. A. Clements, U. S. A.† Dr. Clements (now Surgeon and Brevet Lieutenant Colonel U. S. A.) has had the kindness lately to communicate his "present estimate of the safety and value of the operation of excision of the head of the femur for gunshot wounds," as follows: "My actual experience of the operation is limited to the two cases recorded in the Circular No. 6, from the Surgeon General's Office, 1865, but the same conviction of the superior value of the operation, as compared with amputation at the joint, which induced me to perform those operations, still remains. After an examination of the records of such cases as are accessible to me, and a careful consideration of the operation for several years before I had occasion to perform it, I can say that I regard it as a safe operation, attended with few difficulties in its execution, and that its value as a means of saving life cannot, in my judgment, be diminished by a comparison with amputation at the joint I think it may be performed, in properly selected cases, with a confidence which the slender success thus far gained should not greatly shake, and I indulge the hope that eventually it will materially diminish the mortality of gunshot wounds of the hip-joint:"

FIG. 8.—Comminution of the head and neck of the right femur by a musket ball. Spec. 328, Sect. I, A. M. M.

CASE XXXVII.—Private Charles E. Marston, Co. F, 1st Massachusetts Volunteers, aged 19 years, a pale and delicate boy, was wounded at the 2d battle of Bull Run, August 30th, 1862. He was admitted to the College Hospital, Georgetown, D. C., on September 6th, having laid on the battle-field several days, and then moved in an ambulance wagon thirty miles over very rough roads. An examination revealed a large bullet wound an inch anterior to and on a line with the right trochanter major, with great comminution of the head and neck of the femur. The limb was shortened, and the foot was everted. The circumference of the limb exceeded that of its fellow by half. The pulse was at 112, and of moderate volume. The tongue was rather dry. The patient suffered little pain. The general condition was not promising, and yet not very bad. Excision of the fractured bone was decided upon, and, on the 27th September, Assistant Surgeon B. A. Clements, U. S. Army, assisted by Assistant Surgeon Charles H. Alden, U. S. A., and the surgical staff of the hospital, proceeded to perform the operation. Chloroform was administered, and a slightly-curved incision five inches in length was made on the outside of the thigh, the shot hole in the middle of the incision, and the trochanters and neck were thus exposed. The neck was crushed into about forty fragments, which were extracted. The head was also much broken, and the round ligament was absorbed or destroyed, so that exarticulation was easy. The roughened portion of the neck, at its attachment with the trochanter, was sawn off with a small chain saw. The

---

* The case is cited in the Surgical Report in Circular No. 6, S. G. O., 1865, p. 64, CASE 6.

† The cases are concisely recorded in the Surgical Report in Circular No. 6, S. G. O., 1865, p. 64, and in the Catalogue of the Surgical Section of the Army Medical Museum, pp. 244, 245.

missile, a conoidal musket ball, was found on the inner side of the thigh at the bottom of a large cavity, and was removed with difficulty. After thoroughly syringing the wound and removing the powdered bone, the wound was closed by silver sutures, except at the bullet hole, and sand bags were placed to keep the limb in position. Slight extension was made by a weight to the foot. The patient expressed himself as relieved by the operation, and he slept well that night. On the following day his pulse had risen to 128, and the discharge from the wound was very copious, thin, and brown. On September 29th, his pulse was still quick and feeble, and his tongue dry, and, though he took nourishment well, and was free from distress, he gradually sank. He died on September 30th, 1862, at half past eight o'clock A. M., three days after the operation. At the autopsy, on October 1st, made by Acting Assistant Surgeon G. K. Smith, the wound made by the operation was found to be filled with very offensive pus. The upper end of the shaft of the femur was found to be diseased on its posterior surface near the trochanter minor, and the periosteum was loosened from the bone for some distance above and below this point. The fracture of the ischium, which was noticed at the operation, extended obliquely upwards and backwards from the lower border of the acetabulum, terminating in the sciatic notch, about an inch and a quarter above the spine of the ischium. The lower half of the acetabulum had been broken into several fragments, which were held in position by the cotyloid ligament. The excised portions of the femur were forwarded to the Army Medical Museum, and are numbered 328 in the Surgical Section. They are represented in the preceding wood-cut. (FIG. 8.)

CASE XXXVIII.—Private F. Macblin, 11th Pennsylvania Volunteers, a robust man, was wounded at the second battle of Bull Run, August 30th, 1862. He laid on the field several days, and was then transported thirty miles in a wagon to the Warehouse Hospital, Georgetown, D. C., where he was received on September 8th. A musket ball had entered the right buttock and emerged an inch and a half below and within the anterior superior spinous process of the ilium. The limb was shortened, and the foot was completely everted. Any movement of the limb gave excessive pain. A thorough examination was made under the influence of chloroform, and showed that the neck of the bone was comminuted and the shaft uninjured. The case was considered to be a particularly favorable one for the operation of excision of the broken fragments, and the general condition of the patient was such as to offer some hopes of its success, and it was determined to do the operation on the following day. But the surgeon in charge of the hospital was confined to his bed by illness, and was unable to see the patient for five days. On September 13th, the general condition of the latter was less favorable. His pulse was at 130, quick and weak, and his tongue was dry. He was placed under very careful nursing, and beef essence, brandy, eggs, and milk, and other concentrated nourishment and stimulants, were administered ad libitum. By the 20th of September this treatment had produced a slight improvement, and though he was still in a very unfavorable condition for an operation, it was determined, on consultation, that an excision might afford the patient relief from the constant pain he suffered, and that it could not greatly depress him. At noon on September 20th, accordingly, Assistant Surgeon B. A. Clements, U. S. A., assisted by Dr. George K. Smith, of Brooklyn, and the surgical staff of the hospital, performed the operation. An incision five inches in length was made from a point two inches behind and an inch below the anterior superior process of the ilium, downwards over the prominence of the trochanter major. The incision exposed the parts freely, and the muscular insertions being divided, and several small loose fragments of bone removed, the irregular broken extremity of the shaft, at its junction with the neck and tip of the great trochanter, was excised by a chain saw. The remnant of the capsular ligament and the round ligament were now cut, the joint being opened from below and in front with a probe-pointed bistoury. A blade of a long bullet forceps was then introduced as a lever and the head was disarticulated by gently prying it out of the cotyloid cavity. These steps in the operation were facilitated by rotating the trochanter outwards, and by lifting the extensor muscles by a metallic retractor. A small vessel was tied at the upper end and another at the lower end of the wound. The wound was well washed out by means of a syringe, and a few stitches were applied, the middle portion of the wound being left open. The limb was suspended in a Smith's anterior wire splint. The patient appeared to rally from the operation satisfactorily. At nine P. M. he was free from pain ; his pulse was at 136, and the skin was cool and natural. He had slept tranquilly. On the following day, September 21st, his pulse was 120 and very feeble. The discharge from the wound was dark and thin, and copious. His countenance was placid, the pinched, distressed expression it had worn having disappeared. But in the afternoon he sank rapidly, and died in the evening, thirty-six hours after the operation, September 21st, 1862. The specimen is No. 329 of the Surgical Section of the Army Medical Museum, and is represented in the accompanying wood-cut. (FIG. 9.)

FIG. 9.—Head of right femur, excised on account of a gun-shot fracture, Spec. 329, Sect. 1, A.M.M.

The history of the next case is drawn up from materials furnished in a special report by the operator, Assistant Surgeon J. H. Bill, U. S. Army, and from the memoranda forwarded with the pathological preparation. It is the only instance among the intermediate operations in which the injury was inflicted by a large projectile. From the limited nature of the fracture, the case was particularly well adapted to the operation:

CASE XXXIX.—Private Cornelius Callaghan, Co. G, 2d Delaware Volunteers, was wounded in the left hip by a fragment of shell, at the battle of Antietam, September 17th, 1862. On September 19th, he was admitted to Hospital No. 3, at Frederick, Maryland. He was placed under the influence of chloroform, and an examination of the wound was made by Assistant Surgeons Bill and Colton, U. S. A. The wound being enlarged sufficiently to admit of free exploration, the trochanteric region of the femur was found to be badly comminuted, the great trochanter entirely detached and drawn backwards by the action of the gluteus, while fissures extended up the neck within the capsular ligament. No fissures extended below the trochanter minor. The patient's general condition was good, and all the circumstances being favorable to such an attempt, it was determined, in a consultation of the medical staff, and with the approval of Medical Inspector Coolidge and Surgeon Milhau, U. S. A., that the injured portions of bone should be excised. On September 29th, Assistant Surgeon J. H. Bill made an incision from the wound three inches downwards in the course of the shaft, and another three inches long curving upwards and inwards from the wound to a point a little below the anterior superior spinous process of the ilium. The muscular attachments being dissected aside, a chain saw was passed around the shaft of the femur and made to divide it just below the trochanter minor. The head of the bone was then disarticulated. The edges of the wound were united by six sutures, and adhesive plasters and water dressings were applied. The limb was kept in position by pillows, without the use of splints. A full dose of morphin was given, and light but nourishing food was directed. On the following day the patient was quite comfortable. His pulse, which was at 100 before the operation, now beat 120. He was ordered a diet of beef tea, eggs, and oysters, with a small amount of wine. At midnight he was sleeping quietly. On October 1st he was still cheerful. His pulse was very compressible at 120, and he was sweating profusely. The thigh was swollen and painful. A draught of aromatic sulphuric acid with a little quinia was added to his prescriptions, and the allowance of wine was increased. The prognosis was now very unfavorable. On October 2d the sweating was checked, but diarrhœa had supervened. The pulse was still softer and more frequent, and suppuration had commenced. At midnight the patient was attacked with vomiting and hiccough. On October 3d the vomiting persisted, and the sweating was renewed. This state continued through the day and night, the patient sinking gradually. He died at 3 p. m. of October 4th, 1862. The pathological preparation is deposited in the Army Medical Museum, and is numbered 840 of the Surgical Section. An anterior view of it is presented at page 247 of the catalogue of the Surgical Section, and a posterior view in the accompanying wood-cut. (FIG. 10.)

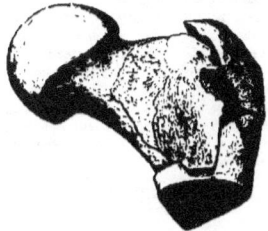

FIG. 10.—Neck and trochanters of right femur shattered by a fragment of shell, and excised.—*Spec.* 840, Sect. I, A. M. M.

In the tabular statement of operations on the monthly report of Hospital No. 21, Nashville, Tennessee, for January, 1863, there is an entry of a fatal secondary excision of the head of the right femur, with the remark by Surgeon Edward Sennet, 94th Ohio Volunteers, who signs the report, that, "In our opinion, the case of resection at the hip might have been successful if an early operation had been resorted to on the field, as the operation was very successfully performed, and accompanied by comparativly little hæmorrhage." After protracted inquiries, some particulars of the case have been learned from Dr. M. M. Hooton, of Centralia, Illinois, formerly Surgeon 56th Illinois Volunteers:

CASE XL.—"I have the honor to report," writes Dr. Hooton, "that I did not take charge of Hospital 21 until March 10th, 1863, and did not treat the case referred to. I was somewhat conversant with its history, however, and will give it to you as I received it from Surgeon Sennet, 94th Ohio Volunteers, whom I relieved. The wound was received at the battle of Stone River or Murfreesboro', the last of December, 1862, or January 1, 1863. Resection of the upper part of the shaft, including the shattered trochanters, was performed on the field. The head of the bone was left in the cotyloid cavity. A few days subsequently the patient was sent by rail to Nashville, a distance of twenty miles. Notwithstanding the fatigue of the long journey, his general condition was very satisfactory, and the wound looked well. Part of the incision healed; but there was a profuse discharge from the lower part, and gradually this discharge became thin and ichorous. About January 21st it was determined to reopen the wound. It was found that the head of the bone was carious throughout its whole extent. It was therefore removed. After this operation, which was accomplished without hæmorrhage or other untoward circumstance, the limb swelled very much, irritative fever set in, and the patient died exhausted in a few days." [The records of the hospital identify this patient as Sergeant Dallas W. Hade, Co. H, 101st Ohio Volunteers, wounded at Stone River, December 31, 1862; died January 31, 1863.]

The following case was concisely recorded on the register of St Louis Hospital, New Orleans, but a full account of it has been communicated by Dr. Francis Bacon, Professor of Surgery at Yale College, who, while Surgeon of Volunteers, was in charge of the hospital. Professor Bacon mentions "that the excised portions of bone, together with a considerable number of other valuable specimens intended for the Surgeon General's Office, were lost through the unaccountable carelessness of the medical officer to whom they were entrusted:"

CASE XLI.—Private John Miller, Co. E, 162d New York Volunteers, aged about 33 years, a robust, phlegmatic German, in good health, was wounded on June 14th, 1863, in the assault on Port Hudson, Louisiana, by a conoidal musket ball, which passed through the upper portion of the thigh, breaking the neck of the left femur transversely, and splitting it longitudinally, but without great comminution. The pelvis was uninjured, and there was no serious damage to the soft tissues. The patient was conveyed to New Orleans on a hospital transport, and was placed in the St. Louis Hospital, on June 16th. It was determined that excision of the injured bone was advisable, and the officer in charge of the hospital, Surgeon F. Bacon, U. S. Vols., being confined to his bed by illness, the operation was performed on July 8th, by Assistant Surgeon George W. Avery, 9th Connecticut Volunteers. The head, neck, and great trochanter were removed in the usual way, through a single straight incision of moderate extent over the trochanter and in a line with the axis of the femur. There was an immaterial loss of blood. The state of the tissues involved and the constitutional condition of the patient were as good as might be. Dr. Bacon remarked that the operation was well and rapidly performed. The patient rallied from it promptly, and afterwards received the most assiduous care. His progress was very favorable until the early part of September. The wound had nearly healed, and Surgeon Bacon and his assistants were very hopeful of the patient's recovery. But in September the weather became most oppressively hot, and the patient steadily declined. The wound assumed a bad appearance, discharging copiously, and despite sustaining measures, the patient sank and died from exhaustion on September 21st, 1862, seventy-five days after the operation. Dr. Bacon examined the fragments of bone removed, and found the periosteum adherent throughout the larger pieces.

The following case is recorded in the Proceedings of the Pathological Society of Philadelphia:*

CASE XLII.—"Private Michael Welsh, Co. H, 10th Regiment, Kentucky Volunteers, aged 40 years, was struck by a conoidal ball in the region of the left great trochanter, at the battle of Chickamauga, September 20, 1863, and at once conveyed to a field hospital, which soon after fell into the hands of the enemy. Ten days subsequently, he was brought to Chattanooga, Tennessee; and during the period of his captivity the only nourishment that he received was a small portion of corn-meal gruel daily. Having lost his blankets, he also suffered much from cold, and had contracted a rather severe bronchial inflammation. On the 1st of October, he was admitted into the general field hospital of the 14th Army Corps, when a conoidal musket ball was removed from among a mass of small fragments of the neck of the femur, the ball having entered just anteriorly to the great trochanter. Two days subsequently, Surgeon F. H. Gross, U. S. Volunteers, carried a curvilinear incision, with its convexity presenting forwards, and including the opening made by the ball, from above downwards, and excised the head of the femur along with the attached greater portion of the lower surface of the neck of the bone. Many fragments were removed with the forceps; but as the trochanters were not involved in the injury, the remaining sharp portions of the neck were trimmed off close to the intertrochanteric lines, which completed the procedure. No ligatures were required. The man bore the operation well, and the limb was placed in a comfortable position. From the date of the operation up to October 20th, the man did very well, in spite of his enfeebled condition and bronchial trouble, when the discharge from the wound became sanious. On the same night he had a chill and was delirious, and the pulse was very feeble and frequent. On the 25th October his condition is thus described: Pulse 125 and very feeble; tongue dry and red; had a natural alvine evacuation. At 9 A. M. hæmorrhage recurred from the wound, which was arrested by injecting a solution of sulphate of iron. Stimulants freely administered, but the man grew more and more feeble, and expired at 2 P. M. No post mortem examination was held, but death was evidently due to pyæmia. The specimen shows that about one-fifth of the head at its upper aspect has been shot away, together with the entire upper surface of the neck, about one-half of the anterior and posterior surfaces and the lower border of the neck remaining." The pathological specimen, illustrated by the accompanying wood-cut, (FIG. 11,) was forwarded to the Army Medical Museum, June 3, 1868, by the operator.

FIG. 11.—Fracture of the neck of the left femur by a musket-ball.—Spec. 5442, Sect. I, A. M. M.

The following account of a successful intermediate excision of the head and upper extremity of the femur in a case of gunshot fracture of the trochanteric region, is compiled from the account recorded in the Case-book of Hospital No. 4, Richmond, Virginia, from the narrative published by the operator,† and from data communicated by Assistant Surgeon Latimer, C. S. A., who treated the case after the operation. The appearance of the patient, seven months after the operation, is exhibited in the accompanying plate, in the figure on the left hand, drawn from a photograph presented to the compiler by Dr. Latimer, and now deposited in the Army Medical Museum. It does not appear that the fracture implicated the joint primarily, but the result was so satisfactory that one is not inclined to inquire if exarticulation was indispensably necessary:

CASE XLIII.—Lieutenant James M. Jarrett, Co. C, 15th North Carolina Confederate Regiment, a spare man, 28 years of age, of medium size, of fair complexion, of temperate habits, and good general health, was wounded at the affair at Bristow Station, Virginia, on October 14th, 1863, by a conoidal musket-ball, which entered in front and a little to the outside of the median line of the left thigh, two inches below Ponpart's ligament, shattered the femur, and made its exit posteriorly at the

---

* *American Journal of the Medical Sciences*, Vol. LV, p. 410. Some remarks on the case may be found in the same Journal, Vol. LVI, p. 130. A memorandum of the case was recorded in the casualty lists forwarded by Surgeon Glover Perin, U. S. A., the Medical Director of Gen. Rosecrans's army. Dr. Perin replied to inquires from this office that no particulars of the operation could be ascertained. Hence, the case was not included in the tabular statement in Circular 6, S. G. O., 1865.

† *Confederate States Medical and Surgical Journal*, Vol. I, p. 5, January, 1864.

outer part of the limb, the wound of exit being on a rather higher level than that of entrance. The fracture was dressed with a straight splint, and the wounded officer was placed in an ambulance and transported over rough roads to Richmond, a distance of one hundred and sixty miles. On October 20th, he was admitted to Hospital No. 4, at Richmond, in an exhausted state, and was placed in charge of Surgeon James B. Read, P. A. C. S. He complained of extreme pain upon any movement of the limb, and was unwilling to submit to an examination of the injury unless insensibility was induced. Chloroform having been administered, the splints and soiled bandages were removed, and the limb was placed in an easy position on pillows. Water dressings were applied to the wounds, and an opiate was administered. For the next three weeks the progress of the case was very unfavorable. The wound of exit discharged copiously unwholesome thin pus, mixed with blood and bubbles of fœtid gas and small bits of dead bone. The pulse was quick and small, the tongue red and dry. There was a tendency to diarrhœa, and night sweats frequently recurred. On November 9th, as the patient was steadily growing worse, a consultation was asked for, and Surgeons C. B. Gibson and M. Michel saw the case with Dr. Reed. It was decided that the circumstances called for operative interference, and that an excision and exarticulation of the head of the femur offered the best prospect of recovery. The patient cheerfully acceded to this proposition, desiring in any way to be relieved of the anguish he endured upon the slightest movement. On the afternoon of November 9th, the patient was anæsthetized, and then placed on his right side on the operating table. A straight incision was commenced two inches below the posterior or exit wound, and was carried through this to the great trochanter, and thence upward for two inches further, thus making a wound about seven inches in length. This incision being carried down to the bone, the upper end of the shaft of the femur was examined and was found to be jagged and pointed, a thin laminæ of bone about three inches long being broken off from its anterior aspect. The lower fragment was projected through the incision by adducting the limb and pushing the knee upwards, and it was sawn about two inches below its upper sharp extremity. The trochanteric portion of the femur was then sought for, and was found drawn upwards by the psoas and iliacus internus. Its extremity was seized by the lion forceps and drawn downwards, and the attachments of these muscles to the lesser trochanter was divided. To dislocate the head of the femur so as to admit of the division of the round ligament was a work of great difficulty. It was finally accomplished, partially by twisting the neck of the bone, and the head was exarticulated. The appearance of the principal portion of this excised bone is shown in the annexed wood-cut. (FIG. 12.) Several large detached fragments and spiculæ were then extracted, and other closely-attached bits of bone were enucleated by the finger-nail. The wound was cleansed and then closed by sutures and adhesive strips. Dry dressings were applied, and the thigh was fixed by a large straight splint. The patient was ordered two grains of opium and a drachm and a quarter of brandy every two hours. At bed-time, the patient was quite comfortable, and could shift his position slightly without pain; his pulse was at 120. On the following day, anodynes were given at greater intervals. On November 11th they were omitted, except at bed-time, and nutritious diet was ordered. The next day the sutures were removed; the wound began to discharge laudable pus in small quantity. The case progressed without any untoward complication. On December 9th the wound was healed, except at two points, connected by sinuses leading to the cotyloid cavity and the upper end of the shaft. The patient had gained flesh and strength; his pulse was full and strong at 76; his appetite and digestion were natural; he slept well, was cheerful, and did not complain of pain. The limb was shortened five inches. The daily discharge of pus was less than half an ounce. Two weeks subsequently the wound was entirely united; the cicatrix was firm; the patient could move about his bed without inconvenience; there was no pain on pressure about the muscles of the injured part. The patient was now removed to his home in North Carolina, and was soon able to move about on crutches. In September, 1864, ten months after the operation, he reported that he was able to bear considerable weight on the limb, and that he had discarded his crutches and walked about in a high-heeled boot with the aid of a cane.

FIG. 12.—Upper third of left femur shattered by a musket-ball.—(From a wood-cut in Dr. Read's paper.)

The next case may be regarded, perhaps, as having furnished the most satisfactory and successful result which has as yet rewarded the efforts of military surgeons to establish excision as a recognized resource in fractures involving the head or neck of the femur. Here the precarious alternative of amputation at the hip-joint could alone be considered, for the neck and trochanter were completely smashed, and the ball remained in the articulation. Temporization, under such circumstances, was simply to abandon the patient to inevitable death. Amputation in the intermediate period would have been a very hopeless enterprise. Fortunately, excision was determined on, and rapidly and skilfully accomplished. The man has survived the operation five years, and is now in excellent health and in the possession of a very useful limb. He walks long distances without a cane or other assistance, ascending lofty ladders, carrying heavy burthens, and performing the various duties of a day laborer. The history of his case settles the question of the alleged absolute inutility of the lower extremity after decapitation of the head of the femur for

INTERMEDIATE OPERATIONS. 41

injury. The history here given is compiled from a great variety of sources, from letters from the operator, Dr. George A. Mursick, now of Harlem, New York, Acting Assistant Surgeon J. B. Garland, Surgeon General L. W. Oakley, of New Jersey, Dr. E. D. Hudson, Dr. C. A. McCall, Dr. W. Pierson, Jr., Dr. Stephen Wickes, and from the register of Stanton Hospital:*

CASE XLIV.—Private Hugh Wright, Co. G, 87th New York Volunteers, aged 28 years, a robust, healthy man, was wounded on May 5th, 1864, at the battle of the Wilderness, by a conoidal musket ball, which entered the right thigh an inch within the track of the femoral vessels and two inches below Poupart's ligament, passed backwards and outwards, shattering the neck and trochanters of the femur, and, having been greatly flattened and distorted by the impact, it lodged amid the fragments of bone. The precise direction of the fracture is indicated in the accompanying wood-cut. (FIG. 13.)† He states that, after being wounded, he was carried to the rear by a number of his companions, and, in the evening, was taken to the field hospital of the 2d Division of the 2d Corps. Here he remained for three days. He was then sent in an ambulance to Fredericksburg, and placed in a temporary hospital. He states that his wound was repeatedly examined by different surgeons, but that no treatment was instituted beyond the application of a compress dipped in cold water to the wound. He was transferred, after a fortnight, on a hospital steamer, to Washington, and on May 25th he was admitted to Stanton Hospital, then under the charge of Surgeon D. B. Wilson, U. S. Vols. He was placed in Ward 6, under the care of Acting Assistant Surgeon J. B. Garland, who has communicated to the Surgeon General a special report of the case. The injured limb was swollen, everted, and shortened. Pus had accumulated in the tissues about the hip. Notwithstanding the gravity of the injury, the patient's constitutional condition is said to have been hopeful. On exploring the wound with the finger, the patient being under the influence of chloroform, detached fragments of bone could be felt. On consultation with Acting Assistant Surgeon George A. Mursick, an operation was decided upon, for the purpose of removing these loose sequestra and the missile, if it could be found. On May 27th, the patient was rendered insensible by the inhalation of sulphuric ether, and Dr. Mursick, assisted by Dr. Garland and others of the hospital staff, made a straight incision, commencing above and behind the trochanter major and carried downwards in the axis of the thigh. It was not in contemplation, at the beginning of the operation, Dr. Garland states, to exarticulate the head of the femur; but when the muscular attachments were divided, and the full extent of the fracture was revealed, and the joint was found distended with pus, it was at once determined to make a formal excision. The fragments of the neck were extracted piecemeal. The ball was found lying behind the neck, and was extracted. The capsular ligament being freely incised, a bistoury was inserted into the cotyloid cavity and the round ligament was severed, and the head of the femur was removed without difficulty. The jagged upper extremity of the shaft of the femur was then turned out of the wound by carrying the limb over the opposite knee, and was smoothed off by a chain saw. There was but trifling hæmorrhage, and no ligatures were required. The wound was carefully cleansed, dressed with dry lint, and left to heal by granulation. To keep the limb in position, long sand bags were laid on either side of it, and moderate extension was made by means of a weight attached to the leg and suspended from the foot of the bed. At night he took a grain of sulphate of morphia in a draught. The operation seemed to depress him very much, and reaction was slow. He passed a restless night, manifesting much nervous excitement. In the morning, his pulse was feeble and frequent; his tongue dry and furred. He was ordered an ounce of brandy every three hours, a grain of opium every four hours, and as much beef tea and concentrated nourishment as he could take. On May 29th his general condition had much improved; the pulse was less frequent and stronger. There was free suppuration. The wound was dressed with a weak solution of permanganate of potassa. On June 1st he continued to improve, the wound looked well, and the character of the suppuration was good. The amount of brandy was reduced to four ounces daily. On August 1st he was still doing well. The wound was filled up with granulations from the bottom, with the exception of a sinus leading to the bone. It continued to suppurate quite freely, and some small pieces of dead bone had come away with the discharges. He had gained in flesh, and his health and spirits were good. On August 22d, he attempted, for the first time, to sit up in bed, but, owing to the rigidity of the parts and the agglutination of the muscles, the pain caused by the sitting posture was so severe that he was compelled to lie down again. Cold evaporating lotions were applied to the thigh. On August 23d, the upper part of the thigh swelled and was painful, and the discharge was increased in quantity. On August 27th, the swelling of the thigh had increased, the discharge from the wound was very free, thin, and flaky, and the surrounding surface was glazed and doughy to the touch. The wound of entrance had re-opened and discharged thin pus. An abscess formed on the inner side of the thigh, and about four ounces of thin flaky pus was discharged. The patient was restless. He was ordered twenty drops of the tincture of the sesquichloride of iron every six hours, with stimulants and nutritious diet. On September 1st, the swelling and inflammation of the thigh continued. He complained of nausea and want of appetite. An abscess formed on the outer side of the thigh. On September 5th, the

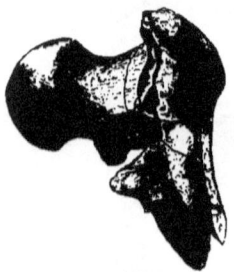

FIG. 13.—Comminution of the neck and trochanters of the right femur by a conoidal musket ball.—*Spec.* 3375, Sect. I, A. M. M.

---

* The case is reported in *Circular* No. 6, S. G. O., 1865, p. 68, and in the *New York Medical Journal*. Vol 1, p. 424.

† An anterior view of this specimen is printed in the surgical report in *Circular* No. 6, S. G. O., 1865, p. 74, and in the *Catalogue of the Surgical Section of the Army Medical Museum*, p. 246.

abscess was incised, and a large quantity of thin, flaky, and offensive pus was evacuated. He had an irritable stomach, and Hoffman's anodyne was administered. On September 6th, the edges of the incision in the abscess were beginning to slough, and nitric acid was freely applied. On September 9th he had diarrhœa. Ten grains of subnitrate of bismuth and a grain of opium were given every six hours. On September 11th the slough had cleaned off, and the edges of the wound were covered with florid granulations. The discharge from the abscess is free, and the patient still has diarrhœa. Five grains of tannic acid and one grain of opium were administered thrice daily. On September 13th the diarrhœa had nearly ceased. The patient's general condition had improved, and the wound looked well, though the suppuration was still copious, and had improved in quality. On September 25th a large ring-shaped exfoliation from the upper end of the femur was removed through the wound of operation. On September 26th another exfoliation was removed. On October 6th, 1864, Private Wright was discharged from the military service of the United States, on account of the expiration of his term of service. On October 7th the swelling of the thigh had subsided; the discharge from the wounds had much diminished in quantity, and presented the appearance of laudable pus; the diarrhœa had ceased, and his general condition was much improved, he being able to sit up in bed. On October 30th a sinus communicating with necrosed bone opened on the outside of the thigh. On November 24th the limb again became inflamed and swollen. Cold applications were made to it, and tonics and stimulants were given internally. In the latter part of December, another abscess formed on the outer side of the thigh. When this was opened the swelling and inflammation subsided. He continued to do well until February 6th, 1865, when another abscess formed in the lower third of the thigh, on the outer side. This was incised and the pus evacuated. Several pieces of dead bone came away with the discharges from the wound of operation. About the middle of March, 1865, he was able to get out of bed, and to walk about the ward on crutches. Soon after, in getting out of bed, "he let his leg fall and hurt it." This accident was followed by inflammation and swelling of the thigh, and an abscess in the lower third of it, on the inner side. This was incised, and a small quantity of pus was evacuated. He was now attacked with erysipelas, which extended from the knee to the hip. This was combated with tonics and stimulants, such as iron and quinine, and rapidly disappeared. From this time he did well, taking daily exercise about the hospital on crutches. On April 17th he was transferred to the Ward Hospital, at Newark, New Jersey. His general health was tolerably good. He could not bear much weight on his limb, and inflammation and abscesses followed any unusual exertion. He remained at this hospital until May 6th, 1865, when it was reported that he "eloped." As a discharged soldier, he was no longer under military authority, and of course was at liberty to go. For many months, though diligent inquiries were made by this office, he could not be traced; but, in July, 1866, Surgeon General L. W. Oakley, of New Jersey, transmitted a letter from Dr. W. Pierson, of Orange, New Jersey, which stated that Wright had entered the almshouse at that place in June, 1865, and had remained there until the following spring, under Dr. Pierson's professional care. At first, the mutilated limb had been enormously swollen from œdema, and there was an ichorous discharge from a sinus near the hip-joint. With careful bandaging, the œdema gradually disappeared. In the spring of 1866, Wright left the almshouse, and engaged himself as a laborer on a farm. He wore, Dr. Pierson reported, a cork-soled shoe of his own manufacture. The limb was shortened precisely five inches. The circumference of the injured thigh at the highest part was one inch less than that of its fellow. He walked well without crutch or cane, bearing his full weight on the mutilated limb. There was quite free motion at the hip, but little at the knee. There were no open fistulæ, and no tenderness about any of the cicatrices. Dr. Stephen Wickes, of Orange, reported, late in the summer, that Wright was in good health, though rather intemperate; that he worked daily at light tasks, and was even able to mow grass. He commonly walked with a cane. According to the measurement of Dr. Wickes, the limb was shortened four and three-quarter inches. About this period, Dr. Mursick, the operator in the case, discovered his former patient, and examined him. He found the resected end of the femur firmly attached to the pelvis by ligamentous tissue an inch and a half long. The agglutination of the muscular sheaths had nearly disappeared. The limb was quite under control. The man could flex and extend it slightly, and adduct to a limited extent; the power of rotating and abducting was lost. Motion at the knee was quite restricted, on account of the thickening and consolidation of the surrounding tissues resulting from inflammation. He stated that latterly the improvement in his limb had been very decided; that when he first commenced to walk, the limb felt like a weight attached to the body; this sensation had entirely disappeared. On October 19th, 1867, Dr. Mursick again examined Wright, and took him to New York, and had his photograph taken. The negative is preserved at the Army Medical Museum. The photograph is No. 183 of the Surgical Series. It is copied in Figure 2, Plate II, of this Report. (Ante, op. p. 38.) At this period, Wright reported that his limb had given him no trouble since the sinuses healed, in May, 1865, and that it sufficed for all purposes of locomotion. He stood on it very firmly, and could move it in any direction with an easy, swinging motion. He had been engaged for a year and a half as a farm hand, and was employed at that time as a wood-chopper. He had for a short time earned larger wages as a hod bearer, and had climbed high ladders with a heavy hod of bricks on his shoulders; but he found this avocation too fatiguing. His general health and physical condition were good. The knee-joint continued quite stiff. It could be flexed to about quarter, perhaps, of the normal extent. When he walked, the rounded upper extremity of the femur played up and down on the dorsum of the ilium over a space of an inch and a half. In November, 1868, Dr. Mursick again examined Wright, and reported on his condition. The utility of his limb had augmented during the twelve months that had elapsed since the last examination. The attachment of the femur to the pelvis was strong; the cicatrices were firm and healthy. All the movements of the thigh were performed with almost as much facility as in the normal state; rotation, even, as well as flexion, extension, adduction, and abduction. His general health was good. He still worked as a day laborer.

Dr. W. F. Richardson remarks that CASES XVII and XIX, included in the category of primary excisions, and the case now to be described, comprise "all the coxo-femoral resections of Dr. Charles Bell Gibson, at the Receiving and Wayside Hospital, No. 9, Richmond, Virginia." The fracture did not involve the articulation:

INTERMEDIATE OPERATIONS.

CASE XLV.—Private Marsella Smith, Co. F, 38th Virginia Infantry, a robust middle-aged man, was wounded near Spottsylvania, early in the morning of May 10th, 1864, by a conoidal musket ball, which entered at the upper posterior part of the left thigh, passed through the perinæum without injuring the urethra, and through the soft parts of the right hip. He was sent to Richmond by rail, and was admitted to Hospital No. 9, otherwise known as the Receiving and Wayside Hospital, on the following morning. On May 12th he was placed under chloroform and the wound was thoroughly explored. The limb was everted and shortened and swollen; there was crepitus on rotation. The fracture appeared to be limited to the great trochanter and neck. It was supposed that the urethra was divided; but this was afterwards proved not to be the case. Surgeon Charles Bell Gibson, C. S. A., decided to excise the injured bone, and the operation was performed forty-eight hours after the reception of the injury, the head, neck, and two inches of the shaft being removed. It is stated that the effects of the chloroform were unfavorable. On the following day "patient commenced sinking at an early hour, and continued growing more and more feeble until three o'clock P. M., when death ended his sufferings." May 13th, 1864. The excised portion of the femur was preserved by Dr. Richardson, and has been presented by him to the Army Medical Museum. It is represented in the accompanying drawing, (FIG. 14,) and shows that the injury to the bone was altogether external to the hip-joint.

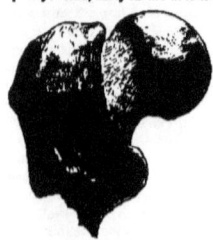

FIG. 14.—Oblique gunshot fracture of the upper portion of the shaft of the left femur. *Spec.* 5500, Sect. I, A.M.M.

The abstract of the next case is compiled from a special report from the operator, Assistant Surgeon Alexander Ingram, U. S. A. That a slender boy, after such a preparation for an operation as a journey in an army wagon from Spottsylvania to Belle Plain, should have speedily succumbed under almost any surgical proceeding, was not surprising. Unhappily, temporization was as fatal as interference.

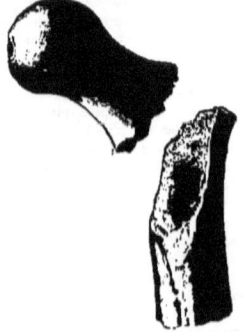

CASE XLVI.—Private Charles C. Cleaver, Co. C, 2d U. S. Infantry, aged 18 years, was wounded at the battle of Spottsylvania on May 12th, 1864, by a conoidal musket ball, which fractured the neck and trochanters of the right femur. He was transferred to Fredericksburg and thence to Belle Plain in a wagon, and thence on a steam hospital transport to Washington. On May 18th he arrived at Washington, and was admitted to Judiciary Square Hospital. The soft parts of the upper and outer part of the thigh were extensively lacerated, and pus of an ill-conditioned character was burrowing in every direction around the wound. The patient's condition was unfavorable; but it was decided that excision of the fractured portion of the femur afforded the only prospect of relief. Accordingly, on the 19th, Assistant Surgeon Alexander Ingram, U. S. A., proceeded to make a curved incision six inches in length, through which the head, neck, and four and a quarter inches of the shaft of the femur were removed. Charpie soaked with permanganate of potash was applied to the wound, and tonics and stimulants were freely given. Pyæmia supervened, and the patient died on May 23d, 1864, four days after the operation. The pathological specimen, figured in the accompanying wood-cut, (FIG. 15,) is in the Surgical Section of the Army Medical Museum.*

FIG. 15.—Fracture of the neck and trochanters of the right femur by a musket ball. *Spec.* 2919, Sect. I, A. M. M.

The next case, the notes of which are taken from the register of Judiciary Square Hospital, is another instance, probably, of what are termed "compulsory" operations. It would be difficult to picture a body of men less capable of undergoing operations than those who filled the ambulance trains that were compelled, by the exigencies of war, to make the horrible journey from the battle-fields of the Wilderness to the landing places on the Potomac. The results of the intermediate amputations and excisions performed on these unfortunates, on their arrival at Washington, are simply appalling:

CASE XLVII.—Private Alexander Ewing, Co. A, 140th Pennsylvania Volunteers, aged 30 years, was wounded at the battle of Spottsylvania, Virginia, on May 12, 1864, by a conoidal musket ball, which comminuted the upper part of the left femur. He was taken to the hospital of the 1st Division of the 2d Army Corps, and on the following day was sent to the rear in a wagon. Arriving at Belle Plain after a three-days' journey over rough roads, he was conveyed on an hospital steamer to Washington, and, on May 18th, he was admitted to Judiciary Square Hospital. There was considerable inflammation and swelling of the soft parts, and the patient was in poor health. On the following day he was anæsthetized, and Acting Assistant Surgeon

---

* In the report in *Circular* No. 6, S. G. O., 1865, Act. Asst. Surgeon J. H. Thomson is named as the operator in this case on the authority of the quarterly surgical report from Judiciary Square Hospital. The investigations of Assistant Surgeon A. A. Woodhull, U. S. A., prove that Dr. Ingram was the operator.—See *Catalogue of Surgical Section* A. M. M., p. 240.

J. F. Thompson made an incision five inches in length over the great trochanter, including in it the wound of entrance. The muscular attachments being divided, it was found that the neck was splintered, that fissures extended within the capsule, that the great trochanter was separated from the shaft, and the upper part of the shaft much comminuted. The head and fragments of the neck and trochanters were removed, and the shaft was sawn just below the trochanter minor. Ice was applied to the wound, and stimulants were freely administered. The wound assumed an unhealthy action, and the patient gradually sank and died from exhaustion on May 24th, 1864, five days after the operation. The pathological specimen was not received at the Army Medical Museum.

The next case is compiled from the quarterly reports from Mount Pleasant Hospital, and from a special report by Assistant Surgeon C. A. McCall, U. S. A. In a former report[1] the operation is ascribed to Acting Assistant Surgeon M. C. Mulford, in accordance with a statement signed by Dr. McCall himself. Dr McCall subsequently announced that this statement was erroneous, and was transmitted through inadvertence, and furnished a detailed account of this very interesting case, which, at one time, gave some promise of a successful issue.

CASE XLVIII.—Captain John Phelan, Co. A, 73d New York Volunteers, aged 22 years, received a compound comminuted fracture of the neck and upper extremity of the left femur at the battle of Spottsylvania Court House, Virginia, on May 14th, 1864. On May 16th, he was admitted to Mount Pleasant Hospital, Washington. The rapid and incessant influx of wounded was such that the attention of the overworked hospital staff was not especially drawn to his case for some time after his admission, and the delay in minutely examining the case was extended by the uncomplaining fortitude of the sufferer, who expressed his wish that the more serious cases should first be attended to, and declared that his own sufferings were comparatively slight. When, however, Acting Assistant Surgeon Mulford, the ward surgeon, proceeded on June 3d to adapt apparatus to what he supposed to be an ordinary gunshot fracture of the upper third of the thigh, he was led to apprehend that the injury extended to the coxo-femoral articulation, and requested the opinion of the surgeon in charge of the hospital, Assistant Surgeon C. A. McCall, U. S. A., as to the diagnosis and treatment. Dr. McCall immediately visited the patient, and found him to be a large, muscular, finely formed man, whose previous health had been excellent. When lying quietly in bed, he suffered but little. His appetite was good; and his strength, so far, had diminished but little. Altogether, his general condition was extraordinarily good, in view of the gravity of the injury he had sustained. The ball had entered in front, just over the point at which the profunda is given off from the left femoral artery. The aperture of entrance was small and characteristic as an entrance wound of a conoidal musket ball. The missile had passed towards the great trochanter and shattered it. Further, its course could not be ascertained at the time. Any movement of the limb caused extreme pain. Though the femur was much comminuted, Dr. McCall was not positive that the hip-joint was implicated, and, with a view to a full exploration of the injury, he directed Dr. Mulford to make a longitudinal incision three inches in length over the trochanter, to explore the parts thoroughly, and to ascertain by digital examination the condition of the articulation. If it was uninjured, Dr. Mulford was instructed to extract detached fragments of bone and foreign matters, to close the upper part of the wound, and to avail of the lower portion for drainage. In the afternoon the patient was etherized, and the exploratory incision was made, and it was found that the fracture extended to the head of the femur. It was then decided to excise the head. The patient was again rendered insensible by the inhalation of sulphuric ether, and Dr. McCall extended Dr. Mulford's incision upwards an inch or more, and then made an oblique incision across its upper extremity, as represented in the accompanying figure. (FIG. 16.) The two flaps thus marked out were reflected, and the joint was readily exposed, the round ligament divided, and the head of the femur exarticulated. The acetabulum was carefully examined, and found to be uninjured. Seven large and numerous small fragments of the neck and trochanter major were then removed, a task requiring much time and patience, many fragments being driven into the gluteal muscles, or deeply retracted by the muscles attached to the great trochanter. The fractured upper extremity of the femur was then brought out at the wound, by adducting and pushing upwards the knee of the injured limb, and all diseased tissue was removed. The periosteum was in a healthy condition quite up

FIG. 16.—Direction of the incisions in case of excision of the head of the Femur.—(From a drawing by Dr. McCall.)

to the end of the bone. The wound was now thoroughly washed out, and approximated by three stitches, and by adhesive strips. A grain of sulphate of morphia was administered, and the patient was put to bed. The operation lasted three-quarters of an hour. Dr. McCall thinks that the ball was removed during the operation; but is not positive on this point. The hospital report, which is quoted at page 69 of Circular No. 6, S. G. O., 1865, states that the patient's pulse was quick and irritable at the time of the operation, that he had a furred tongue and diarrhœa, and was reduced by suppuration. But Dr. McCall (Letter of February 11th, 1868) thinks that this report exaggerates the gravity of the constitutional symptoms, and is quite sure that the general condition was favorable. The patient rallied well from the operation. For two days, the wound was dressed

---

[1] In *Circular No. 6, S. G. O.*, 1865, p 68.

with lint. Suppuration then commencing, the limb was placed in Fergusson's apparatus for excision of the head of the femur, the counter extension straps being left off. The wound was freely syringed with cold water containing a little permanganate of potassa. A nourishing diet was ordered, with tonics and stimulants. For a week or ten days subsequently, the case progressed favorably. Suppuration was moderate in amount, and of a healthy character. About the middle of June, the weather became intensely hot. The atmosphere of the wards, in which nearly every bed was occupied by a patient with suppurating wounds, became intensely oppressive. About this time, the patient began to grow worse. The cheerful resolution and hopefulness he had hitherto evinced, gave way. Diarrhœa supervened, and he lost strength rapidly. The fatal event was thought to have been delayed by the plan which was pursued of daily removing the patient in his bed at nine in the morning to a spot beneath the shade trees near the hospital, where he had pure air and escaped the distressing scenes of the ward; he remained each day until five in the afternoon. He died on June 21st, 1864. The portion of bone excised was forwarded at the time of the operation to the Army Medical Museum. The preparation is No. 2618 of the Surgical Series. It is represented in the adjacent wood-cut. (FIG. 17.)

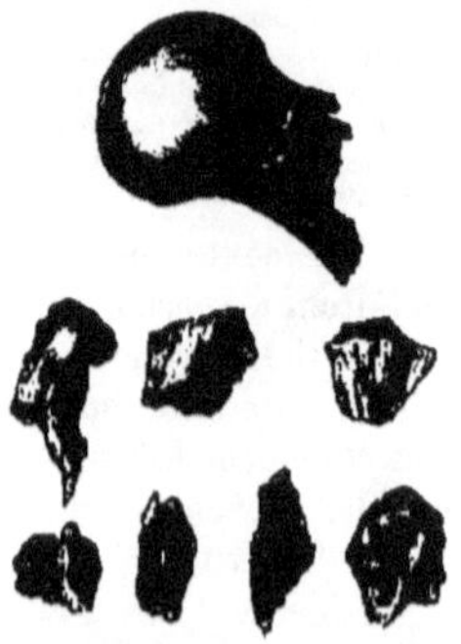

The abstract of the next case is derived from the returns of the field hospitals, and the report of the operator, Assistant Surgeon W. Thomson, U. S. A. The patient was the oldest of any of the wounded men subjected to the operation, and incurred the further disadvantage of prolonged transportation:

FIG. 17.—Head and fragments of the neck of the left femur, excised on account of gun shot injury.—*Spec.* 2618. Sect. I, A. M. M.

CASE XLIX.—Private Peter Doyle, Co. D, 59th Massachusetts Volunteers, aged 60 years, was wounded in front of Petersburg, Virginia, July 30th, 1864. A conoidal musket ball entered the left hip and passed antero-posteriorly through the soft parts and surgical neck of the femur, and fractured the trochanter major. His entry at the base hospital at City Point and transfer to Washington are recorded on August 1st. He was conveyed to Washington on an hospital steamer, and on August 3d was admitted to Douglas Hospital. His constitutional condition, on admission, was poor. The wound, however, had an healthy aspect, and a thorough exploration showed that the injury to the soft tissues involved no important part, and that the fracture, at the junction of the neck and trochanter major, was not accompanied by much longitudinal splintering in either direction.

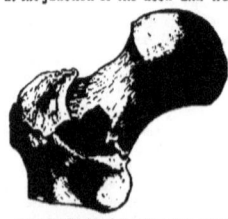

Accordingly, Assistant Surgeon William Thomson, U. S. A., in charge of Douglas Hospital, decided that an excision of the injured bone was expedient, and that the operation should be done as soon as the patient had rallied from the fatigue and irritability induced by his long journey. On August 5th, ether having been administered, Assistant Surgeon Thomson proceeded to excise the head, neck, and trochanters of the left femur, through a straight incision of sufficient length made over the trochanter major. The rotator muscles and the tendons of the psoas and iliacus being divided, the round ligament was readily cut and the head exarticulated. The section of the shaft was made by a chain saw at the level of the trochanter minor. The operation was rapidly accomplished, and there was no hæmorrhage of moment. The wound was dressed with an antiseptic solution of one drop of creasote to each ounce of water on charpie, and the limb was supported by a sand bag on either side. The patient reacted well after the operation; but at night there was profuse sweating and some nausea. On the following morning he ate a good breakfast. He still had a cool, sweating skin, and his pulse was at 123. He gradually sank, and died from exhaustion on August 7th, 1864. The pathological preparation presented by Dr. Thomson to the Army Medical Museum, is No. 3593, Sect. I, and is figured at page 245 of the Surgical Catalogue of the Museum, and also in the accompanying wood-cut. (FIG. 18.)

FIG. 18.—Perforation of the base of the neck of the left femur by a musket ball. *Spec.* 3593, Sect. I, A. M. M.

The very extensive fractures of the pelvis, which existed in the case next to be described, would probably have prevented any operative interference, had they been recognized. The history of the case is compiled from the reports of Armory Square Hospital, and from a special report by Acting Assistant Surgeon George K. Smith. It is one of the very few cases complicated by secondary hæmorrhage. Unavoidable venous hæmorrhage was here the cause of death:

CASE L.—Lieutenant D. N. Patterson, 46th Virginia Regiment, aged 31 years, was wounded at an engagement on the Boydton Plank Road, near Petersburg, Virginia, March 29, 1865, and was captured and sent to City Point, and immediately conveyed on an hospital transport to Washington, and placed in Armory Square Hospital on April 2d. On examination, it was found that a conoidal ball had entered the left thigh behind the trochanter major, and had passed inwards and forwards, fracturing the trochanter, neck, and head of the femur, and the anterior border of the acetabulum. On the day after the patient's admission, five days subsequent to the reception of the injury, Surgeon D. W. Bliss, U. S. Vols., in charge of the hospital, decided that the case was one in which excision of the upper extremity of the femur was applicable. The wounded man was anxious that an operation should be performed, and his general condition was very satisfactory. On April 3d he was placed under the influence of chloroform, and Surgeon Bliss exposed the fractured bone by a curvilinear incision with its convexity forwards. The shattered fragments of the neck were extracted, the rent in the capsular ligament was enlarged and the round ligament was divided, and the head of the femur was exarticulated. It was found that the ball had not only comminuted the head, neck, and great trochanter, but that fissures extended down the shaft of the femur. The bone was divided by the chain saw two inches below the trochanter minor. The deep wound was now washed out, and small fragments of bone were removed, and search was made for the ball. It was finally detected by means of a Nélaton probe deeply buried in the obturator muscle near the posterior margin of the obturator foramen. The operation was accomplished with little loss of blood, and the patient reacted satisfactorily. The wound was lightly dressed, sufficient outlet for discharges being left, and the limb was extended and supported by pillows. Careful nursing was provided, and such stimulants and concentrated nourishment as seemed best adapted to the patient's condition. For three days he progressed very satisfactorily. Notwithstanding the extensive lesions of the pelvis, which were not detected until after death, there was no indication of peritonitis or disturbance of the urinary organs. On April 7th, however, profuse hæmorrhage took place, which could not be controlled, and the patient died on the morning of that day. At the autopsy it was found that the bleeding had proceeded from the internal iliac vein, gradually worn away by a sharp bit of bone forced inwards by the ball. The pathological specimen, represented in the adjoining wood-cut, (FIG. 19,) is also figured at page 246 of the Catalogue of the Surgical Section of the Army Medical Museum, where another view is given.

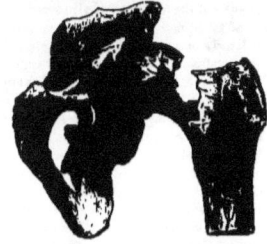

FIG. 19.—Extensive fractures of the shaft, trochanters, head and neck of the left femur, and of the os innominatum. *Spec.* 4048, Sect. I, A.M.M.

In the next case the acetabulum was chipped by the ball, and the shock to the pelvis appears to have been communicated to the contained viscera, from the grave abdominal symptoms which promptly supervened:

CASE LI.—Corporal Henry C. Sennett, Co. F, 122d New York Volunteers, aged 27 years, was wounded in front of Petersburg, Virginia, March 27th, 1865, by a conoidal musket ball, which entered midway between the anterior superior spinous process of the ilium and the trochanter major, and lodged in the head of the left femur. The patient was removed to Washington, and, on April 2d, was admitted to Mount Pleasant Hospital. He was feverish and fretful, and his tongue was furred; but the wound had a healthy aspect, and there was but little swelling or deformity of the limb. But exploration with the finger proved that the ball had penetrated the hip-joint. On April 4th, the patient being anæsthetized by an equal mixture of chloroform and ether, Assistant Surgeon H. Allen, U. S. A., made a T-shaped incision, four inches by six inches, over the trochanter major, and excised the head and neck of the femur. The head was fractured into three pieces, and the ball was embedded in it. Violent hiccough came on immediately after the operation and continued through the night, but was finally arrested by the persistent use of antispasmodics. On April 5th and 6th there was great tympanitis, the bowels being obstinately constipated. An enema of castor oil was administered without effect, and in two hours another of molasses and water and salt, which induced a slight evacuation. Singultus again recurred. On April 7th the bowels moved freely. A chill occurred, lasting half an hour. There was great abdominal tenderness on pressure, and other well marked symptoms of peritonitis. On the 8th, the hiccough continued; the abdomen became greatly distended; the countenance became pinched and ghastly, and the patient died at ten at night. At the autopsy, made twelve hours after death, the lungs were found healthy; the liver greatly hypertrophied; the lower fifth of the ilium inflamed and injected. The tissues surrounding the hip-joint were in a sloughing condition, and were infiltrated with fœtid pus, which had burrowed several inches under the gluteal muscles and two inches below the trochanter minor. The acetabulum was denuded and slightly fractured at its upper and posterior border. Two inches of the upper extremity of the shaft of the femur was denuded of periosteum. The specimen was presented by Dr. Allen to the Army Medical Museum, and is represented in the adjoining wood-cut, (FIG. 20.) The innominate bone was not removed; but the upper fourth of the femur was sawn off after death, and mounted with the excised head to show how completely the injury to the femur was limited to the epiphysis. Had it not been for the fracture of the pelvis, it would have been difficult to have found a case better adapted for the operation of primary excision.

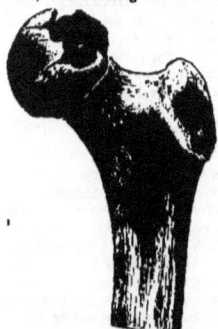

FIG. 20.—Upper extremity of left femur from which the head, with a conoidal ball lodged in it, has been excised. *Spec.* 153, Sect. I, A. M. M.

The next case is compiled from the field reports and from a careful record in the Casebook of Douglas Hospital, Washington, drawn up under the supervision of Assistant Surgeon W. F. Norris, U. S. A.:[1]

---

[1] The case is recorded as Case 29 in the Tabular Statement in the Surgical Report in *Circular* No. 6, S. G. O., 1865, and in the *Catalogue of the Surgical Section of the Army Medical Museum*, p. 246.

INTERMEDIATE OPERATIONS.

CASE LII.—Private Henry Phillips, Co. I, 146th New York Volunteers, a robust man, aged 34 years, was wounded at the South Side Railroad, near Petersburg, Virginia, on April 1st, 1865, by a conoidal musket ball, which entered the left thigh and lodged against the anterior surface of the neck of the femur. The patient was conveyed to the field hospital of the 2d Division of the 5th Corps, and thence by ambulance and rail to City Point, where he arrived on April 4th, and was transferred by steamer to Washington, and, on April 6th, was admitted into Douglas Hospital. He was much exhausted and had considerable fever, though in frequency the pulse and respiration were nearly normal. The wound was painful, and the beginning of its grave constitutional aspect was becoming manifest. There was no shortening or deformity of the limb. At a preliminary exploration the ball was found impacted near the anatomical neck, and was extracted. The limited nature of the fracture was also ascertained, and excision was decided on. On April 8th the patient was placed under the influence of ether, and Assistant Surgeon Wm. F. Norris, U. S. A., excised the head, neck, and trochanter major through a curved incision six or seven inches in length with its convexity forwards. About twelve or sixteen ounces of blood were lost during the operation. One small artery required a ligature. The ball had crushed the laminated structure of the anterior face of the neck, and from this cup-shaped cavity a small fissure ran up the articular surface and a deep fissure nearly around the neck; but the separation between the head and neck was incomplete. During the operation this fracture was converted into a complete one in rotating the bone to facilitate the exarticulation. The operation concluded, the limb was supported by pillows, and the patient was ordered beef tea and milk punch every three hours, and a full dose of opium at midnight. He had another dose of laudanum at four o'clock the next morning. On the 9th, 10th, and 11th, there was little pain, and anodynes were not required, but concentrated nourishment and stimulants were assiduously administered. It was thought the nurse exceeded his instructions in the amount of whiskey given, for, on April 12th, the patient had hiccough and nausea, and his breath was redolent of alcoholic fumes. He was now transferred to the immediate charge of Acting Assistant Surgeon C. Carvallo. A laxative enema was administered, and, when the bowels were unloaded, a sinapism at the epigastrium, and small doses of creasote allayed the irritability of stomach. On April 13th, the stomach was quiet, pulse 120, rather weak, and there was profuse perspiration. The patient was ordered a cupful of beef tea every two hours, one of milk punch every four hours, milk toast and soft boiled eggs at breakfast and dinner. No change the next day. On the 14th the wound looked well. Some shreds of disorganized connective tissue were removed by the dressing forceps. There was some pain and difficulty in micturition. Small doses of tincture of the sesquichloride of iron were directed thrice daily, and chicken broth was added to his dietary. No entry of importance appears on the 16th. On the 17th the pulse was 120, respiration 32. Slight pain on right side, and signs of pleurisy on auscultation. There was an erythematous blush about the wound, and, in the evening, there was diarrhœa, which was checked by pills of opium and nitrate of silver. On the 18th the pleurisy was worse, the breathing more rapid, and there was retention of urine, so that it was necessary after this to use a catheter. April 19th the countenance was sunken, and the wound was flabby. There was a sore on the sacrum. The patient was moved to a Crosby Invalid Bed. He had a draught containing ammonia and sugar, and a blister on his side.

FIG. 21.—Fracture of the head and neck of the left femur by a conoidal musket ball. Spec. 3235, Sect. I, A. M. M.

April 20th the nurse reported a chill during the night. The breathing was labored. There were sordes on the teeth. At the next morning visit the patient was very low. He died before noon, April 21st, 1865, eleven days after the operation. The autopsy revealed dry pleurisy on either side; lungs healthy, somewhat congested posteriorly; heart and liver not abnormal. A large sub-peritoneal abscess in the course of psoas and internal iliac muscles, which appeared to originate in the obturator foramen and ascend along the left iliac fossa, denuding the bone of its periosteum. No evidences of pyæmia were found, though it was strongly suspected after the occurrence of the chill on April 20th. The pathological specimen, contributed by the operator to the Army Medical Museum, is represented at page 246 of the Catalogue of the Surgical Section, and also by the accompanying wood-cut. (FIG. 21.)

The abstract of the next case is compiled from the quarterly report of surgical operations at the St. Louis, and the Marine Hospitals at New Orleans, Louisiana:[1]

CASE LIII.—Private T. E. Foulke, Co. D, 2d Alabama Regiment, aged 17 years, was wounded and captured at Fort Blakely, Alabama, April 9th, 1865. A conoidal musket ball entered posteriorly at the middle third of the left thigh, fractured the upper third of the femur, including the trochanters and neck, and was removed from above the anterior superior spinous process of the left ilium. The patient was then transferred to New Orleans, and, on April 15th, he was admitted to the St. Louis Hospital. On admission he was very much exhausted by profuse suppuration, the soft parts about the hip-joint being filled with unhealthy pus. On April 27th Surgeon A. McMahon, U. S. Volunteers, excised the head, neck, trochanters, and two inches of the shaft of the left femur, the patient being under the influence of chloroform. He was placed on nourishing diet, with two bottles of porter daily, eggs, beef tea, and everything he desired. On May 26th he was transferred to the Marine Hospital at New Orleans. He was then doing well; but he gradually failed, and died June 5th, 1865, of exhaustion, thirty-nine days after the operation.

The history of the fifty-fourth case is taken from a quarterly report of surgical operations at the St. Louis Hospital, New Orleans, Louisiana. It has been impracticable to learn more minute particulars in regard to it:[2]

CASE LIV.—Private G. W. Brantley, Co. C, 2d Alabama Regiment, aged 18 years, was wounded and taken prisoner at Fort Blakely, Alabama, April 9th, 1865. A conoidal musket ball had passed through the left groin, fractured the neck of the

---

[1] The case appears in the Surgical Report in Circular No. 6, S. G. O., 1865, p. 74, Case 31.
[2] The case is reported in Circular No. 6, S. G. O., 1865, p. 74, Case 32.

femur, and emerged posteriorly at the apex of the left buttock. He also received a gunshot fracture of the external condyle of the right humerus. He was conveyed to New Orleans, and, on April 15th, he was admitted to the St. Louis Hospital. The thigh, groin, and surrounding parts were infiltrated with unhealthy pus, and the patient was very much exhausted. On April 28th, the patient was anæsthetized by chloroform, and Surgeon A. McMahon, U. S. Volunteers, proceeded to excise the head, neck, and trochanters of the left femur. No arteries required ligation. The patient did not rally very well. Stimulants were freely administered; but the patient sank, and died on May 2d, 1865, of capillary hæmorrhage. The condition of the patient did not admit of any operation on the elbow-joint.

Of the twenty-two intermediate excisions, it may be remarked that the local lesions fully justified the operations except in the cases numbered XLIII and XLV, in which the fractures were entirely external to the joint, and in the cases numbered L and LI, in which the pelvis was fractured. Case XL, in which the trochanters and a portion of the neck were first excised, and the carious head was subsequently enucleated, would appear to teach that when the joint is opened, and fragments of the neck are extracted, no advantage accrues from leaving the head, even if it is uninjured, since it is unlikely that its vitality will be maintained. In almost all of the cases, the constitutional condition of the patients was very unsatisfactory; and the operations were only performed under the conviction that unless some operative interference was undertaken an inevitably fatal issue was imminent.

SECONDARY OPERATIONS.

The list of secondary excisions at the hip is brief, and it is probable that this will always remain an infrequent operation; for, unhappily, the number of patients with gunshot injuries involving the hip-joint, who survive until the secondary period, without some operative interference, is very small. Of the nine operations included in this category, one was successful. This last was an example of partially consolidated fracture below the trochanters, with necrosed fragments and diseased callus; the hip-joint was apparently not involved until the exarticulation was undertaken. Of the eight unsuccessful operations, two were performed for secondary involvement of the hip-joint without fracture, the ball having grazed the cervix femoris in the one case, and, in the other, having grooved the cylinder of the shaft in front of the trochanter minor, and lighted up a secondary traumatic arthritis. In a third case, almost all hope of a successful result was precluded by the discovery of a fracture of the pelvis. The average duration of life after the eight unsuccessful operations was sixteen days, or rather longer than in the intermediate operations. One of the patients survived one hundred days, the remainder less than a week. One patient died from secondary venous hæmorrhage, two died from shock, and five from exhaustion. The interval between the reception of the injury and the date of the operation averaged two and a half months. The shortest interval was thirty-three days, and the longest two hundred and four days, or nearly seven months, this being the period which elapsed from the reception of the injury to the operation in the successful case. The injuries in all of the secondary cases were inflicted by musket balls.

Surgeon General Dale, of Massachusetts, Surgeon J. Simpson, U. S. A., Surgeon Lavington Quick, U. S. V., and Dr. Edmund G. Waters, of Baltimore, have contributed the materials from which the history of the first case is compiled. Dr. Waters conducted the after-treatment in the case, and assisted at the operation and autopsy:[1]

CASE LV.—Private John W. Nelling, Co. K, 1st Massachusetts Volunteers, aged 25 years, was wounded on June 30th, 1862, at the engagement at White Oak Swamp, by a musket ball, which entered his right groin, passed horizontally backwards,

---

[1] See Report in *Circular* No. 6, S. G. O., 1865, p. 62, and *Catalogue of Surgical Section of Army Medical Museum*, p. 244.

comminuted the neck of the femur, and emerged posteriorly. He was abandoned with other wounded in the retreat of General McClellan's army. Being made a prisoner, he was confined in Richmond for three weeks, and was then released and sent by water to Baltimore, where he was admitted to the National Hospital, on July 25th, in a very depressed condition. There was copious suppuration, and through the large orifices of entrance and exit it was easy to explore the extent of the injury to the bone, and to determine that the comminution was limited to the epiphysis. It was deemed advisable to excise the shattered extremity of the bone as soon as the patient could acquire, by a tonic treatment, strength to undergo such an operation. In a few weeks his general condition was much improved, though he was still anæmic and feeble. On August 21st, Assistant Surgeon Roberts Bartholow, U. S. A., in charge of the hospital, proceeded to perform the operation. The patient being placed under chloroform, Dr. Bartholow made a vertical incision, commencing a little above and behind the great trochanter, continued downwards into the axis of the limb four inches, and carried it down to the bone. The head of the femur was found to be entirely separated from the neck, and was retained in the acetabulum only by the round ligament. This was divided, and the

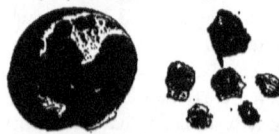

Fig. 22.—Head and fragments of neck of right femur, excised for gunshot fracture.—*Spec.* 400, Sect. 1, A. M. M.

head was removed. Several necrosed fragments were then extracted, and the jagged extremity of the neck was smoothed by an osteotome. But little blood was lost in the operation, and the patient rallied promptly from the effects of the anæsthetic. After he was put to bed, the limb was suspended by Smith's anterior splint, and the patient's condition was rendered comparatively comfortable. The case progressed favorably and without an untoward symptom till the afternoon of August 25th, when a sudden and very profuse gush of dark blood from the wound of incision and the entrance bullet wound took place, and the patient expired before the hæmorrhage could be controlled. At the post mortem examination, the soft parts in the vicinity of the wound were found to be in a softened and semi-gangrenous condition. The end of the excised neck was denuded of periosteum and was necrosed. The external iliac and femoral arteries were traced some distance above and below Poupart's ligament, and were found to be in a normal condition. The femoral vein was softened, and near the track of the ball appeared to be broken down so as not to be distinguished from the surrounding tissues. A quantity of dark fluid blood was found under the integuments. The excised head and neck were deposited in the Army Medical Museum by the operator, and are represented in the annexed figure.

The particulars of the next case were recorded in a Frederick Hospital case-book, by Assistant Surgeon R. F. Weir, U. S. A.:

CASE LVI.—Private Edward Huut, Co. D, 71st Pennsylvania Volunteers, aged 24 years, was wounded at the battle of Antietam, September 17, 1862, by a conoidal musket ball, which entered about two and a half inches above the trochanter major, and grazing the neck of the right femur, passed out at the nates. Shortening and eversion were not present, and it was thought that there was not a complete fracture. After the battle, he was removed to a barn near the battle field, where he was treated with cold water dressings for eleven days. He was then removed to the City Hotel at Frederick, Maryland, and the cold applications were continued up to November 19th. During this period the wound was discharging healthy pus profusely. As pus was burrowing in the muscles, a seton was run through the wound and six inches down the thigh. On November 19th, he was removed to Jail Street Hospital, and about the last of December he was transferred to the U. S. Hotel Hospital, and thence, on January 20th, 1863, to Hospital No. 5, at Frederick, Maryland. On January 31st, an abscess formed on the anterior internal aspect of the thigh, which was opened, and discharged nearly a quart of laudable pus in twenty-four hours. On February 2d, the opening on the posterior aspect of the thigh was enlarged and the wound syringed out with warm water. Erysipelas attacked the wound, but it was not of an intense character, and by February 10th, it had subsided, and the patient was in good condition comparatively. On February 23d, he had become more emaciated and had night sweats, but his strength continued good. On exploring for the ball with the finger, in the opening on the inner side of the thigh, a round, smooth surface was felt, which was thought at first to be a piece of a conoidal ball, but was ascertained to be the head of the femur just outside of the acetabulum. Excision of the hip-joint was now decided upon, and Assistant Surgeon Henry A. Dubois, U. S. A., operated, the patient being under chloroform, by enlarging the opening on the inner side of the thigh, cutting the capsular ligament, and removing the head of the femur. The neck was divided by the lion-jawed cutting forceps. A small quantity of pus was found behind the head of the bone. But very little blood was lost during the operation, and the patient rallied partially, but he never fully recovered from the shock, and died February 25th, 1863, two days after the operation. At the autopsy, the neck of the femur was found rounded off, and formed a false centre of motion on the inner side of the acetabulum. The rounded extremity of the neck and the acetabulum were carious. The annexed figure imperfectly represents the specimen, which was sent to the Army Medical Museum.

Fig. 23.—Caries of the acetabulum and neck and trochanters of the right femur, following an excision for gunshot fracture.—*Spec.* 3307, Sect. 1, A. M. M.

The following example of a secondary excision of the head, neck, trochanters, and several inches of the shattered shaft of the femur for a gunshot fracture below the trochanters, has the happy distinction of having been completely successful. An extract from his monthly hospital report, containing an account of the case, was published by the operator four months subsequent to the operation.* The recovery then announced has proved permanent, and, after the lapse of more than five years, the subject of the operation continues to enjoy robust health:

CASE LVII.—Private Joseph Brown, Co. I, 3d Michigan Volunteers, aged 38 years, was wounded at the second battle of Bull Run, August 29, 1862, by a musket ball, which passed through the left thigh, fracturing the femur just below the trochanter minor. He laid on the battle-field three days, and was then removed to Centreville. On September 11, 1862, he was admitted to Fairfax Seminary Hospital, near Alexandria. The limb was kept in position by appropriate apparatus; but suppuration was profuse, and, on two occasions, fragments of bone were removed from the wound. Early in March, 1863, there was great swelling of the thigh, the discharge became scanty and fœtid, and pus burrowed amid the muscles. On March 21st, an exploratory incision was made from three inches above to five inches below the prominence of the great trochanter. The neck and upper extremity of the shaft of the femur were found to be extensively diseased, and excision was decided on. Surgeon D. P. Smith, U. S. Volunteers, performed the operation. Difficulty was experienced in separating the muscular attachments from the trochanters, on account of the foliaceous masses of callus that had been thrown out. When this dissection was accomplished, many necrosed fragments were extracted, and the periosteum and new bone were separated by the handle of the scalpel and preserved as far as practicable. The shaft of the femur was then divided by powerful cutting bone forceps, about six inches below the tip of the great trochanter. A screw was driven into the mass of callus, below the trochanters, to be used as a lever in disarticulating the head, but it would not hold, and the bone was seized with large forceps and rotated, so as to facilitate the division of the capsular and round ligaments. The head, and neck, and trochanters, and the masses of callus adhering to the trochanters, were then removed. The operation was accomplished with but very trifling hæmorrhage, yet great prostration followed, and the patient rallied slowly. As the anæsthesia passed off, he had much nausea and vomiting. As soon as this subsided, he was given a very full allowance of concentrated nourishment, such as strong beef-tea, eggs, milk, etc., with half an ounce of brandy every two hours. The wound was partially closed; the limb was supported on pillows until the third day, when it was dressed in a Smith's anterior splint. About forty-eight hours after the operation, an erysipelatous blush pervaded the limb, and the constitutional symptoms assumed a typhoid character. A female catheter was passed through the middle of the wound and another at its lower extremity, through which much offensive decomposed serum and grumous blood escaped. The wound was thoroughly washed out through the catheters with warm water impregnated with chlorinated soda. On the fifth day there was a rigor, and hæmorrhage to the extent of six ounces. As the anterior splint did not permit convenient access to the limb, it was removed, and the leg and thigh were suspended in a canvas hammock, the leg being horizontal and the thigh in an almost vertical position. A piece of soft towelling extending from the perinæum to the popliteal space, and, connected by cords with an upright post at the head of the bed, supported the muscles on the sides and under surface of the thigh. The wound freely discharged synovia, bloody serum, and thin pus, until the seventh day, when healthy suppuration was fairly established. During April, 1863, the patient's progress was satisfactory. He was supplied with very nutritious diet, with porter, and cod-liver oil. He took for a time as much as half a pint of oil daily. During May, the case continued to progress favorably. It was necessary to keep a tube in the wound until June 1st. Previously, whenever it was removed pus would accumulate and burrow. A mesh of suture wire was finally substituted for the tube. This was retained until June 20th, when the patient began to get about on crutches. In the latter part of July the wounds closed. In August, Brown was reported as "well," and on August 23d, 1863, he was discharged from the hospital and from the military service of the United States. On March 21st, 1864, he wrote from his home in Coopersville, Michigan, that his health was good; that he could get about and attend to home business; could saw and split a little wood for fuel, though his knee was stiff and his leg painful. On the whole, there had been steady improvement. In September, 1865, he again wrote,† and stated that his general health was good; that he had some control over the movements of the thigh, being able, when standing on the right foot, to swing the left backwards and forwards, and to adduct the thigh enough to carry the injured limb across the

FIG. 21.—Head, neck, and trochanters of left femur, with masses of foliaceous callus, excised by a gunshot fracture below the trochanter minor.—Spec. 1192, Sect. 1, A. M. M.

other. He could bear some weight on the limb, and used but one crutch, with a stirrup for the foot. There had been no fistulous orifices since March, 1864, and there was no soreness about the cicatrices. In November, 1865, in accordance with a request from the Surgeon General's Office, Mr. Brown had a photograph taken to represent the amount of deformity in his limb. This picture is numbered 110 in the Photograph Series of the Army Medical Museum. It is carefully copied in the accompanying lithograph. The excised bone is preserved at the Museum, and is numbered 1192, Section I. It is repro-

---

* American Medical Times, Vol. VII, p. 13, July 11th, 1863. Further particulars of the case were published at pp. 66–75 of the Surgical Report in Circular No. 6, S. G. O., 1865. See, also, Catalogue of the Surgical Section of the Army Medical Museum, p. 246.

† See Circular No. 6, S. G. O., 1865, p. 75.

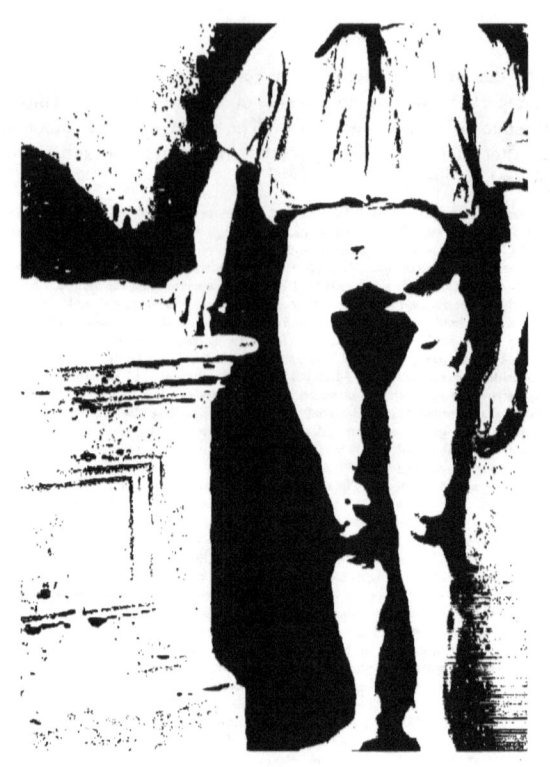

sented in the preceding wood-cut." (See FIG. 24.) The reports of the Pension Office for 1866 and 1867 indicated that Mr. Brown continued to enjoy tolerable health. On February 12th, 1868, he wrote to the compiler of this report as follows: "I take pleasure in informing you that my limb is in as good condition as when I last wrote you; but think there is no improvement, except that it is not as tender. There have been no abscesses, nor any pain in the limb, excepting slight pains about the knee, just before storms. About two years ago, I slipped and fell upon the ice, injuring the limb severely about the knee, and was thereby confined to the house for about three weeks. In March last I had a severe attack of ague. The limb swelled quite badly at this time, and was much inflamed for about ten days. I applied cold water and a bandage to reduce the swelling. I had to keep it bandaged about two weeks after the inflammation was removed. Since that time the limb has given me no more trouble than usual. Since I was discharged I cannot see that there is any lengthening of the limb. I have to use a crutch and cane all the time when moving about, and I think I shall always have to do this. The injured limb has wasted away somewhat since I last wrote. The circumference of the well limb at the upper extremity is 22 inches, and the injured limb measures at the same place 19½ inches. The knee of the well limb measures around the centre of the knee-pan 15½ inches; the injured limb measures at the same place 17 inches. The above measurements were made in the evening; I think that in the morning the measurements of the injured limb would be less. The knee still remains quite stiff, and gives me about all the pain there is anywhere in the limb. I have been troubled during cold weather by coldness of the outer side of the leg, and I have to warm it by the fire before going to bed nearly every night when I have been out." On November 19th, 1868, another letter was received from Mr. Brown, from which the following extract is made: "I take pleasure in informing you that my limb is in as good condition as it has been at any time since it was entirely healed, and, if anything, in better condition. It does not pain me about the knee as much as it did one year ago. It does not have any spell of swelling at the knee as it did for the first two years after my discharge, and there is less soreness about the limb than there was even one year ago. I can get around without hurting it as much as formerly. I can bear some weight upon it. I have walked across a room without the aid of crutch or cane, by stepping very quick with the well limb; but it is more like hopping than walking. There have been no abscesses in the limb. I think that it is gradually improving, and hope that I may yet see the day that I can go without a crutch. My general health is good. I have not been sick a day for a year and a half, and then only a few days with ague. My weight now is 167¼ pounds. Before I entered the army my weight was never quite up to those figures, but within a few pounds of it. I have been postmaster at this office over a year, and have attended to all the business of the office almost entirely without assistance, and it gives me pretty good exercise."

The following case, remarkable for the delay in the apparition of grave symptoms after a fracture of the acetabulum, is recorded in the Confederate States Medical and Surgical Journal,† and in one of the Confederate hospital registers deposited in this office:

CASE LVIII.—Private Alfred Toney, Co. A, 16th North Carolina Regiment, aged 43 years, a farmer by profession, was wounded June 30th, 1863, and admitted to Hospital No. 4, Richmond, Virginia, on the same day. A conoidal musket ball had entered the left buttock and lodged. No particular attention was called to this case for some time. The patient seemed to be doing well. On August 11th, however, he complained of great pain in the knee and ankle; the slightest touch caused great anguish. The foot was œdematous. Chloroform was administered, and digital examination of the wound was made. The finger could pass but half an inch into the wound until the limb was carried forward; it then could be passed into the cotyloid cavity, and the ball was found in the acetabulum. The round ligament was severed and the head of the femur was ascertained to be slightly fractured and deprived of its cartilage. Excision of the head of the femur was decided upon, and on August 12th, Surgeon James B. Read, P. A. C. S., proceeded to operate. The patient was laid on his face, and his buttocks were brought to the edge of the table. A straight incision was commenced two inches below the anterior superior crest of the ilium and carried downwards to one inch below the trochanter major. The muscles were then separated, and the joint exposed. The head was then dislocated by forcibly bringing the leg under the table. The soft parts were protected by a spatula and the head was sawn off. The ball was removed from the cotyloid cavity, which was found to be broken across and the cartilage loosened. The wound was then closed by sutures and the patient was removed to his bed. He suffered no pain, and in twenty-four hours the swelling had subsided. His general condition was very feeble, and he was freely stimulated during the after-treatment. He died on August 19th, 1863, eight days after the operation, exhausted by hectic fever. There is no account of any abdominal disturbance or pyæmic symptoms resulting from the fracture of the acetabulum.

The particulars of the fifty-ninth case are derived from the casualty lists of the Sixth Army Corps, and from the quarterly surgical report from Harewood Hospital. The patient was young and slender, and was one of those who had made the exhausting journey in the ambulance trains through the Wilderness. There was nothing like a complete fracture in the case. The ball flattened its apex against the shaft in front of the lesser trochanter, grooving the bone, but without any splintering. The joint was involved secondarily. It was a case for primary *débridement* of the entrance wound:

---

* Another view of the specimen is given at p. 246 of the *Catalogue of the Surgical Section of the Army Medical Museum*, and in *Circular* No. 6, S. G. O., 1865, p. 75.

† READ, *Confederate States Med. and Surg. Journal*, 1864, Vol. I, p. 6.

52   EXCISIONS AT THE HIP.

CASE LIX.—Private Henry Woodworth, Co. A, 4th Vermont Volunteers, aged 18 years, was wounded at the battle of Spottsylvania Court House, Virginia, on May 11th, 1864, by a conoidal musket ball, which entered the left thigh, just below the trochanter major, passed inwards and forwards, grooving the femur anteriorly at the level of the lesser trochanter, and lodging under the sartorious muscle. The patient was conveyed to the field hospital of the Second Division of the Sixth Corps, where the ball was removed through an incision at the edge of the sartorius. A week subsequently, he was placed on one of the trains for the Rappahannock, and was transferred from Fredericksburg to Washington, where, on May 25th, he was admitted to Harewood Hospital. His condition on admission was very unpromising; his pulse was quick and feeble; he was anæmic, and without appetite. He was placed upon a tonic regimen, but he did not improve. The wound discharged profusely; there was much pain in the joint, pain aggravated by the slightest movement, and pus had burrowed in every direction about the articulation. Surgeon R. B. Bontecou, U. S. V., in charge of Harewood Hospital, decided that an excision of the head of the femur offered the only possible chance of saving life, and, on July 1st, the patient having been anæsthetized by sulphuric ether, Dr. Bontecou proceeded to perform the operation. A curved incision, with its concavity forwards, embracing the trochanter, readily exposed the joint. The muscular attachments were divided, and the head was easily disarticulated, the joint being disorganized and the round ligament destroyed. The continuity of the bone being uninterrupted, the upper extremity was readily turned out and sawn just below the point of impact of the ball. On examination of the portion of bone removed, it was found that much of the head had been absorbed, and that the remainder was carious.[1] The specimen is represented in the accompanying wood-cut. (FIG. 25.) The neck and trochanters are covered with traces of the effects of periostitis. The cotyloid cavity was ulcerated. The wound was drawn together by adhesive strips, and the limb was dressed in a fracture apparatus with moderate extension. Every means of supporting the patient's strength was adopted, but he did not rally from the operation, and, sinking gradually, expired on July 2d, 1864.

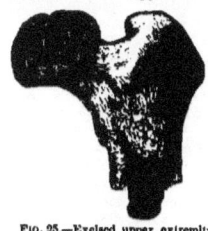

FIG. 25.—Excised upper extremity of the left femur with a conoidul ball which contused and grooved the bone. Spec. 3049, Sect. I, A.M.M.

The particulars of the next case are derived from the register and case-books of the Confederate Hospital No. 4, at Richmond, Virginia:

CASE LX.—Ensign W. J. Henry, 21st Mississippi Infantry, aged 22 years, was wounded near Petersburg, Virginia, on June 21st, 1864, by a conoidal musket ball, which entered at the upper third of the left thigh anteriorly and made its exit at the opposite side of the limb, having badly comminuted the femur, without injuring the principal vessels or nerves. The wounded man was conveyed to Richmond, and was admitted to Hospital No. 4, on June 28th. The injured limb was suspended by Smith's anterior splint, the wound was dressed with water dressings, a nourishing diet was ordered, with an opiate at bed-time. Under this treatment the case progressed satisfactorily until August 2d, when the copious suppuration and wasting of the patient excited so much anxiety that a consultation was held, at which Surgeons C. B. Gibson, M. Michel, and J. B. Read assisted. After a careful exploration of the injury under chloroform, it was decided that an excision of the upper extremity of the femur was expedient, and the operation was at once performed by Surgeon J. B. Read, P. A. C. S. An incision seven inches in length was made, commencing above the trochanter and carried downward in the axis of the thigh. The joint was opened and the head of the bone exarticulated. The shaft was sawn below the trochanter minor, about six inches of the bone being removed. The wound being thoroughly cleansed, was approximated and kept in position by sutures and adhesive strips. A long straight splint was then applied on the outside of the limb. An ounce of brandy and thirty drops of tincture of opium was ordered to be administered every hour until reaction should be fully established. On August 3d, the patient had reacted and had passed a comfortable night. The pulse was at 136; there was no pain, except in the left knee, which was swollen, but without increased heat or redness. The appetite was poor but improving. A nourishing diet was directed, and half an ounce of brandy every two hours with a grain of opium every three hours. On August 4th, the patient had rested well, had a good appetite, the tongue was clean, the skin moist, the bowels had been moved naturally, the pulse was at 129, the countenance was cheerful. The treatment was continued with the addition of porter thrice daily. On August 5th, the pulse was stronger at 120, the countenance was cheerful; the treatment was continued. On August 6th, the patient was reported to have passed a bad night. He complained of acute pain in his left knee and in the right side of his chest. The pulse was 136 and weak. Incipient pneumonia was detected in the right lung. He had vomited the porter, and it was discontinued; the brandy and opium were continued as before. The wound was suppurating profusely. The sutures were clipped, and the wound was kept together by adhesive strips. The bowels were constipated. An enema of warm soap and water was administered, which procured a normal dejection in a few hours. Sinapisms were applied to the chest. On August 7th he had rested tolerably well, but breathed badly. He was sweating profusely, and complained of much pain in the chest. The pulse was at 140, and was very weak. He was ordered an ounce of brandy every hour and a grain of opium every four hours. On August 8th he was reported as having passed a very restless night. He was too weak to expectorate; the pulse was very feeble at 148. He was evidently sinking. He died at 3 P. M., August 8th, 1864, six days after the operation, and forty-nine days from the reception of the injury. The report gives no account of the post mortem appearances.

---

[1] See *Catalogue of the Surgical Section of the Army Medical Museum*, 1866, p. 245, and Surgical Report in *Circular* No. 6, S G. O., 1865, p. 70, CASE 21.

SECONDARY OPERATIONS. 53

In a special report of the following case, the operator, Assistant Surgeon C. Wagner, U. S. A., remarks that he believes that, "had the operation been performed earlier, the result would have been more favorable." The following abstract is compiled from Dr. Wagner's report, the memorandum accompanying the pathological specimen, and the register of the Beverly Hospital:[1]

CASE LXI.—Private John Zaborouski, Co. H, 7th Connecticut Volunteers, aged 33 years, was wounded at the engagement at Deep Bottom, Virginia, August 16, 1864, by a conoidal musket ball, which entered just below the right trochanter major, and, passing upwards and inwards, fractured the neck and slightly injured the head of the femur. The patient was conveyed to the field hospital of the Tenth Corps, where his wound was dressed, and he was then sent to City Point and transferred to the hospital transport steamer De Molay, and conveyed to the U. S. Hospital, at Beverly, New Jersey, where he was admitted on August 22d. His condition at this period is not described, and it is not stated whether the ball had been extracted. From the subsequent history, it would appear that there was profuse suppuration about the joint, since a free transverse incision was practised to give free escape to the discharge. On September 27th it was determined to excise the head of the femur. The patient had greatly lost in flesh and strength since his admission, and seemed to be rapidly failing from the exhaustive suppuration; extensive sloughing of the soft parts had supervened, and, upon the whole, the case was unfavorable and unpromising.

FIG. 26.—Head of the right femur excised for caries following a gunshot fracture.—Spec. 3716, Sect. I, A. M. M.

Chloroform having been administered, and the patient being placed on his sound side, Assistant Surgeon C. Wagner, U. S. A., made an incision four and a half inches in length, extending from just below the anterior superior spinous process of the ilium towards the tuberosity of the ischium, crossing the transverse incision previously made over the trochanter major to permit free exit of pus. The thigh was flexed and rotated inwards, the tendons of the muscles were divided, and a chain saw was passed under and between the head of the femur and the trochanter major, and the bone was sawn through the neck, the soft parts being protected by spatulas. About one inch of the trochanter major was found to be necrosed, and was removed by a small saw. No blood was lost during the operation. The patient sank rapidly, and died September 23th, 1864, of exhaustion. A post mortem examination revealed nothing of note, except slight caries of the acetabulum. The carious head of the femur was sent to the Army Medical Museum, and is represented in the adjacent wood-cut. (FIG. 26.) The fragments of the neck and the necrosed trochanter major were lost. The reports do not state whether the ball was extracted at the time of the operation or previously.

The following case is briefly mentioned in the surgical report in Circular No. 6, S. G. O., 1865, as a partial excision, or extraction of fragments, in accordance with a statement of Acting Assistant Surgeon Liebold. Subsequently, Surgeon John Vansant, U. S. A., made a special report of the operation, describing it as a formal decapitation of the femur. The injury was limited to the head and anatomical neck of the bone, and the case seemed for a long time to promise a successful issue:

CASE LXII.—Private Joseph Roth, Co. B, 188th New York Volunteers, aged 25 years, a large, robust man, was wounded in the engagement at Hatcher's Run, near Petersburg, Virginia, on February 6th, 1865, by a round musket ball, which entered a little below Poupart's ligament, an inch external to the course of the vessels, and lodged in the neighborhood of the hip-joint. He was received at the field hospital of the First Division of the Fifth Corps, and thence conveyed to the base hospital at City Point, and, as there was no pain or deformity, the case was regarded and treated as a flesh wound, and a week subsequently, the patient was sent in the hospital transport steamer State of Maine to the General Hospital at Point Lookout. After a short time, Roth began to complain of great pain in the knee and leg of the wounded limb, which aroused suspicion that the hip-joint was implicated. The symptoms becoming aggravated, and pointing clearly to some injury of the joint, Surgeon John Vansant, U. S. A., in charge of the hospital, determined to make an exploratory incision, and to ascertain the true condition of things. The patient being anæsthetized by a mixture of chloroform and ether on March 9th, an incision was made, commencing at the wound of entrance, and continued downwards three inches or more. The ball was now found impacted in the head of the femur, the anterior part of which was shattered, while the posterior two-thirds of the head was intact. The muscular attachments being divided, and the capsular ligament freely opened, the round ligament was severed, and the head was exarticulated. A chain saw was then passed around the neck, which was divided near to the head, and the latter, with the ball inserted in it, was removed. Some sharp projecting portions of the neck were smoothed off with bone forceps. There was very little hæmorrhage. The wound was drawn together and treated by water dressings, and the limb was kept in a suitable position by pads and pillows. The patient seemed to do well for several weeks, but gradually became feeble and emaciated, losing all appetite. There was but a slight discharge. In May, the patient became quite yellow, and apparently suffered from malarial complications. In spite of a careful tonic and sustaining regimen, and the bracing, wholesome salt air of Point Lookout, he gradually declined, and died

---

[1] A history of the case was printed by Dr. Wagner in a pamphlet entitled: *Report of Interesting Surgical Operations performed at the U. S. General Hospital, Beverly, New Jersey,* by C. WAGNER, Assistant Surgeon, U. S. A., Commanding Hospital, p. 14. See, also, the Surgical Report in *Circular* No. 6, S. G. O., 1865, p. 70, CASE 25.

June 17th, 1865. The pathological specimen was not forwarded to the Museum, and no account of an autopsy was rendered, an annoying omission, since it would have been interesting to have learned what reparative action had taken place during the three months after the operation which this patient survived.

The particulars of the last case to be recorded are transcribed from the registers of the St. Louis Hospital, at New Orleans. As in the intermediate operations by Surgeon McMahon, it has been impracticable to learn any further details. The pathological specimens were not forwarded to the Army Medical Museum. This case is of peculiar interest as an instance of excision for caries resulting from secondary involvement of the hip-joint:

CASE LXIII.—Private Hugh Train, Co. G, 31st Massachusetts Volunteers, aged 22 years, was wounded February 1, 1865, by a conoidal musket ball, which entered the anterior surface of the middle third of the left thigh, ranging upwards, and making its exit above the left gluteus, fracturing the neck of the femur in its course. He was on horseback, his regiment serving at the time as mounted infantry. He was treated in the regimental hospital for several days, and then, on February 13, he was admitted to the St. Louis Hospital, at New Orleans. When admitted, he stated that the surgeon of his regiment had given him chloroform, and examined the wound, and that the bone was not touched. He had walked upon the limb, and there was then no evidence of fracture. His general health became poor. He had night sweats. His tongue was clean and moist, and his appetite was good. The whole thigh gradually became dissected with pus of an unhealthy character. Abscesses discharged through fistulous openings in the groin. The limb was inverted and shortened, and was drawn over to the right. It was decided that an excision of the head of the femur was expedient. On March 24th, Surgeon A. McMahon, U. S. Vols., proceeded to perform the operation. An incision four inches in length was made over the great trochanter, the soft parts were dissected up, and the femur was divided by the chain saw just below the trochanter minor. The ligamentum teres was softened, and the head of the femur was removed without difficulty. The wound was filled with lint, and the patient was ordered porter, chicken broth, eggs, stimulants, and everything necessary to sustain the drain upon his system. He felt easier for a few days after the operation, but he gradually sank, becoming very much emaciated, and died March 30th, 1865, six days after the operation. The ball had injured the neck of the femur, and the subsequent caries had caused the destruction of the head, and the disorganization of the surrounding tissues.

In six of the nine secondary incisions, the local lesions fully justified operative interference. Comminuted intra-capsular fractures existed in four of these, and in the other two the hip-joint was disorganized by secondary traumatic arthritis. In case LX, the location and extent of the fracture is not described; in case LVIII, the acetabulum was fractured; and in case LVII, the injury was altogether extra-capsular.

UNCLASSIFIED OR DOUBTFUL CASES.

Besides the sixty-three abstracts of cases recorded in the three preceding sections, a certain number have been reported and excluded as unfit for statistical purposes, either on account of the indefiniteness or inexactness or contradictory character of the reports, or because the absence of names and dates involved the hazard of duplicating cases. It is proper, however, to notice these cases briefly, with the hope that farther information may be elicited.

Dr. Edward Batwell, of Ypsilanti, Michigan, late Surgeon 14th Michigan Volunteers, records[1] the following:

CASE.—"Before the evacuation of Corinth, I witnessed the operation of excision of the head of the femur at Farmington, May 9th, 1862, performed on a soldier of a Missouri regiment. I did not know the operator, nor the result of the case. I regarded the operation as one undertaken to show the skill and fortitude of the operator, rather than with a view to benefit the patient."

Assistant Surgeon Thomas F. Azpell, U. S. A., reports[2] that he took notes of two cases of excision of the head of the femur in patients on board the hospital river transport R. C. Wood. It is Dr. Azpell's impression that one was embarked after the battle

---
[1] Letter from Dr. Batwell, of October 26, 1868.
[2] Letters of February 27, 1867, February 12th, and September 22d, 1868.

UNCLASSIFIED OR DOUBTFUL CASES.   55

of Helena, and the other after the battle of Franklin. One was conveyed to an hospital at Memphis, and the other either to Mound City or Jeffersonville. In one, the operation was performed through a longitudinal incision; in the other, by raising a semi-circular flap. The records of the hospital steamer and of the hospitals at Memphis, Mound City, and Jeffersonville, throw no light on these cases, and inquiries addressed to surgeons who were on duty at those places at the periods referred to, have been unavailing.

Dr. J. W. Brock, formerly Surgeon 66th Ohio Volunteers, states[1] that he excised the head of the femur primarily for gunshot injury on two occasions: after Gettysburg, and at Resaca, Georgia; but remembers no particulars of either case, and knows nothing of the results. The very full returns from Gettysburg and Resaca contain no allusion to these operations.

Dr. Benjamin Woodward, of Galesburg, Illinois, formerly Surgeon 22d Illinois Volunteers, relates[2] two cases which he regards as examples of the operation in question:

CASE.—"Private Joseph Derusha, Co. F, 14th Ohio Volunteers, was sent to the hospital at Tullahoma, two days after the battle of Lookout Mountain, [November 25th, 1863], wounded in the right hip. I found a small hole of entrance over the greater trochanter, and with a Nélaton probe detected a ball firmly lodged in the bone. As I could not remove it with any instrument without doing greater damage to the bone, I made a crucial incision, the ball hole in the centre, and, reflecting the flaps, sawed out with a Hey's saw a wedge-shaped piece of bone with the ball in it. There was little hæmorrhage, and the case did well, so that by April he was acting as nurse in the hospital." [The hospital register reports Private Derusha as treated for "inflammation of the kidneys," and makes no mention of any wound or operation.]

CASE.—"The other was that of a negro railroad hand, wounded by guerillas, but, as he was not a soldier, his case, like those of others at the same time, did not go on the hospital reports. This man was shot in the left hip and the head of the bone with two inches of the shaft shattered. He was chloroformed and the head of the bone with two and a half inches of the shaft excised. There was no hæmorrhage requiring any ligation. The cavity was filled with lint, and water dressings used. Suppuration was profuse, and he sunk in about a month." [The hospital register states that Jenaway Engle, colored laborer on military railway, aged 19, was admitted March 16th, 1864, with a gunshot fracture of the femur by a conoidal musket ball. March 17th, ball and splinters of bone removed. Death, March 24th, 1864.]

From the great number of negative responses to a circular letter from this office requesting medical men to communicate any facts within their knowledge in relation to unpublished cases of excision of the head of the femur, it may be concluded that the record of these operations in the United States armies is very complete. For obvious reasons, similar completeness can hardly be hoped for in the reports from the Confederate armies. I am informed by Professor T. G. Richardson,[3] of New Orleans, that the operation was not frequently performed in the Confederate service, and this opinion is corroborated by the records that have been preserved, and by the statements of Professor J. F. Heustis,[4] of Mobile, of Dr. S. M. Bemiss,[5] Director of Hospitals, of Dr. R. H. Taylor,[6] of Memphis, and of Dr. A. C. Crymes,[7] of Alabama. I have recorded but fifteen operations by Confederate surgeons, and have felt compelled to exclude others on account of the insufficiency of the data. Among these are three cases reported by Dr. C. A. Rice,[8] of Brandon, Mississippi, as follows:

---

[1] Letters from Leavenworth, Kansas, dated October 17th, November 9th, 26th, and December 15th, 1868.
[2] Letter of October 23, 1868.
[3] Letter of October 20, 1868.
[4] Letter of October 12, 1868.
[5] Letter of October 15, 1868.
[6] Letter of October 22, 1868.
[7] Letter of November 27, 1868.
[8] Having learned through Dr. P. F. Whitehead, of Vicksburg, that Dr. Rice had excised the head of the femur for gunshot injury in the case of a stout man wounded at the siege of Vicksburg, I addressed a letter of inquiry to him, and he politely furnished me with the subjoined particulars in a letter from Brandon, Mississippi.

CASES.—' In regard to a case of excision at the hip-joint, in the case of a stout, robust man, during the siege of Vicksburg, * * I can only state from memory. I do not remember his name or command. He was some distance in rear of the works, cooking. The oven was minus a leg. He substituted a spherical case shot or bomb, and the fire soon exploded it. A fragment of the shell and one of the balls struck the neck or periphery of the femur, shattering it very badly, and producing dislocation. After making the incision, (and I do not now remember what kind I made, as we were compelled so often to disregard surgical instructions, and improvise our cuts to suit the lacerated condition of the parts,) I found the head and neck very badly shattered or comminuted. I resorted to excision, without the loss of more than two or three tablespoonfuls of blood, dissected out the synovial sac carefully, and coaptated the parts with sutures and adhesive plaster and a light roller. The patient did well until the fourth day, when meningitis supervened, which speedily terminated the case. The wound was doing well, and an examination of it after death proved that the reparative process was going on finely. The cause of the meningeal disturbance I cannot tell. I certainly do not regard it as a metastasis to that point. I think it was purely idiopathic, and not dependent in any respect upon the wound or operation. I have had two more cases of excision of hip-joint since then, and both were successful, as far as I know. They were doing well at the end of fifteen days. I have never seen or heard of them since. Properly managed, I believe the hip-joint can be excised, if it is a primary operation, with almost as much success as the shoulder-joint."

At the meeting of the American Medical Association in Washington, in May, 1868, Dr. J. M. Keller, of Memphis, Tennessee, informed a number of the delegates that he had twice excised the head of the femur for gunshot injury. Recently, Dr. Keller writes from St. Louis, whither he has removed, that his notes having been destroyed in a conflagration at Mobile, it is impossible for him to furnish the particulars of these cases.

Dr. A. M. Fauntleroy,[1] of Staunton, Virginia, regrets the loss of his notes of a case of excision of the head and neck of the femur, treated in the military hospital at Staunton:

"I cannot," Dr. Fauntleroy writes, "give you the man's name, nor his exact condition when he was transferred to an interior hospital. I remember well that the case was tedious or slow in its progress; the man becoming much reduced at one time in his physical condition, and demanding the most nutritious regimen; the limb being suspended two months or more in Smith's anterior splint. During this time an exfoliation of bone occurred from the surface of the great trochanter, which I removed instrumentally."

On the register of the Jackson Hospital in Richmond, there is recorded the case of Private J. M. Pucket, Co. I, 27th Georgia Regiment, admitted in July, 1862, for a "gunshot wound of the right hip." Then follows the entry "resection of bone" and "furloughed November 4, 1864, for sixty days." This man is still living, and it is of the highest interest to determine whether he underwent a formal excision of the head of the femur. Dr. Thomas H. Williams, formerly Medical Director of General Lee's army, and Professor E. S. Gaillard, have kindly endeavored to clear up the obscurity which envelopes the case, but unsuccessfully. It appears most probable that it was an instance of extraction of fragments of a shattered trochanter.[2]

The Transactions of the American Medical Association for 1867, contain an article by Professor Paul F. Eve,[3] in which the author has sought to collect the instances of excision of the head of the femur for gunshot injuries in the Southern armies, and has prepared a

---

[1] Letter dated Staunton, November 14, 1868.

[2] Dr. Gaillard transmits a letter from Pucket, who now lives in Monroe county, Georgia. He says: "I was wounded in the right hip. It was broke. There were twenty-nine pieces of bone extracted by J. M. Dannell, Surgeon 27th Georgia Infantry. It was done at Malvern Hill, Virginia, July 1st, 1862, while charging a zouave battery. The wound still discharges; it has never healed. There are continually pieces of bone coming out." Dr. Williams writes: "I regret to state, that after having made diligent inquiry of several medical officers residing in Richmond, some of whom were on duty at Jackson Hospital at the time referred to, I can obtain no information in regard to the operation or the proper address of the medical officer who performed it. Dr. S. P. Moore, formerly Surgeon General. C. S. A., says he does not remember any medical officer by the name of Dannell or Daniel."

[3] *A Contribution to the History of the Hip-joint Operations performed during the late Civil War.* In *Trans. Am. Med. Assoc.*, Vol. XVIII, pp. 256, 263.

tabular statement of thirteen alleged examples of this operation. I have elsewhere[1] analyzed this table, and pointed out its mistakes and the fallacy of its conclusions. Only seven of the cases, and but one of the five alleged successes, can be accepted.[2]

Dr. D. D. Saunders, of Memphis, Tennessee, has published[3] a paper on excisions containing the following account of excisions of the head of the femur for gunshot injury:

CASES.—"Only three cases of excision of the head of the femur have fallen under my observation. Two of them were primary, and one was operated on twenty-eight days after the injury. The two primary operations proved fatal—one the twenty-second day, and the other the twenty-fifth day succeeding the operation—from exhaustive suppuration. In the case operated on twenty-eight days after the receipt of the wound, I was not present at the operation, which was performed after the battle of Gettysburg, by Dr. Asch, of Philadelphia. The head of the femur, neck, and trochanter were comminuted by a grape shot, which knocked off a portion of the acetabulum. About four and a half inches of the bone was removed, including the head. The subject of the operation was a healthy young man, of twenty-three years of age. I last saw this case eighteen months after the operation. He was walking about with a crutch, and could even walk pretty well with a stick; the wound had entirely healed, and the limb was three inches shorter than the sound one. The upper end of the femur, with the new bony tissue which had been thrown out, was drawn up in contact with the acetabulum, and considerable mobility at the seat of the joint was retained, sufficient, at any rate, to make the limb very useful, much more so than commonly exists after this operation. Excision of this joint, when the blood vessels, nerves, and soft parts are not torn, is greatly to be preferred to amputation."

It is certain that Dr. Asch did not excise the head of the femur after Gettysburg. He did, however, in a case of fracture near the trochanters, in the person of Private Durbin, 9th Alabama Regiment, extract detached fragments, and the patient recovered with a shortened limb, and was exchanged, and may, with the common exaggeration of wounded prisoners, have represented that he had undergone a formal excision. As to the other two cases mentioned by Dr. Saunders, it is impracticable to determine whether they are reported elsewhere in detail or not. These three cases are included in Dr. Eve's table. It also includes a case reported by Surgeon B. W. Avent, as follows:

CASE.—"An officer received a gunshot fracture of the neck of the femur; two fingers at entrance and exit fully explored the wound; date of injury unknown. The wound was dilated at entrance and spiculæ removed with saw and forceps. He lingered for months, but finally recovered with a limb four inches short, but able to walk with the assistance of a cane."

---

[1] *Buffalo Medical Journal*, Vol. VIII, p. 21, et seq., and *American Journal of the Medical Sciences*, Vol. LVI, p. 131.

[2] The operations numbered in the Table 1 and 2, (Read), 4 (Ladd), 5 (Grant), 7 (Cullen), 8 and 9 (Gilmore), are authentic, and correspond with CASES XLIII, LVIII, XXII, XXIII, XII, IX, and XVI, of this report. Case 3 of the Table was an example of consolidated gunshot fracture of the trochanteric region; it is proved by the report of Surgeon H. S. Hewit, U. S. V., and by a letter from Dr. Miltenberger, who is cited as the operator by Dr. Eve, that no operation whatever was performed. In Case 6 there is nothing in the report to indicate that the hip-joint was implicated. Assistant Surgeon M. J. Asch, U. S. A., cited as the operator in Case 10, positively disavows any knowledge of the operation. No names or dates are given in connection with Cases 11 and 12, and there is no reason to believe that they are not identical with cases reported elsewhere in detail. Case 13 is pronounced not to have been an excision of the head of the femur by the authority cited by Dr. Eve, as will be seen from the following letter:

"NEW ORLEANS, LA., *October 5th*, 1868.

"COLONEL: I have delayed answering your communication in consequence of a press of business for the last two or three days; but it gives me pleasure to furnish the information which Assistant Surgeon Otis desires.

"When Professor Eve wrote to me on the subject of gunshot injuries of the hip, I somehow conceived the impression that his object was to ascertain merely the percentage of mortality in cases where surgical interference had been attempted. One of my cases I thought illustrated this, though not falling strictly under the head of 'excisions of the head of the femur,' but as his printed slip for analysis was small, I put it under this head as most convenient. The circumstances of the case were these: The patient had received a gunshot wound, from a minié ball, which had fractured the neck of the femur, the ball passing out near the ischiatic notch. An attempt was made at the time to remedy the injury as far as possible, but, after resecting the lower end of the femur, I desisted from further interference, as the hæmorrhage was very great, and I had no competent assistant at hand. Some ten days or two weeks after, as well as I can now recollect, the patient was laid on the table for the purpose of excising the head of the femur, for the examination had showed that it was extensively split up. After the primary incisions had been made, and several fragments of the neck had been removed, the operation was suspended, as the patient was suffering from the effects of the chloroform administered, and as the free explorations then made showed that the condition of the soft parts was such (from sloughing, burrowing of pus, &c.) that death was inevitable.

"As I am not a surgeon, and have never practised surgery, except for the exigencies of war, I had no vanities to gratify in the case, which was not a flattering one, and should not have reported it if I had entertained any idea that a strict classification was desired.

"I am, very respectfully, your obedient servant,

(Signed)   "J. DICKSON BRUNS."

"Bvt. Lt. Col. J. F. RANDOLPH,
  Surgeon U. S. Army, Med. Dir. Dept. Louisiana."

[3] SAUNDERS, *Excision or Resection of the Bones and Joints of the Lower Extremity.* Memphis Medical Monthly, April, 1866. Vol. I, No. 2.

It is sufficient to state the case to indicate that it was not an excision of the head of the femur. The apocryphal operations ascribed to Drs. Bruns and Miltenberger constitute the fifth and sixth entries that must be rejected from Dr. Eve's table of thirteen cases.

The record in the much disputed case of Private J. G. Durbin,[1] 9th Alabama Regiment, is as follows:

CASE.—Private J. G. Durbin, Co. I, 9th Alabama Regiment, aged 23 years, received at Gettysburg, July 3, 1863, a fracture of the upper third of the left femur by a grape shot. He was left on the field, and was made a prisoner and placed in hospital. The case is recorded in the register of Confederate wounded as a gunshot wound of left thigh. The wound was examined by Assistant Surgeon M. J. Asch, U. S. A., who found the femur comminuted just below the trochanters, and extended the wound by an incision and extracted a number of large detached fragments. The injured limb was then placed in a fracture apparatus, and the patient was soon after sent to the General Hospital, at Chester, Pennsylvania, where he was admitted on July 18th, 1863. The case is here recorded as a fracture of the upper third of the left femur by a grape shot. No details of the progress or treatment of the case are recorded in the register of this hospital. Six months subsequently he was transferred to the McClellan Hospital, at Philadelphia, where he is reported as admitted on January 18th, 1864, with a partially united fracture of the upper third of the left femur. On February 12th, 1864, he was moved to Broad and Prime Streets Hospital, a depot for transfers, and on the following day, February 13th, he arrived at the West Buildings Hospital, at Baltimore, a depot for convalescent Confederate prisoners. He is registered at this hospital as a case of fracture of the upper third of the left femur by grape shot, with resection of fragments. On April 25th, 1864, he was sent to Fort Monroe for exchange. His name appears on the register of the Howard Grove Confederate Hospital, at Richmond, Virginia, in the following entry: "Private J. G. Durbin, Co. I, 9th Alabama, *vulnus sclopeticum*, with excision of the head of the left femur, was examined by the medical examining board at the Howard Grove C. S. A. General Hospital, October 11, 1864, and granted a furlough of sixty days. Destination and post office address Montgomery, Alabama."

This is believed to be the case regarded by Dr. Saunders and the Richmond Examining Board as a successful excision of the head of the femur by Dr. Asch.

Another doubtful case, of a Confederate prisoner, is recorded in the case-books of the Chattanooga General Hospital No. 1, and in the books of the Medical Director of the Department of the Cumberland, Surgeon and Brevet Colonel George E. Cooper, U. S. A. The patient declared that he had undergone an excision of the head, neck, and trochanters of the femur; but the surgeon in charge of the hospital, Surgeon John H. Phillips, U. S. Volunteers, describes the operation as an excision of the trochanters and upper portion of the shaft:

CASE.—Private John W. Brewer, Co. C, 4th Tennessee Cavalry, aged forty years, was wounded in a skirmish in front of Dalton, Georgia, on August 14th, 1864, by a musket ball, which shattered the right trochanter major. Five hours after the reception of the injury, he was placed under the influence of chloroform, and his regimental surgeon, Dr. Sevenson (also recorded as Swanson and Swainson) excised four and a half inches of the femur, including the trochanter major, which was badly smashed. The wounded man subsequently fell into the hands of the U. S. troops, and was placed, on October 25th, 1864, in a hospital train, and sent by rail to Chattanooga, a distance of thirty miles. He was there admitted to Hospital No. 1, on October 25th, 1864, and entered on the register as a case of "gunshot fracture of the right femur near trochanter; exsection, prior to admission, of four and a half inches of the bone, including great trochanter." The case-book of the hospital further states that the general condition of the patient was excellent, that there was very little inflammation of the soft parts about the wound, and that the incision made over the trochanter was nearly healed. The subsequent entries regarding the case refer to the ordering of tonics and stimulants and nourishing diet and occasional laxatives, and to the transfer of the patient, cured, on March 13th, 1865, to a military prison. On March 20th he was returned to hospital, on account of the opening of an old sinus, and he was sent back to the prison depot to be forwarded to his home on June 8th, 1865. His wound was then firmly healed, and the limb was shortened two and a half inches. He was released on June 16th, 1865, at Louisville, Kentucky, under the provisions of General Order 109, A. G. O.

RECAPITULATION.

The record of the surgery of the war as relates to excisions of the head, or of the head and neck, or head, neck, and trochanters of the femur for gunshot injury, comprises the histo-

---

[1] Also registered as J. W. Dulin, J. G. Derbin, and John Dribben.

RECENT ADDITIONAL CASES. 59

ries of sixty-three authenticated cases, and the reports of twenty-one[1] other cases, some of which are described incorrectly, and the remainder so indefinitely as to be useless for statistical purposes.

The results of the sixty-three authenticated cases are set forth in the following table:

EXCISIONS OF THE HEAD OF THE FEMUR FOR GUNSHOT INJURY.		CASES.	SUCCESS-FUL.	FATAL.
PRIMARY	U. S. Armies	21	1	20
	Confederate Armies	11	1	10
INTERMEDIATE	U. S. Armies	20	1	19
	Confederate Armies	2	1	1
SECONDARY	U. S. Armies	7	1	6
	Confederate Armies	2	0	2
TOTAL	U. S. Armies	48	3	45
	Confederate Armies	15	2	13
AGGREGATE		63	5	58

ADDITIONAL EXCISIONS AT THE HIP FOR GUNSHOT INJURY.

Dr. C. Heine mentions[2] that in the second war in the Danish Duchies, in 1864, Dr. J. Neudörfer, of the 8th Austrian Army Corps, performed two (intermediate?) excisions of the head of the femur with fatal results. Dr. Heine adds that he believes that no excisions of the head of the femur were performed in the Prussian hospitals during the Seven Weeks' War, and Dr. J. Heyfelder,[3] after a minute inspection of the hospitals of Berlin, Dresden, Prague, and the numerous depots for wounded in Bohemia, makes the same report. But in their recent works on the surgery of that war, Dr. Stromeyer[4] and Dr. Bernhard von Langenbeck[5] record five examples of the operation, with one complete recovery, and Dr. Bernhard Beck,[6] adds an account of a sixth operation, which resulted unsuccessfully. The following are concise abstracts of these six cases:

CASE.—"Sigismund von Kurcharsky, aged 18 years, was wounded in a skirmish at Olschowa, in Poland, March 22, 1863. The ball entered behind the left trochanter major and lodged. Profuse suppuration followed, and led to great exhaustion, with hectic fever. In July, 1863, excision of the head of the left femur was performed by Dr. B. Von Langenbeck. A semi-circular incision was made behind the trochanter major; the bone was sawn immediately below the trochanter major; the head of the femur was then exarticulated, and the ball and fragments of the ilium were removed. The operation produced but slight shock. Profuse suppuration persisted, and the patient sank and died fourteen days after the operation. The acetabulum was found to be fractured and perforated, and denuded of cartilage."

CASE.—"Private K., an Austrian, was wounded July 3, 1866, at the battle of Sadowa, by a musket ball which entered behind the prominence of the trochanter, fractured the neck of the femur, and made its exit at the inner side of the thigh. He was

---

[1] Cases of Drs. Batwell (1), Azpell (2), Brock (2), Woodward (2), Rice (3), Keller (2), Fauntleroy (1), Dannell (1), Eve (6), Sevenson (1). Total 21.

[2] *Die Schussverletzungen der unteren Extremitäten, nach eigenen Erfahrungen aus dem letzen Schleswig-Holsteinschen Kriege*, in Langenbeck's Archiv., B. VII, S. 655.

[3] *Gazette Médicale de Paris*, 1867, pp. 496, 509, 539, 557.

[4] STROMEYER, *Erfahrungen ueber Schusswunden im Jahre 1866*. Hannover, 1868, S. 52.

[5] B. LANGENBECK, *Ueber die Schussfracturen der Gelenke und ihre Behandlung*. Berlin, 1868, S. 15.

[6] BERNHARD BECK, *Kriegs-Chirurgische Erfahrungen während des Feldzuges 1866 in Süddeutschland*, Freiburg, 1867, S. 351.

placed in a hospital at Horic. Profuse suppuration come on, and, on August 4th, there were rigors. On August 5th, the head, neck, and trochanters were excised by Dr. B. von Langenbeck, by raising a semi-circular flap over the trochanter. The operation was bloodless, and produced but little depression. The patient died on August 14, 1866, of pyæmia."

CASE.—"Emil Dauer, 10th Battalion of Saxon Infantry, was wounded at the battle of Gitschin, in Bohemia, on June 29th, 1866, by a musket ball, which entered half an inch behind the right trochanter major, and lodged about the hip-joint. The patient was received in the hospital at Zittau. The fracture seemed to be limited to the epiphysis. The wound gave rise to a copious suppuration with burrowing of pus, and denudation of the periosteum of the innominatum. On August 20th, 1866, excision of the head of the femur was performed by Dr. von Langenbeck, and the ball and fragments of the neck were extracted. The neck was sawn off smoothly at its base; the trochanter was uninjured. The patient died early in September, of exhaustion."

CASE.—"Private Maxim Glutschak, aged 24 years, 8th Austrian Lancers, was wounded at the battle of Sadown, or Königgrätz, July 3d, 1866, receiving a gunshot fracture of the head of the left femur. On August 22d, 1866, Dr. Schönborn excised the head of the femur, making a curved incision behind the great trochanter, disarticulating, and sawing the bone close below the trochanter minor. In May, 1867, the patient was entirely cured. The limb was shortened nine centimetres; but the patient could walk very well. The head of the bone was shattered into fragments, and it is quite probable that the acetabulum was fissured."

Dr. Bernhard Beck describes his case as follows:

CASE.—"O. F. S., drummer in the King of Wurtemburg's Battalion, was wounded at Tauberbischofsheim, July 24th, 1866. The ball entered to the outer side of the anterior crest of the ilium, and, passing downwards, shattered the base of the neck and the trochanter major, and, splitting, one portion passed into the gluteal region, and the other fractured the trochanter minor. On August 2d the patient was admitted to hospital, and the limb was placed on air pillows on a double inclined plane. The discharge from the wound was fœtid and profuse. Nutritious diet, beef-essence, wine, etc., were ordered, with the hope of placing the patient in a fit condition to undergo an operation; but, on August 5th, a copious hæmorrhage rendered immediate interference unavoidable. A slightly curved incision was made on the outside of the thigh, the muscular attachments were separated, and one by one the shattered fragments of bone were removed, the larger one being peeled from the periosteum with the handle of the scalpel. The shaft was then sawn across, and the bend of the bone was exarticulated and removed. The wound being cleansed, and no divided vessels being found, the bleeding was supposed to have come from the lacerated muscles. The angles of the wound were now united by sutures, the middle portion being left open and stuffed with charpie to absorb the secretions. The limb was then placed in a straight splint. The patient reacted well. The excised portions of bone consisted of twenty-three fragments. In twenty-four hours the wound suppurated freely, but there was a good deal of blood mingled with the discharges. The patient gradually grew very weak, the pulse became small and frequent, delirium supervened, and on the following day he died, August 7th, 1866."

Dr. Beck states that he was convinced that had it been in his power to perform a primary operation, the result would have been different.

In his account of the treatment of the wounded after the battle of Langensalza, June 27, 1866, Dr. L. Stromeyer reports the following case:*

CASE.—"In a very debilitated subject, who was suffering excessive pain, with intra-capsular gunshot fracture of the neck of the femur, excision was performed on the twentieth day, (July 17th,) through a crucial incision, with very little hope of a successful result. In an anatomical point of view, the case was a very favorable one for the resection, as the head of the femur was completely separated from the neck, so that it was only requisite to exarticulate the former and to even off the latter with a saw. Death took place on the twenty-second day, (July 19th, 1866)."

An intermediate excision at the hip, performed in 1867, by Surgeon Glover Perin, Brevet Lieutenant Colonel, U. S. A., may be included with the eight operations above referred to. The abstract is compiled from the quarterly report of surgical operations from Newport Barracks, Kentucky, October, 1867:

CASE.—Private Francis Ahearn, aged 30 years, U. S. General Service, was wounded at Newport Barracks, Louisville, Kentucky, on July 31, 1867. He was a prisoner in the guard-house, and was shot by a sentinel while attempting to escape. The ball entered behind and below the prominence of the right trochanter major and passed inwards and upwards, emerging on the anterior part of the thigh, two inches below Poupart's ligament, a little to the outside of the course of the femoral artery, having shattered the upper part of the femur, the fissures extending within the joint. The wounded man was immediately taken to the Post Hospital, and was examined by Colonel Perin, the surgeon in charge. The patient had been an habitual drunkard for years, and had *mania a potu* when shot. The shock of the injury was so great that an operation was not considered advisable. It was determined to adopt a supporting treatment, and to endeavor to build up the general health, with a view of operating at the first favorable moment when a good result could be reasonably anticipated. On August 26th, 1867, the patient was in a

---

* STROMEYER, *Erfahrungen über Schusswunden im Jahre 1866 als nachtrag zu den Maximen der Kriegsheilkunst.* Hannover, 1867, S. 52.

RECENT ADDITIONAL CASES.   61

better condition than at any time subsequent to the reception of the injury. The pulse was at 90; there had been troublesome diarrhœa, but it was somewhat abated; the injured limb was much wasted, except at the upper part of the thigh, where it was greatly swollen; the discharge from the wound was very copious, and there was extreme pain on the slightest movement. There were abscesses about the joint communicating with its cavity. Excision having been decided upon, Surgeon Perin, assisted by Assistant Surgeon T. E. Wilcox, U. S. A., proceeded with the operation. The patient being rendered insensible by a mixture of chloroform and ether, the entrance wound was enlarged by a straight incision downwards three inches in length. The head of the bone was disarticulated, and the shaft was sawn several inches below the lesser trochanter. The wound was then cleansed and approximated. Scarcely any hæmorrhage took place, no ligatures being required. On recovering from the anæsthetic, the patient complained of great pain and nausea. Brandy was administered, and half a grain of sulphate of morphia; but there was such irritability of stomach that everything was rejected. A quarter of a grain of sulphate of morphia was then administered hypodermically, and this relieved the pain. But there was no decided reaction, and, sinking gradually, the patient died from the shock of the operation twenty hours after its completion. No autopsy was made. The shattered excised bones were sent to the Army Medical Museum, and are represented in the adjoining wood-cut. (FIG. 27.) Many of the fragments were carious.

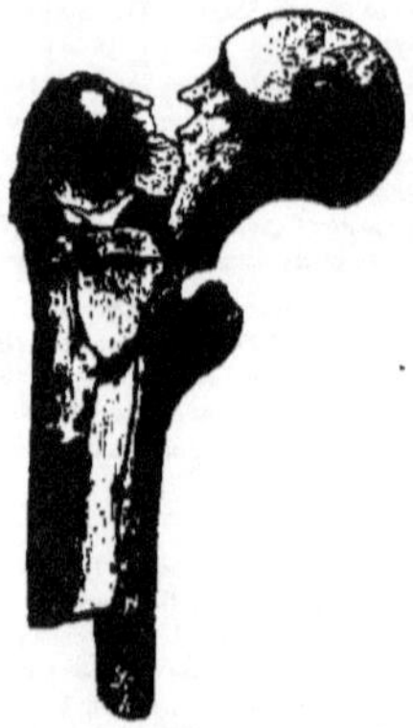

FIG. 27.—Shattered upper extremity of the right femur, excised for caries following gunshot fracture.—*Spec.* 5499, Sect. I, A. M. M.

Of the nine recent cases above enumerated,* eight, including the successful operation by Dr. Schönborn, were performed, it is believed, during the intermediate stage. In one of Dr. von Langenbeck's cases, the operation was deferred until the secondary period, having been performed nearly four months subsequent to the reception of the injury.

The results of the sixty-three authenticated cases of the American War, of the twelve operations recorded prior to that war, and of the nine later operations, are consolidated in the following table:

EXCISIONS AT THE HIP FOR GUNSHOT INJURY.		CASES.	SUCCESSFUL.	FATAL.
CASES PRIOR TO 1861	Primary	7	1	6
	Intermediate	3		3
	Secondary	2		2
CASES IN THE AMERICAN WAR	Primary	32	2	30
	Intermediate	22	2	20
	Secondary	9	1	8
RECENT CASES	Intermediate	8	1	7
	Secondary	1		1
AGGREGATE		84	7	77

* I have been informed by a surgeon who visited the Austrian hospitals after the battle of Sadowa, that he witnessed an unsuccessful intermediate excision at the hip, at the hospital at Brünn, by Professor Klopach, of Breslau, who had performed the operation in several cases of gunshot injury. But I can find no published account of these cases.

TEMPORIZATION AND AMPUTATION AT THE HIP COMPARED WITH EXCISION.

It is difficult, if not impossible, to obtain statistics of the results of gunshot injuries involving the hip-joint, and treated on the expectant plan, which shall be complete, and, at the same time, accurate. The difficulties of diagnosis in these cases are notorious. The best surgeons are sometimes at fault in determining the extent of the lesions produced by a ball passing towards the head or neck of the femur. The problem, then, would be sufficiently complex, if only precise reports from skilled observers were to be dealt with, and cannot be solved if complicated by vague statements or unfounded assertions by incompetent witnesses. If all the alleged recoveries without operative interference after gunshot fractures implicating the coxo-femoral articulation were cited, the results of temporization would appear in a far too favorable light. But, if any cases are rejected, the advocates of the expectant method object to the incompleteness of the returns. It is to be remembered, however, that, in statistical inquiry, a large series of unselected instances, or of cases impartially selected, may often be as valuable as a complete series; and hence there is encouragement to group together the examples of gunshot injury of the head, neck, or trochanters of the femur, or of wounds involving the hip-joint primarily or secondarily, which have been submitted to expectant treatment, and to endeavor to ascertain the average results. In so doing, I shall exclude, of the cases recorded on the registers of the Office, only those in which the diagnosis is wanting in precision, and shall present also a number of cases in which, although there is a possibility that fissures may have extended within the capsule, or that extra-capsular fracture may have induced traumatic arthritis, it is improbable that the hip-joint was injured primarily.

Three modern writers on military surgery—Dr. Hermann Demme,[*] of Berne, Professor N. Pirogoff,[†] and Dr. S. W. Gross[‡]—have taught latterly that the expectant plan of treatment of gunshot injuries of the hip-joint is inadequately appreciated by surgeons, and have collected instances to prove that the commonly received opinion that such injuries are uniformly fatal when abandoned to the resources of nature, is altogether fallacious.[§]

Professor Pirogoff states his opinions, as formed from his Crimean experience, as follows. It must be premised that Pirogoff divides gunshot injuries about the hip-joint into three categories, viz: 1. Fractures complicated with injury to the acetabulum; 2. Wounds from balls entering through the trochanter or through the groin; 3. Cases followed by immediate symptoms of fracture of the neck of the femur, as shortening, eversion, etc.:

---

[*] *Specielle Chirurgie der Schusswunden*, Würzburg, 1864, S. 348.
[†] *Grundzüge der Allgemeinen Kriegschirurgie*, Leipzig, 1864, S. 814.
[‡] *The American Journal of the Medical Sciences*, Vol. LIV, p. 417.
[§] "Those surgeons who pointedly condemn all efforts to save the limb after gunshot fracture of the surgical or anatomical neck of the thigh bone, exercise, in our judgment, but little discrimination, when they declare, as does the surgical historian of the late American War, that 'experience has demonstrated the uniform fatality of gunshot fractures of the head and neck of the femur, when abandoned to the resources of nature.' *No statement can be farther from the truth.*" Dr. S. W. Gross, *loc. cit.* The compiler of the Surgical Report in *Circular* No. 6, S. G. O., 1865, should have said "*almost* uniform fatality;" but his statement, if it did not represent the exact truth, was assuredly not very far from it.

"The summary of my observations made during my inspections of hospitals, together with observations derived from other surgeons, although too few in number after all, would lead me to conclude that it is preferable to subject cases of gunshot fractures of the upper epiphysis and articular extremity of the femur, in the early period, to an expectant conservative treatment. Of twenty cases of complete recovery which I found in hospitals, more than half belonged to the second and third categories. An exact diagnosis was at first left entirely out of view in most of them, the nature of the injuries being only determined by the subsequent profuse suppuration, the separation of splinters, considerable shortening of the limb, and the formation of callus in the region of the trochanters. The healing process, therefore, occurred under the most unfavorable circumstances, the wounded men being placed among other patients, and having been transported a long distance in a very uncomfortable way, and entrusted to the care of physicians who had no idea of the gravity of their injuries. I have already cited three successful cases (out of six) of expectant conservative treatment of gunshot fractures of the hip-joint observed by Legouest. Hyrtl cites in his anatomy one case observed by himself, in 1848, a fracture of the neck of the femur caused by a musket ball. After profuse suppuration and the elimination of splinters, the patient recovered in thirteen months. Demme mentions two cases observed by himself in the Italian war, and also the case of Brandish. In one of Demme's cases a piece of the cartilaginous covering of the acetabulum became exfoliated, and the limb, though shortened, was well in five months. In the other, anchylosis took place after nine months. In the case of Brandish, total exfoliation of the head of the femur took place. Out of twelve cases of excisions of the hip-joint that have been performed for gunshot injuries up to the present time, only one, that of O'Leary, in the Crimea, was successful."

Dr. Demme (*loc. cit.*) claims that the experience of the Italian War of 1859 speaks loudly in favor of expectant treatment in gunshot injuries of the hip-joint, and in opposition to operative measures; declares that it should be regarded as a maxim in military surgery to exhaust all curative conservative means in such injuries before having recourse to an operation; and states that these opinions are sanctioned by the leading military surgeons of the present day. He also reports the two cases of gunshot fracture of the neck of the femur cited by Pirogoff, in which recovery took place after the extraction of detached fragments.

Dr. S. W. Gross (*loc. cit.*) quotes the observations of Pirogoff and Demme, and adds three alleged cases of recovery from gunshot injuries of the hip under the expectant treatment from the reports of the American War. It is observable that he derives no illustrations from his own extended observations of the surgery of that war. He claims to present a series of thirty-one recoveries from gunshot injuries of the hip-joint treated on the expectant plan, a much larger number of successes than have been obtained by amputation or excision.* The best means of dispelling this grave misapprehension, will be to examine this series of cases in detail.

Had Professor Pirogoff observed, as Dr. Gross reports, "not less than twenty cases of unquestionable injury of the hip-joint by projectiles of war, in which conservative treatment was successful," one might inquire why he amputated at the hip in the latter part of the Crimean War in not less than eight cases, with unvarying fatality, upon patients whom he describes as "in almost every instance anæmic and unfit to undergo so grave a mutilation." But Pirogoff's statement really is, that more than half of the recoveries from gunshot wounds about the hip, which he saw or heard of, belonged to his second and third categories. Now, his third category alone embraces cases of intracapsular fractures, revealed by unequivocal signs. From such a general statement, no precise deductions can be drawn.

Dr. S. W. Gross proceeds to state that "Legouest has witnessed six cases of fracture of the neck of the femur in which neither amputation or resection were resorted to, and of these three recovered." To refute this extraordinary assertion, it is only necessary to quote M. Legouest's report of these six cases:†

---

* The cases are cited from Demme (2), Larrey (1), Hyrtl (1), Pirogoff (20), Legouest (6), Hamilton (1), Miles (2), Brandish, (1); total 34. Dr. Gross, by a misprint, probably, refers to them as a series of 31 cases. But with his generous way of dealing with statistics, such discrepancies are of little moment.

† *Recueil de Mémoires de Médecine, de Chirurgie et de Pharmacie Militaires.* Deuxième Série, T. XV, p. 237.

1. "M. X., présenté en 1854, à la clinique chirurgicale du Val-de-Grâce, reçut en 1812 un coup de feu dans *les trochanters* du côté gauche; * * * sa cuisse est consolidée, etc." He was afterwards able to walk with a limb shortened five inches.

2. "M. X., lieutenant d'artillerie en 1843 fut blessé un duel d'un coup de feu *dans les trochanters*, côté droit. Après quelque hésitation, M. Sédillot conserva le membre et traita la fracture par l'extension." * * * He recovered and was able to walk with a shortening of a little more than an inch.

3. "Le nommé Tanguel, chasseur au 2ᵉ bataillon d'infanterie légère d'Afrique, reçut dans une sortie en Janvier, 1841, à Cherchell, une balle qui lui fractura le fémur dans *les trochanters* du côté droit." * * * He recovered, and, in 1855, pursued the trade of a weaver at Schlestadt.

4. "Garakowski (Simon) prisonnier Russe, * * âgé de 40 ans, reçut à la bataille d'Alma, 20 Septembre, 1854, un coup de feu qui lui fractura le fémur gauche *dans le trochanters*." * * * He died a fortnight afterwards. M. Legouest states that he did not amputate at the hip-joint in this case, because the patient was suffering from severe chronic diarrhœa.

5. "Mouchard (Pierre) clairon au 3ᵉ régiment de Zouaves, * * * âgé de 24 ans; blessé le 5 Novembre, à Inkermann, entré à l'hôpital le 13. Coup de feu ayant pénétré à la racine de la cuisse gauche par le côté interne; sorti à la hauteur de la partie supérieure du grand trochanter: le doigt, introduit dans la plaie, reconnaît que le col du fémur *tout à la base*, à été traversé par le projectile, de bas en haut et de dedans en dehors; il est fendu en deux moitiés, l'une postérieure, l'autre antérieure; à cette dernière est resté attaché presque tout le grand trochanter; immédiatement au *dessous* de cette éminence, fracture comminutive avec esquilles nombreuses. Nous proposons à ce blessé de l'opérer en lui conservant le membre, c'est-à-dire de lui faire une résection. * * * Le 19 * * nous proposâmes * * l'amputation comme dernière ressource; il nous fut impossible de le convaincre, et ce malheureux succomba le 24, avec une gangrène de tout le membre."

6. "Delos (Jacques), fusilier au 6ᵉ de ligne, * * * âgé de 26 ans, fut blessé le 5 Novembre, à Inkermann, et entra à l'hôpital de l'érn le 21. Il à reçu un premier coup de feu qui lui à fracturé comminutivement l'avant-bras gauche; un second coup de feu lui a traversé le flanc gauche; pénétrant à trois travers de doigt en dehors de l'ombilic, sortant vers le milieu du carré des lombes; enfin un troisième coup de feu lui à *brisé la cuisse gauche dans les trochanters.*"

Throughout his memoir M. Legouest refers to these six cases as "fractures in the trochanters." In the fifth case alone, the neck was injured, "quite at its base;" and, in this case, M. Legouest proposed to excise the head of the femur; but the patient having refused this, and a subsequent proposition of amputation at the hip-joint, died on the nineteenth day, with gangrene of the entire limb. Neither Professor Pirogoff nor Dr. Gross would misrepresent intentionally, yet here are six cases which they record as illustrations of the advantages of expectant treatment in gunshot injuries of the hip-joint, when no injury of the joint was suspected except in a single case, and in that excision was proposed, and subsequently amputation, and the patient finally died. Thus a proper estimate may be placed upon the value of the evidence which it is not practicable to examine in detail, adduced by these authors.

Another case is cited from the elder Larrey, and purports to be that of "an officer who was struck, in Egypt, by a ball, which entered the neck of the femur. He recovered from the injury, but died twenty years subsequently of disease of the chest, when the ball was found to be impacted in the neck of the bone." To prove the correctness of the diagnosis in this case, Dr. Gross refers to the specimen, presented by Larrey to Seutin, now deposited in the Museum of Val-de-Grâce, and figured at p. 649 of M. Legouest's *Chirurgie d'Armée*. What is M. Legouest's description of this specimen? "Le projectile s'est arrêté et enclavé superficiellement dans le milieu de *la ligne réunissant le petit et le grand trochanter du fémur*," and the drawing indicates that the ball lay quite without the capsule.

The case of Brandish, already referred to, (*ante* p. 8), is as irrelevant to the question immediately under discussion, as could be any of the numerous examples of spontaneous exfoliation of the head of the femur destroyed by the various forms of coxo-femoral arthritis.

Dr. Gross quotes a case of recovery after gunshot fracture of the neck of the femur, under expectant treatment, from Dr. F. H. Hamilton,[1] and falls into the error of the latter author in ascribing reports of *two* such cases to Dr. B. B. Miles.[2]

Two cases are quoted from Demme, who states that he observed fourteen examples of gunshot injuries of the hip-joint in the Italian War of 1859, treated on the expectant plan, and that he saved two patients, Borelli, from whom a portion of the acetabulum was extracted, and an Austrian foot-soldier, Franz Veter. Dr. Rocco Gritti, chief surgeon of the great hospital (Ospedale Maggiore) at Milan, after pointing out a number of erroneous quotations and other blunders by Dr. Demme, cites this allegation regarding the fourteen hip-joint fractures with amazement, remarking that but three such injuries had come under the observation of his colleagues and himself.[3] It is certain, that unless the Swiss surgeon's description of these cases is not more accurate than his account of hip-joint amputations and resections, his unsupported assertions should be allowed very little weight.

One other case is adduced by Drs. Gross and Demme: that of a guardsman, treated by Hyrtl.[4] It was an example of recovery after extensive removal of fragments, for Hyrtl speaks of having in his cabinet the pieces forming the neck of the femur.

That the imposing array of thirty-one recoveries under expectant treatment in gunshot injuries of the hip has not been fairly demonstrated by Drs. Pirogoff, Demme, and Gross, will probably be admitted. It is not intended, however, to deny that, in very rare instances, life may be preserved after such injuries, without operative interference. The authors reviewed above might have referred to Cole,[5] and to M. Boinet,[6] for illustrations of such a result more satisfactory than most that they have offered.

To ascertain the average result of expectant treatment in gunshot injuries of the hip-joint, it is proposed to analyze the histories of the cases recorded on the registers of this Office, beginning with those in which the diagnosis is unquestioned. The cases are arranged in alphabetical order:

CASE 1.—"Private Amos Baker, 5th Maine Battery, aged 28 years, was wounded at the battle of Cedar Creek, Virginia, October 19th, 1864, by a conoidal musket ball, which passed through the left hip. The wounded man was conveyed by ambulance, a distance of about fifteen miles, to Winchester, whence he was taken by rail to Baltimore, and admitted, on October 24th, into the National Hospital. An exploration of the wound revealed a fracture of the neck of the femur. The limb was supported by pillows, and cold water dressings were applied. On November 4th, pyæmia supervened. Tonics and stimulants were freely administered. The patient died November 11th, 1864." The case is reported by Surgeon Z. E. Bliss, U. S. V.

CASE 2.—"Corporal James A. Baker, Co. C, 49th New York Volunteers, was wounded at the battle of Fredericksburg, Virginia, December 13th, 1862, by a gunshot missile, which penetrated the hip near the joint. In a few days he was conveyed by steamer to Washington, and was admitted, on December 18th, into Armory Square Hospital. A fracture implicating the joint was recognized. An expectant treatment was pursued, comprising no therapeutic measures save a sustaining regimen with tonics and stimulants, and the support of the limb by pillows. Abscesses formed about the hip, and the patient died exhausted by the profuse suppuration, January 20th, 1863." The case is reported by Surgeon N. H. Ballou, 2d Vermont Vols.

CASE 3.—"Private Joseph Balinea, Co. C, 26th Wisconsin Volunteers, at the battle of Gettysburg, Pennsylvania, July 1st, 1863, received a gunshot fracture of the neck of the femur involving the right hip-joint. He was conveyed to the Seminary Hospital, and treated on the expectant plan. Profuse suppuration and irritative fever soon set in, and the patient died on July 14th, 1863." Report by Surgeon H. Janes, U. S. V.

---

[1] HAMILTON, *A Treatise on Military Surgery.* New York, 1865, p. 397.

[2] *American Medical Times,* Vol. VII, p. 14. The successful cases related by Professor Hamilton and Dr. Miles are those of Private James Vanderbeck and Captain W. A. Bugh. The histories of these cases are detailed in the *Report on Amputations at the Hip-Joint,* in *Circular No. 7, S. G. O.,* 1867, and again in this report. Dr. Miles's paper reports four cases of fractured femur, in only one of which is it alleged that the joint was implicated.

[3] *Delle Fratture del Femore,* etc., Milano, 1866, pagina 75.

[4] HYRTL, *Handbuch der Topographischen Anatomie.* 5. Aufl. Wien, 1865. B. II, S. 534.

[5] J. J. COLE, *Military Surgery; or Experience of Field Practice in India,* London, 1852, p. 139.

[6] *L'Union Médicale,* Juin 23, 1860.

EXCISIONS AT THE HIP,

CASE 4.—"Private William Bancho, Co. H, 3d New Hampshire Volunteers, aged 36 years, was wounded in the action at Weir Bottom Creek, Virginia, June 16th, 1864, by a conoidal musket ball, which penetrated the region of the left hip-joint. On the following day he was conveyed to the Base Hospital of the Army of the James, at Point of Rocks, and, on June 19th, he was transferred to Hampton Hospital, at Fort Monroe. A few days subsequently, he was conveyed by a hospital steamer to Philadelphia, and, on June 21st, he was admitted into Broad and Cherry Streets Hospital. It was now ascertained that the neck of the femur was fractured. The patient was treated by rest, position, and simple dressings. Tonics and stimulants were freely administered. He died, exhausted, on July 6th, 1864." The case is reported by Assistant Surgeon T. C. Brainerd, U. S. A.

CASE 5.—"Corporal Marshal Barden, Co. K, 10th Massachusetts Volunteers, was wounded at the battle of Fair Oaks, Virginia, May 31st, 1862, by a fragment of shell, which struck the hip and comminuted the neck of the femur. Surgeon A. B. Crosby, U. S. V., dressed the limb, and sent the patient, in an ambulance wagon, to the White House Landing, and thence by steamer to David's Island, New York Harbor. He was admitted into De Camp Hospital on June 8th. Death took place on June 13th, 1862." The case is reported by Acting Assistant Surgeon E. Lee Jones.

CASE 6.—"Private Joshua Barnes, Co. B. 52d Indiana Volunteers, was wounded in the hip at the assault on Fort Blakely, Alabama, April 9th, 1865, by a gunshot missile. He was conveyed to his regimental hospital, where it was ascertained that the head and neck of the femur were badly comminuted. The patient died from shock a few hours after the reception of the injury, April 9, 1865." Surgeon E. H. Abadie, U. S. A., reports the case.

CASE 7.—"Private Charles S. Bates, Co. G, 27th Michigan Volunteers, aged 18 years, was wounded in front of Petersburg, Virginia, July 27th, 1864, by a conoidal musket ball, which entered in front and above the left great trochanter, and, passing downwards, lodged about the left hip-joint. The missile was extracted on the field. The patient was conveyed by rail to City Point, and thence by steamer to New York, and was admitted into the McDougal Hospital, at Fort Schuyler, on August 7th. Pyæmia had already supervened, and the patient died a few hours after his admission to hospital, August 7th, 1864." The case is reported by Assistant Surgeon S. H. Orton, U. S. A.

CASE 8.—"Private Anthony B———, Co. C, 12th Illinois Volunteers, aged 23 years, was wounded at the battle of Corinth, Mississippi, October 3d, 1862, by a musket ball, which penetrated the left thigh, directly over the great trochanter. An exploration of the wound revealed a perforation of the left femur. He was taken in an ambulance, a distance of twenty-five miles, to the Tennessee River, and thence by hospital steamer to St. Louis, where he was admitted, on October 15th, into the City General Hospital. The limb was not placed in splints. The patient died on October 23d, 1862. The missile, a round ball, was discovered, post mortem, beneath the integuments in the inner aspect of the thigh. The pathological specimen, with the missile attached, is No. 466, Sect. I, A. M. M. The neck of the femur is perforated obliquely from the anterior portion of the great trochanter. The head of the femur is eroded, and a slight fissure runs into its articular portion." See the adjoining wood-cut. (FIG. 28.) The case is reported by Surgeon J. T. Hodgen, U. S. V.

FIG. 28.—Perforation of the neck of the left femur by a round musket ball.—Spec. 466, Sect. I, A. M. M.

CASE 9.—"Private Chancellor Benjamin, Co. F, 118th Pennsylvania Volunteers, aged 19 years, was wounded in the action at Shepherdstown Ford, Maryland, September 20th, 1862, by a conoidal musket ball, which penetrated just above the right great trochanter, and, traversing through the hip-joint, emerged behind the inner edge of the adductor longus muscle. There was also a severe contusion of the front and middle of the thigh. The patient was conveyed to Sharpsburg, where for several days he was treated at a private house. He was then conveyed to Frederick City, thence by rail to Philadelphia, and, on September 27th, was admitted into Broad and Cherry Streets Hospital. The wounds were in a healthy condition, but the patient was much debilitated from diarrhœa. Astringents, tonics, and stimulants were administered, nourishing food was provided, and warm water dressings were applied; but remedies were of little avail. The diarrhœa continued, and the patient died, exhausted, on November 22d, 1862. The autopsy showed a comminution of the head of the femur. There was a large abscess at the sent of the contusion, filled with a mixture of pus and broken down coagula. The mucous membrane of the rectum and of the lower portion of the colon was much thickened and at several points ulcerated." The case is reported by Dr. John Neill. The pathological specimen was sent to Dr. F. Hartshorne for the Army Medical Museum, but, if received, it was not catalogued.

CASE 10.—"Private Charles R. B———, Co. E, 16th Maine Volunteers, aged 26 years, was wounded at the battle of Fredericksburg, Virginia, December 13th, 1862, by a conoidal musket ball, which entered the anterior and outer aspect of the left thigh and lodged about the hip-joint. He remained in a field hospital, treated only by rest, position, and cold water dressings, until December 19th, when he was conveyed by steamer to the Third Division Hospital, at Alexandria. On the following day Surgeon Edwin Bentley, U. S. V., made an exploration of the wound under chloroform, and it was determined beyond a doubt that the projectile had penetrated the capsular ligament. The patient died from pyæmia, December 25th, 1862. The autopsy revealed the bullet deeply imbedded in the upper anterior portion of the neck; fissures extended from it upon the articular surface and outwards upon the trochanter major." The specimen is numbered 598, Sect. I, A. M. M., and is figured in the accompanying wood-cut. (FIG. 29.)

FIG. 29.—Conoidal musket ball impacted in the neck of the left femur.—Spec. 598, Sect. I, A. M. M.

CASE 11.—"C. F. Beyland, Q. M. D., aged 26 years, was wounded December 8th, 1861, by a conoidal musket ball, which fractured the great trochanter and neck of the right femur. Excision was proposed and refused. The fracture was treated by Hagedorn's apparatus. After protracted and profuse suppuration, the patient recovered with a limb shortened two inches. In July, 1863, he had dispensed with crutches, and walked quite well." Assistant Surgeon C. K. Winne, U. S. A., reported the case.*

CASE 12.—"Sergeant George Bond, Co. A, 137th Illinois Volunteers, aged 19 years, was wounded at the affair at Memphis, Tennessee, August 21st, 1864, by a conoidal musket ball, which penetrated in the region of the right hip-joint. On August 22d he was admitted to Overton Hospital, Memphis. He was treated by rest, position, and simple dressings, and died on August 23d, 1864. The autopsy revealed a fracture of the neck of the right femur." The case is recorded by Assistant Surgeon J. C. G. Happersett, U. S. A.

CASE 13.—"Corporal William Bowen, Co. K, 56th Pennsylvania Volunteers, aged 32 years, was wounded near Spottsylvania, Virginia, May 12th, 1864, by a conoidal musket ball, which produced a fracture of the upper extremity of the femur, extending into the hip-joint. He was conveyed to the field hospital of the 4th Division, 5th Corps. Smith's anterior splint was applied. He died on May 13th, 1864." Assistant Surgeon A. B. Haines, 19th Indiana Volunteers, reports the case.

CASE 14.—"Private Frederick Bowman, Co. F, 1st U. S. Sharpshooters, aged 33 years, received at the battle of Cold Harbor, Virginia, June 3d, 1864, a gunshot fracture of the head of the left femur, the missile striking three inches below the anterior superior spinous process of the left ilium. He was conveyed, June 4th, by ambulance, a distance of about fifteen miles, to the base hospital of the 3d Division, 2d Corps, at White House Landing, and was examined by Surgeon Welling, 11th New Jersey Volunteers, Surgeon Jones, 63d Pennsylvania Volunteers, and Assistant Surgeon Brandt, 110th Pennsylvania Vols., who reported the case as a wound of the *right* hip. He was transferred by steamer to Washington, and on June 11th, was admitted into Carver Hospital. He was treated by rest, position, and simple dressings, with tonics and stimulants. He died exhausted, on June 29th, 1864." Surgeon O. A. Judson, U. S. V., reports the case.

CASE 15.—"Private Thomas B———, 14th North Carolina Infantry, was wounded at the battle of Antietam, Maryland, September 17th, 1862, by a musket ball, which penetrated the right thigh anteriorly, about four inches below Poupart's ligament, and emerged about two inches above the great trochanter. He was presently conveyed in an ambulance, a distance of twenty-five miles, to Frederick City, and admitted into General Hospital, No. 1. The injured limb was placed upon a double inclined plane, suitable dressings were applied, and tonics and stimulants were administered as required. The patient died on October 20th, 1862. The pathological specimen is No. 548, Sect. I, A. M. M. (FIG. 30.) The upper third of the femur is split in three large fragments, and the trochanter major is deeply grooved by the missile. A portion of the fracture is intracapsular. There was no attempt at repair, and the inferior portion of the shaft is roughened and carious." The history and specimen in this case were contributed by Acting Assistant Surgeon W. W. Keen, jr.

CASE 16.—"Captain William A. Bugh, Co. G, 5th Wisconsin Volunteers, aged 35 years, was wounded at the engagement at Williamsburg, Virginia, on May 5th, 1862, and after lying a few hours on the field, he was removed to a temporary hospital, and thence to a hospital transport in the York river and sent to Baltimore, where he was received at the Camden Street U. S. A. General Hospital, on May 10th, 1862. A conoidal musket ball had entered the right groin, passed slightly downward, traversed the line of union between the thigh and trunk, fractured the neck of the femur in its transit, and emerged posteriorly at the fold of the buttock. On flexing or rotating the thigh, crepitus was plainly distinguished. His limb was suspended by Smith's anterior splint, and this treatment was continued for two months. The case progressed without a single untoward symptom, and in the middle of July, 1862, consolidation of the fracture was sufficiently firm to permit the patient's removal to the house of a friend. The limb was shortened one and a half inches. In October, Captain Bugh was able to move about on crutches, and the wounds were entirely healed. About this time he took a journey to Washington, and was promoted to a lieutenant colonelcy in the 32d Wisconsin Volunteers, and placed on recruiting service. He served until April 25, 1863." His recovery was so rapid and uninterrupted that he reluctantly assented to the assurance of his surgeon, Dr. Edward G. Waters, that he would be incapable of active duty in the field. A letter was received at this office from Lieutenant Colonel Bugh, dated June 12th, 1867, more than five years subsequent to his injury, in which he stated that he had partial anchylosis of the hip-joint, and was unable to perform any labor in a stooping posture. Otherwise his condition was satisfactory, though he was more readily fatigued and debilitated than before he was wounded. He suffered no inconvenience from the slight shortening of the femur.

FIG. 30.—Upper third of right femur shattered by a musket ball.—*Spec.* 548, Sect. I, A. M. M.

CASE 17.—"Private W. C———, Co. C, 159th New York Volunteers, aged 37 years, was wounded at the battle of Irish Bend, Louisiana, April 14th, 1863, by a musket ball, which penetrated the left trochanter major, traversed the neck of the femur, and lodged near the hip-joint. He was conveyed by steamer to New Orleans, and admitted, on April 17th, into the University Hospital. The patient was treated only by rest, position, and simple dressings. He died on April 21st, 1863. At the post mortem, the missile was found lodged in the deep structures beneath the central portion of the perinæum. The pathological specimen, with the missile attached, is No. 1291, Sect. I, A. M. M. There is a complete oblique

---

* For a detailed history, see *Circular* No. 7, S. G. O., 1867, p. 73.

fracture through the neck of the femur, with loss of substance at the upper back portion of the anatomical neck, where the missile impinged." (See FIG. 31.) Assistant Surgeon P. S. Conner, U. S. A., reports the case.

CASE 18.—"Private Peter C———, Co. G, 24th Alabama Regiment, aged 57 years, received at the battle of Chickamauga, Georgia, September 19th, 1863, a compound comminuted fracture just below the head of the femur. The wounded man was admitted into the field hospital of General Hindman's Division, Army of the Tennessee, under the care of Surgeon Carlyle Terry, P. A. C. S. The hospital was established in a stable, and was much exposed to the inclemencies of the weather. Profuse suppuration ensued, and there was no disposition to the union of the fracture. The patient died, exhausted, November 5th, 1863." The case is recorded in the *Confederate States Medical Journal*, Vol. I, p. 76.

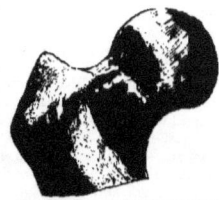

FIG. 31.—Upper extremity of left femur, with a conoidal ball, which lies against the groove it has made in the anatomical neck of the bone.—*Spec.* 1291, Sect. I, A. M. M.

CASE 19.—"Private James Connelly, Co. C, 1st Louisiana Colored Cavalry, aged 25 years, was shot by the Provost Guard at Camp Baton Rouge, Louisiana, on February 2d, 1865. A conoidal musket ball produced a comminuted fracture of the neck of the left femur. The patient was immediately removed to the General Hospital, but succumbed in a few hours to the shock of the injury." Acting Assistant Surgeon J. Roberts reports the case.

CASE 20.—"Private John C———, Co. I, 10th New Jersey Volunteers, aged 19 years, was wounded on August 21st, 1864, at Charlestown, Virginia, by a musket ball, which entered the upper third of the right thigh two inches outside of the course of the femoral artery, fractured the trochanter major, and lodged in the gluteal muscles. He was treated at a temporary hospital at Sandy Hook, Maryland, until September 14th, when he was conveyed to Frederick, Maryland, in a very debilitated condition. Here he was plied with stimulants and tonics, and simple dressings were used. On September 17th, no improvement having taken place, an incision was made in the middle third of the thigh to give exit to pus, which was burrowing beneath the extensor muscles. On October 1st, a further examination of the wound was made, and the missile was removed from midway between the trochanter major and the tuberosity of the ischium. It had probably injured the sciatic nerve, as there was slight paralysis of the limb and intense pain. On November 14th, a piece of cloth was discharged from the wound. He had occasionally fever of a hectic character. On December 16th, there was a decided rigor, followed by high fever, excessive irritability of stomach, and an harassing cough. Symptoms of pulmonary tuberculosis supervened. The wound made little, if any, improvement, evincing at times a tendency to slough. On January 4th, 1865, a piece of necrosed bone was removed. On January 6th, he had involuntary discharges, subsultus, and jactitation. He died, January 9th, 1865. His pulse had become dicrotic two weeks previously, and continued so uninterruptedly until his death. The autopsy revealed vomicæ in the upper lobe of the lungs, while the lower lobes were studded with tubercles in various stages. The hip-joint was disorganized, and the femur was denuded of periosteum nearly to the middle. The trochanter and anterior portion of the neck showed a considerable loss of substance." The specimen is numbered 3806, Sect. I, A. M. M., and is represented in the annexed cut. (FIG. 32). The case was reported and the specimen contributed by Acting Assistant Surgeon F. A. Gove.

FIG. 32.—Secondary lesions resulting from fracture of the trochanter and neck of the right femur by a musket ball.—*Spec.* 3806, Sect. I, A. M. M.

CASE 21.—"Private Daniel Curran, Co. A, 5th Kentucky Volunteers, was wounded at the battle of Chickamauga, Georgia, September 19th, 1863, by a conoidal musket ball, which penetrated at a point near the hip-joint. The patient was soon afterwards conveyed by ambulance to Chattanooga, Tennessee, and admitted, on September 27th, into the General Hospital. He died within a few hours, on September 27th, 1863. The autopsy revealed a fracture of the head and of the femur, with the missile lodged near the cotyloid cavity." The case is reported by Surgeon J. T. Woods, 99th Ohio Volunteers.

CASE 22.—"Private James C———, Co. C, 5th New York Cavalry, aged 21 years, was wounded in the affair at Snicker's Gap, Virginia, July 19, 1864, by a musket ball, which struck the right femur below the surgical neck, passed through both trochanters, and shattered the entire upper third. In a day or two he was conveyed by ambulance, a distance of about thirty miles, to Sandy Hook, Maryland, and thence by rail to Baltimore, and admitted into the Jarvis Hospital. The cellular tissue for eight inches around the place of entrance was gangrenous and sloughing. He died on August 2d, 1864, *of pyæmia*. (!) At the autopsy the missile was found deeply imbedded in the muscles outside of the femur at the middle third of the thigh. The liver and spleen were very soft and easily torn, but the other viscera were normal. The specimen is No. 3189, Section I, A. M. M. A fissure runs in front through the upper third of the shaft, and to the anatomical neck of the femur. The head has been injured, probably post mortem and accidentally." The case is recorded by Acting Assistant Surgeon B. B. Miles.

CASE 23.—"Private Henry Dambach, Co. I, 17th Ohio Volunteers, aged 20 years, in the engagement at Vining's Station, Georgia, on July 9th, 1864, was struck by a musket ball in the right hip. The soldier was conveyed to the hospital of the 3d Division, 14th Corps. A fracture of the neck of the femur involving the joint was ascertained. On July 11th the patient was conveyed, by rail, to Chattanooga, Tennessee, about eighty miles distant, and on July 17th he was received into Hospital No. 1. Death took place on July 24th, 1864." The case is reported by Surgeon C. W. Jones, U. S. V.

CASE 24.—"Private Michael D———, Co. H, 14th New York Heavy Artillery, aged 23 years, was wounded at the assault on Petersburg, Virginia, June 26th, 1864, by a musket ball, which perforated the right thigh and femur from before backwards an inch below the trochanter minor. Fissures extended about three inches down the shaft, and upwards into the neck and trochanter major. D——— was conveyed to the Base Hospital at City Point, and thence transported by steamer to Washington.

where, on July 2d, he was received into the Finley Hospital. Cold water dressings were applied, and the limb was kept in position by pillows, without an apparatus. Death took place July 4th, 1864." The specimen is No. 3261, Section I, A. M. M., and was contributed, with a report of the case, by Acting Assistant Surgeon R. Westerling.

CASE 25.—"Private James A. Deyo, Co. B, 20th Indiana Volunteers, aged 25 years, was wounded at the battle of Petersburg, Virginia, September 10th, 1864, by a conoidal musket ball, which, having perforated the calf of the leg as this was flexed, entered the middle third of the thigh, and lodged somewhere about the hip-joint. He was conveyed at once to the 2d Corps Hospital, where an exploration of the wound failed to detect the position of the missile, but left little doubt of the existence of a fracture. Cold water dressings were applied. Tonics and stimulants were administered as required. Near the close of the month the patient was transferred by steamer to Alexandria, where he was admitted, on October 1st, into the 3d Division Hospital. He was much exhausted from the suppuration and intense pain of the wound. He afterwards experienced great difficulty in voiding his urine and fæces. He lived, however, until November 3d, 1864. The autopsy disclosed the position of the missile, by which the head of the femur had been fractured. No displacement existed. The viscera were healthy." The case is reported by Surgeon E. Bentley, U. S. V.

CASE 26.—"Private James Dice, 43d Missouri Infantry, aged 50 years, was wounded accidentally, early in September, 1864, at a camp near St. Joseph, Missouri, by a conoidal musket ball, which entered the left thigh in front. The neck of the femur was shattered near the base, though incompletely, the fracture not extending entirely through the bone. He was admitted, September 5th, into the post hospital at St. Joseph, too late," the report states, "for the performance of an operation. Simple dressings were applied. Death took place on September 13th, 1864." The case is reported by Acting Assistant Surgeon J. F. Bruner.

CASE 27.—"Private Charles G. Dodson, Co. C, 13th West Virginia Volunteers, aged 46 years, was wounded at the affair at Halltown, Virginia, August 26th, 1864, by a musket ball, which entered at a point over the neck of the left femur, and lodged about the hip-joint. He also received a slight wound of the left wrist. He was conveyed by rail a distance of forty miles to Frederick City, Maryland, and admitted, August 29th, into the General Hospital. The limb was swollen but not painful. Milk punch and full diet was directed. Death occurred suddenly on the morning of August 31st, 1864. The autopsy disclosed the missile embedded in the neck of the femur. There was no lesion of any important viscus." Acting Assistant Surgeon R. W. Mansfield reports the case.

CASE 28.—"Private Alexander D———, Co. B, 43d New York Volunteers, was wounded at the battle of Cedar Creek, Virginia, October 19th, 1864, by a conoidal musket ball, which entered anteriorly in the upper part of the left thigh and lodged somewhere about the hip-joint. On October 22d, he was admitted to the Sheridan Hospital, at Winchester, under the care of Assistant Surgeon J. G. Thompson, 77th New York Volunteers. The patient's general health was good. For several days he suffered comparatively little from the wound, and sat in a chair much of the time very comfortably. On the 30th, he experienced considerable pain in the region of the joint, suppuration was copious, and the thigh was swollen. The pain steadily increased, and on November 4th, was referred in part to the knee-joint. Chills and fever now began to recur daily, and the discharge of pus was sanious. The salts of quinia, with other tonics and stimulants, were administered, and nourishing food. The patient died, exhausted, November 8th, 1864. The autopsy showed extensive burrowing of sanious pus among the congested muscles of the thigh. The ligaments of the hip were so much absorbed that the head of the femur was removed from the acetabulum without aid from the knife. No evidence of the existence of pyæmia was obtained. The pathological specimen is No. 3797, Sect. I, A. M. M. It exhibits an oblique fracture in the upper third, with comminution of the neck, in which the missile is embedded. The bone adjacent to the fracture is carious, and the upper portion of the shaft shows evidence of periosteal inflammation." The specimen is represented in the annexed wood-cut. (FIG. 33.)

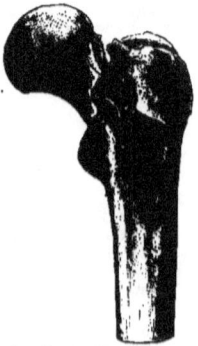

FIG. 33.—Conoidal musket ball impacted in the neck of the left femur.—Spec. 3797, Sect. I, A. M. M.

CASE 29.—"Private E. H. Dunkleberg, Co. F, 11th U. S. Infantry, aged 19 years, was wounded at the battle of Cold Harbor, Virginia, June 3d, 1864, by a musket ball, which, entering the gluteal region, fractured the neck of the left femur. He was captured and conveyed to Richmond, and placed in Hospital No. 21. Death took place on June 19th, 1864." Surgeon G. W. Semple, C. S. A., reported the case.

CASE 30.—"Private N. H. Eisenbower, Co. F, 103d Ohio Volunteers, aged 27 years, was wounded near Allatoona, Georgia, June 19th, 1864, by a canister shot, which entered at a point 'about the neck of the left femur, and lodged.' The patient was conveyed by ambulance to the General Hospital at Allatoona. An exploration of the wound revealed a fracture of the neck of the femur and the lodgement of the missile in the vicinity, whence it was extracted. Cold water dressings were applied. On July 1st, the patient was transferred by ambulance and railroad to the General Hospital at Knoxville, Tennessee. The particulars in the treatment and progress of the case are not known. Private Eisenhower died on July 18th, 1864." Case reported by Brevet Colonel H. S. Hewit, Surgeon U. S. V.

CASE 31.—"Private C. H. Elliott, Co. D, 61st Pennsylvania Volunteers, was wounded at the battle of Fair Oaks, Virginia, May 31, 1862, by several musket balls. One entered near the right acromion process, and emerged at the insertion of the deltoid muscle; another at the gluteal fold; a third passed through the upper part of the thigh; a finger was also shot away. He was conveyed to Portsmouth, and, on June 5th, was admitted into the Balfour General Hospital. A fracture of the head of the humerus was recognized, and, on June 14th, Assistant Surgeon Sheldon, U. S. A., excised the head and about three inches of the shaft. Abscesses formed in the left thigh. Pyæmia supervened, and the patient died on June 17th, 1864. At the autopsy, the head of the left femur was found to be shattered. There was a collection of pus near the wound in the hip." Assistant Surgeon H. L. Sheldon, U. S. A., reports the case.

CASE 32.—"Private J. T. Elliott, Co. H, 22d Georgia Regiment, of Sorrell's Brigade, aged 24 years, received, at the engagement at Deep Bottom, Virginia, August 16th, 1864, a gunshot fracture of the neck of the left femur. He was conveyed to Richmond, and admitted to the Jackson Hospital. He recovered, with three and a half inches shortening of the limb, and, being permanently disabled, and totally disqualified for any military duty, he was retired from the C. S. service on February 17th, 1865." The case is reported by Surgeons A. J. Semmes, Thomas F. Maury, and W. D. Hoyt, members of the Medical Examining Board of the Jackson Confederate Hospital at Richmond, Virginia, in February, 1865.

CASE 33.—"Private David Elmer, Co. M, 14th New York Heavy Artillery, was wounded at the battle of Petersburg, Virginia, June 18th, 1864, by a conoidal musket ball, which fractured the neck of the left femur. He had also a flesh wound of the left forearm. He was conveyed to the hospital of the 3d Division, 9th Corps. The treatment was of an expectant nature. Death took place on June 24th, 1864." The case is reported by Surgeon M. K. Hogan, U. S. V.

CASE 34.—"Private James F———, Co. H, 1st Massachusetts Cavalry, aged 40 years, was wounded in the affair near New Hope Church, in Virginia, November 27th, 1863, by a conoidal musket ball, which entered at a point about four inches below the anterior superior spinous process of the left ilium, passed near the neck of the femur, and lodged. Private Ferguson was immediately conveyed to the General Field Hospital, where an exploration of the wound revealed a fracture of the trochanter major, and the lodgement of the missile near the coccyx, whence it was removed through an incision. Cold water dressings were applied. Within a few days the patient was conveyed by ambulance and steamer to Alexandria, where, on December 4th, he was received into the 3d Division Hospital, under the care of Acting Assistant Surgeon Samuel B. Ward. The limb was placed in an easy position. Cold water dressings were continued. On December 10th, he experienced constant pain at the hip-joint, which was increased by any motion. There was a copious discharge of unhealthy and offensive pus. The wounds were syringed with a solution of three grains of chloride of zinc to an ounce of water. A sustaining treatment was perseveringly followed. Falling gradually, he died December 13th, 1863. At the autopsy, it was found that the tip of the trochanter major had been carried away by the ball, which had grooved the upper surface of the neck, where a portion of the ball had chipped off and lodged under the laminated part of the bone." The specimen, No. 2704, Sect. I, A. M. M., exhibits the upper portion of the femur.

CASE 35.—"Private Alexander E. Fields, Co. B, 6th Maine Volunteers, aged 25 years, was wounded at the affair of Rappahannock Station, Virginia, November 7th, 1863, by a musket ball, which entered behind and just above the left trochanter major, and lodged near the hip-joint. He was conveyed, a distance of fifty miles, to Washington, and, on November 9th, was admitted to Armory Square Hospital, under the care of Surgeon D. W. Bliss, U. S. Volunteers. The injured limb was abducted and rotated outward, and was shortened three-fourths of an inch. In the region of Scarpa's triangle there was an ecchymosis several inches in diameter, beneath which the missile could be felt. On November 11th, an incision was made by Dr. Bliss through the integument and superficial fascia, and a battered conoidal ball, to which fragments of bone and periosteum were adherent, was extracted from its lodgement in front of the femoral artery, one and a half inches below Poupart's ligament. An intracapsular fracture of the neck of the femur was now evident. On November 23d, secondary hæmorrhage took place, which was arrested by compression. The patient died November 25th, 1863." The missile is No. 2932, Sect. I, A. M. M.

CASE 36.—"Private Simon Fleig, Co. E, 45th New York Volunteers, was wounded at the battle of Chancellorsville, Virginia, May 3d, 1863, by a musket ball, which entered the right hip and lodged. A fracture of the neck of the femur was diagnosticated. On May 15th, he was admitted into the field hospital of the 1st Division, 11th Corps. The patient died on June 12th, 1863." The case is reported by Surgeon George Suckley, U. S. V.

CASE 37.—"Private James Forbes, Co. C, 31st Massachusetts Volunteers, was struck, at Port Hudson, Louisiana, June 14th, 1863, by a fragment of shell, which smashed the upper third of the femur, opening the hip-joint. He was taken to a field hospital, and died about twenty-four hours after the reception of the injury, on June 15th, 1863." The case is reported by Surgeon E. C. Bidwell, 31st Massachusetts Volunteers.

CASE 38.—"Private James Foreman, Co. E, 5th Alabama Regiment, aged 27 years, was wounded at the assault on the lines before Petersburg, Virginia, March 25th, 1865, by a conoidal musket ball, which penetrated the left hip. He was conveyed to the Hospital of the 2d Corps, at City Point, and thence was transferred by steamer to Washington, and, March 30th, was admitted into the Lincoln Hospital. The existence of a fracture at the anatomical neck of the femur was ascertained. He died on March 31, 1865." The case is reported by Surgeon J. Cooper McKee, U. S. A.

CASE 39.—"Private Samuel Fowler, 4th Michigan Battery, was struck, June 26th, 1863, at Hoover's Gap, Tennessee, by a shell fragment, which lacerated the soft parts of the left thigh and crushed the femur, the comminution extending upwards to the hip-joint. He was taken to a field hospital, and died four hours and a half after the reception of the injury, on June 26th, 1863." The case is reported by Acting Assistant Surgeon Samuel J. Mills.

CASE 40.—"Corporal William Franks, Co. G, 24th Iowa Volunteers, aged 25 years, was wounded at the battle of Cedar Creek, Virginia, October 19th, 1864, by a conoidal musket ball, which penetrated in the region of the right hip-joint. After a few days, he was conveyed, by ambulance, some fifteen miles, to Winchester, and admitted, on the 23d, into the Sheridan Field

Hospital, under the care of Surgeon L. P. Wagoner, 114th New York Volunteers. The patient was treated by rest, position, and cold water dressings. Pyæmia ensued, and death on November 3d, 1864." The pathological specimen is No. 3793, Sect. I, A. M. M. The head of the femur is shattered and split off from the neck. The missile is lodged against the posterior border of the acetabulum, and has slightly crushed its rim at the point of impact. Absorption of cartilage is observed at the bottom of the cotyloid cavity.

CASE 41.—"Private Samuel F―――, Co. E, 111th New York Volunteers, aged 21 years, was wounded at the battle of Gettysburg, July 3d, 1863, by a musket ball which penetrated the right thigh, about two inches inside of the trochanter major, and lodged in the region of the hip-joint. He was at once conveyed to the regimental hospital, but on the following day was sent to the Seminary Hospital, at Gettysburg, and on the 15th to the Jarvis Hospital, at Baltimore. Pyæmia supervened, and he died July 18th, 1863. The autopsy revealed the missile flattened against the neck of the femur. The head of the bone and the surrounding tissues were disorganized. The pathological specimen is represented in the accompanying wood-cut, (FIG. 34.) The posterior lobe of the liver was much softened, giving way beneath the pressure of the finger. About six ounces of bloody serum was found in the pericardium." The specimen was contributed and the case reported by Assistant Surgeon D. C. Peters, U. S. A.

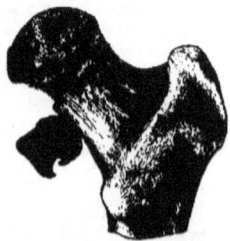

FIG. 34.—Caries of the head of the right femur from gunshot injury.—*Spec.* 1462, Sect. I, A. M. M.

CASE 42.—"Corporal J. E. G―――, Co. I, 2d South Carolina Regiment, aged 22 years, was wounded at the battle of Gettysburg, Pennsylvania, July 2d, 1863, by a musket ball, which entered in front of the right trochanter major and lodged about the hip-joint. He was conveyed to the hospital for prisoners at Gettysburg, and the missile was extracted and cold water dressings were applied. Abscesses formed about the hip-joint and extended up the dorsum ilii and between the gluteal muscles. On October 6th, the wound was still discharging copiously, and diarrhœa existed. The patient died, exhausted, on October 12th, 1863." The pathological specimen is No. 1963, Sect. I, A. M. M. The head of the femur was slightly fractured by the missile, and was found to be almost entirely absorbed. The articular surface of the acetabulum was carious, and on its anterior and outer surface a slight fringe of callus appears.

CASE 43.—"Private John W. Galyean, Co. E, 10th Indiana Volunteers, aged 28 years, received, in the engagement near Atlanta, Georgia, August 6th, 1864, a severe injury near the left hip by a conoidal musket ball, involving, however, it was thought, only the soft tissues. He was treated in the hospital of the 23d Corps, and afterwards at Marietta, and, about the middle of September, was conveyed to Nashville, and admitted into Hospital No. 1. He was furloughed in October, and, near the 1st of December, was transferred to Ekin's Barracks, at Indianapolis, Indiana, and February 17, 1865, to the Madison Hospital, where he was discharged from service, May 24th, 1865. Early in June, 1865, Surgeon Charles Hays, of Warsaw, Indiana, examined Galyean as an applicant for pension. Dr. Hays, in his report, received May 28th, 1866, states that the neck of the femur had been shattered by the missile; that spiculæ were found at dressing of the wound, and others for several months were eliminated in the discharges." The limb was atrophied and to a great extent deprived of muscular power. It seems probable that the patient's statements were accepted by Dr. Hays; for none of the various surgeons who treated the injury report any elimination of bone fragments, or eversion or shortening, or other indication of partial or complete fracture.

CASE 44.—"Private John G―――, Co. B, New Hampshire Heavy Artillery, aged 18 years, was wounded accidentally at Washington, D. C., May 10th, 1864, by a pistol ball, which entered at a point in front of and a little below the neck of the left femur. Private G――― was at once received into the Douglas Hospital, under the care of Assistant Surgeon William Thomson, U. S. A. An exploration of the wound failed to discover the missile. Pyæmia was presently suspected, from the icteroid hue of the skin, the rapid and feeble pulse, and the wandering delirium, and was soon decidedly declared. The first rigor took place on the 26th, and was followed by profuse perspiration. On the 31st, a hæmorrhage to the amount of sixteen ounces took place, probably from the femoral artery, upon which compression was made, temporarily arresting the bleeding. On the following day, however, nearly an equal amount of blood was again lost. Death took place June 1st, 1864. The autopsy revealed a disorganized state of the left hip and knee-joints. An extensive abscess existed in the muscles of the thigh. On a longitudinal section of the femur the medulla was observed to be of a dark red color, and in the early stage of that change which is seen in well-marked cases of pyæmia after amputation. No metastatic abscesses were found in any part of the body." The pathological specimen is No.

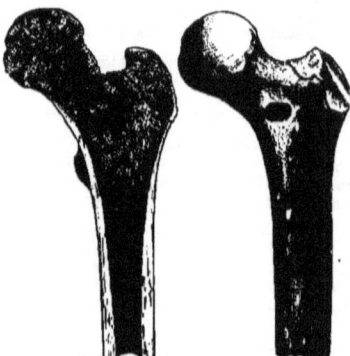

FIG. 35.—Impaction of a pistol ball in the neck of the left femur. *Spec.* 3540, Sect. I, A. M. M.

3540, Section I, A. M. M., and exhibits the upper third of the femur, which has been sawn longitudinally. The pistol ball, which penetrated the base of the neck from before, is impacted in its cancellated structure. (FIG. 35.)

CASE 45.—"Major John J. G———, Co. G, 47th Pennsylvania Volunteers, at the battle of Cedar Creek, Virginia, received a comminuted fracture of the neck and head of the left femur by a conoidal musket ball. He was admitted, October 23d, into the Sheridan Field Hospital at Winchester, where Surgeon L. P. Wagoner, 114th New York Volunteers, dressed the limb in Liston's straight splint. He died, exhausted by irritative fever, on November 5th, 1864." The pathological specimen, No. 3789, Sect. I, A. M. M., shows the head of the femur completely broken off by the missile, which has gouged out its course on the superior border. A complete fracture, with a fissure extending through the depression for the ligamentum teres, separates the posterior third of the head. The specimen is represented in the adjacent wood-cut. (FIG. 36.)

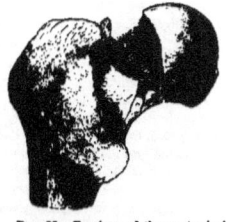

FIG. 36.—Fracture of the anatomical neck and head of the left femur by a musket ball.—*Spec.* 3789, Sect. I, A.M.M.

CASE 46.—"Private Francis G———, Co. A, 1st Louisiana Infantry, aged 42 years, was wounded before Port Hudson, Louisiana, May 27th, 1863, by a conoidal musket ball, which penetrated at a point in front of and just below the base of the left great trochanter, and lodged. He was conveyed by steamer to New Orleans, and, on May 29th, he was admitted to the University Hospital. His general condition was good, and no shortening or displacement of the injured limb was observed, and its motions were so free and voluntary, that only an exploration of the wound under chloroform, revealed the existence of an extensive fracture about the trochanters. He was treated by rest, position, and simple dressings. Nothing of special interest marked the case. The patient died on June 4th, 1863. At the autopsy, a flattened conoidal ball was found lying loose near the seat of fracture." The pathological specimen is No. 1300, Sect. I, A. M. M. The upper part of the shaft of the femur is obliquely fractured and comminuted. Fissures run through the neck, nearly to the head, and also into the great trochanter. In one of the fissures a fragment of the missile still remains. (See FIG. 37.) The specimen was contributed, and the case reported, by Asst. Surgeon P. S. Conner, U. S. A.

FIG. 37.—Gunshot fracture of the neck, trochanters, and upper portion of shaft of the left femur.—*Spec.* 1300, Sect. I, A. M. M.

CASE 47.—"Private Benjamin F. Green, Co. E, 125th New York Volunteers, aged 28 years, was wounded in the engagement at North Anna River, Virginia, May 18th, 1864, by a conoidal musket ball, which entered near the right trochanter major, fractured the neck of the femur, and emerged from the buttock in a line a little in front of the anus. He was taken to the field hospital of the 1st Division, 2d Corps, and thence was conveyed to Washington, and admitted, May 28th, 1864, into Armory Square Hospital. He died, exhausted, on June 30th, 1864." The case is reported by Surgeon D. W. Bliss, U. S. V.

CASE 48.—"Private Thomas Green, Co. L, 12th Tennessee Cavalry, aged 18 years, was wounded at the battle of Franklin, Tennessee, November 30, 1864, by a conoidal musket ball, which entered the right hip. He was conveyed, December 1st, to Nashville, and admitted into Hospital No. 15. Cold water dressings were applied. On December 13th, the patient was transferred, by rail, to the Crittenden Hospital in Louisville, Kentucky. A fracture at the neck of the femur was now discovered. Death took place on December 17th, 1864." The case is reported by Surgeon W. M. Chambers, U. S. V., and by Surgeon R. R. Taylor, U. S. V.

CASE 49.—"Private George H———, Co. D, 26th Michigan Volunteers, aged 40 years, was wounded at the battle of Cold Harbor, Virginia, June 3d, 1864, by a conoidal musket ball, which entered at a point just outside of the left trochanter major, and lodged. Soon after the reception of the wound the patient was admitted into the hospital of the 1st Division, 2d Corps. He was afterwards conveyed by steamer to Washington, and admitted, on June 11th, into the Lincoln Hospital. The limb was rotated outwards and shortened. A search for the missile was unsuccessful. Buck's apparatus was applied, with sand-bags for lateral support. The suppuration was copious. The patient also suffered from colliquative diarrhœa. Bed sores formed over the sacrum. Tonics, stimulants, and astringents were administered, and a nourishing diet was provided. A lotion of creasote was applied to the excoriations. He died, exhausted, July 12th, 1864. At the post mortem examination a conoidal musket ball, much battered, was found in the soft parts just beneath the femoral vein; the femoral vessels, however, were uninjured. The capsular ligament was opened on its anterior surface. Firm adhesions secured the right lung to the thoracic parietes; the lower lobe of this lung was congested and loaded with viscid mucus; the remaining lung tissue was healthy. The liver weighed seventy-nine ounces." Specimen No. 2839, Sect. I, A. M. M., is the upper half of the femur. There is great comminution about the trochanters, and fissures extend half way up the neck anteriorly. The fractured extremities are carious and partly absorbed. The missile is attached near the seat of fracture. Acting Assistant Surgeon H. M. Dean contributed the specimen and history.

CASE 50.—"Private Daniel Haley, Co. B, 57th Massachusetts Volunteers, aged 36 years, was wounded, in front of Petersburg, Virginia, July 6th, 1864, by a conoidal musket ball, which was believed to have fractured the anatomical neck of the right femur. He was conveyed to the field hospital of the 1st Division, 9th Corps. Splints were applied. On July 10th, he was conveyed to City Point, and thence was taken by steamer to New York, and admitted on July 13th, into De Camp Hospital. He died on August 10th, 1864." The case is reported by Surgeon M. K. Hogan, U. S. V.

CASE 51.—"Sergeant William D. H———, Co. A, 6th Iowa Volunteers, aged 24 years, was wounded in the action at Kenesaw Mountain, Georgia, June 27th, 1864, by a conoidal musket ball, which shattered the neck of the left femur. The soft parts were but slightly injured. He was received, on July 1st, into the hospital of the 15th Army Corps, at Barton's Iron Works. Stimulants and concentrated nourishment were administered, but failed to effect any favorable reaction. Death resulted on July 3d, 1864." The pathological specimen is No. 3488, Sect. I, A. M. M., and exhibits a large longitudinal fragment of the neck, with the trochanter minor split off by the missile. There is also a fissure extending completely through the neck at its upper extremity. The case was reported and the specimen contributed by Surgeon A. Goelin, 48th Illinois Volunteers.

CASE 52.—"Corporal Benjamin H——, Co. C, 9th West Virginia Volunteers, aged 19 years, was wounded at the battle of Winchester, Virginia, July 20th, 1864, by a conoidal musket ball, which penetrated in front of and a little below the left hip-joint. He was conveyed a distance of twenty-five miles, in an ambulance, to Martinsburg, and thence by rail to Cumberland, Maryland, where he was received, on July 23d, into the General Hospital. For several days he was treated only by rest and position. On July 25th, an exploration of the wound was made under ether, and the ball was extracted from its lodgement in the base of the neck of the femur. The limb was then suspended in an anterior splint. Gangrene soon invaded the wound, and the patient died on August 3d, 1864." The pathological specimens is figured in the adjoining wood-cut. (FIG. 38). The femur is fractured longitudinally through the anterior portion of the head, the centre of the neck, and great trochanter, the fracture extending several inches down the shaft. There is a large conical cavity in the base of the neck anteriorly, from which the ball was extracted. The specimen and history were contributed by Surgeon J. B. Lewis, U. S. V.

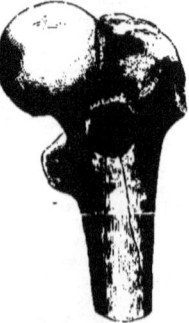

CASE 53.—"Private Jerry Harrigan, Co. K, 36th Massachusetts Volunteers, at the assault on the lines of Petersburg, Virginia, July 18, 1864, was struck in the left hip by a conoidal musket ball. He also received a gunshot wound of the left arm. He was conveyed to the field hospital of the 2d Division, 9th Corps. A fracture of the head of the femur was discovered. Simple dressings were applied. The patient lived only a few hours, death taking place on July 18th, 1864." Assistant Surgeon A. H. Bryant, 35th Massachusetts Volunteers, reports the case.

FIG. 38.—Fracture at the base of the neck of the left femur by a musket ball. Spec. 4207, Sect. I, A.M.M.

CASE 54.—" Private Ladd P. Harvey, Co. B, 11th New Hampshire Volunteers, at the battle of Fredericksburg, Virginia, December 13th, 1862, received a compound fracture of the neck of the femur. He was conveyed to the field hospital of the 2d Corps. The limb was placed in extension, and cold water dressings were applied. Death took place within a few hours, on December 13th, 1862." Surgeon J. S. Ross, 11th New Hampshire Volunteers, reports the case.

CASE 55.—" Private William Herold, Co. B, 6th Alabama Regiment, aged 21 years, was wounded at Fort Sedgwick, Petersburg, Virginia, April 2d, 1865, by a musket ball, which entered one inch below Poupart's ligament, right side, and, passing backwards and slightly downwards, comminuted the neck and the upper part of the shaft of the femur, and emerged from the gluteal muscles. He was captured and conveyed to the field hospital of the 2d Division, 9th Corps. A flesh wound of the hip was reported. The next day he was sent to the Depot Field Hospital of the 9th Corps, at City Point. Here it was only stated that he had a gunshot wound of the right thigh. On April 9th, he was transferred by steamer to Washington, and April 10th was admitted to Lincoln Hospital. The course of the missile was now accurately given, but no mention was made of the existence of a fracture. At the autopsy the neck and upper part of the shaft were found badly comminuted. The muscles of the thigh were very much disorganized." The case is reported by Surgeon J. C. McKee, U. S. A.

CASE 56.—" Musician James B. H——, Co. A, 41st Ohio Volunteers, was wounded at the battle of Chickamauga, Georgia, September 19th, 1863, by a conoidal musket ball, which entered over the right trochanter major, and lodged. He was conveyed in an ambulance a distance of thirty miles, to Chattanooga, Tennessee, where he was admitted, on September 20th, into Hospital No. 4. A fracture of the trochanter was detected. The missile was extracted, and simple dressings were applied. On January 26th, 1864, the patient was transferred to Hospital No. 19, at Nashville. On February 1st, the wound was enlarged and several small fragments of the trochanter were removed. An erysipelatous inflammation immediately afterwards attacked the parts, and rapidly extended to the scrotum and down the thigh. On February 10th, the soft structures upon the dorsum of the foot were gangrenous. The patient died on February 13th, 1864. At the autopsy the capsular ligament was found to have been opened by the fracture. The ensuing inflammation resulted in a partial anchylosis of the joint. The knee-joint contained three ounces of pus." The pathological specimen is No. 2178, Sect. I, A. M. M. The anterior third of the trochanter has been destroyed by the fracture and subsequent caries. The specimen and history were contributed by Surgeon J. W. Foye, U. S. V.

CASE 57.—" Private Robert N. H——, Co. D, 1st New Jersey Cavalry, was wounded at the battle of Spottsylvania, Virginia, May 12th, 1864, by a conoidal musket ball, which penetrated the right thigh, over the trochanter major, and lodged in the region of the hip-joint. He was conveyed in an ambulance to Fredericksburg, and thence by steamer to Washington, and admitted, on May 24th, into the Carver Hospital; but was sent thence, at the close of the month, to the Satterlee Hospital, in Philadelphia. Much constitutional disturbance had by this time arisen, and there was profuse suppuration from the inflamed wound. All attempts to discover the lodgement of the missile were futile. The further treatment consisted mainly in the exhibition of tonics and stimulants, and the provision of nourishing food. He died, exhausted, on June 21st, 1864. The missile was found, post mortem, lodged in the gluteal muscles." The pathological specimen is No. 3636, Sect. I, A. M. M., and is represented in the annexed wood-cut. (FIG. 39). Much of the head has been absorbed, and the remaining portion presents a depression as though chipped out by the missile. The projectile has been placed at this spot. The acetabulum is carious. The specimen and history are contributed by Acting Assistant Surgeon George Kerr.

FIG. 39.—Caries of the head of the left femur and of the acetabulum from gunshot wound of the hip-joint.—Spec. 3636, Sect. I, A. M. M.

CASE 58.—"Private William Huger, Co. D, 7th Virginia Regiment, received, at the battle of Bull Run, August 29th, 1862, a gunshot fracture of the neck of the left femur, involving the hip-joint. He was conveyed to Georgetown, D. C., and, on September 1st, was admitted into the Presbyterian Church Hospital. He died on September 10th, 1862." The case is reported by Assistant Surgeon B. A. Clements, U. S. A.

CASE 59.—"Private J. Hughes, Co. G, 1st Alabama Cavalry, near Hartsville, Tennessee, April 11th, 1863, received a gunshot fracture of the head of the femur. He recovered, and applied in person for a suitable apparatus for his limb. Mr. Hughes resides in Crawford County, Arkansas." The case is recorded in the register of a Confederate relief association.

CASE 60.—"Private William Jackson, Co. B, 106th New York Volunteers, aged 29 years, was wounded at the assault on Petersburg, Virginia, April 2, 1865, by a conoidal musket ball, which entered at a point just behind the left trochanter major, passed downwards and forwards, and emerged about the middle of the anterior surface of the thigh. Private Jackson was at once admitted into the 3d Division, 6th Corps, hospital, where the existence of a fracture was ascertained. Splints were applied. Soon afterwards the patient was transferred to the Depot Field Hospital, at City Point, whence he was sent by steamer to Washington, and received into Armory Square Hospital on April 8th. Cold water dressings were applied. For several days the patient did well. Pyæmia supervened on April 19th, and, in spite of the free administration of tonics and stimulants, and the provision of a nourishing diet, the patient died on April 23d, 1865." Surgeon D. W. Bliss, U. S. V., reports this case.

CASE 61.—"Private Herman J———, Co. F, 14th New York Volunteers, aged 26 years, was wounded at the battle of Spottsylvania, Virginia, May 10th, 1864, by a conoidal musket ball, which entered at the left groin and emerged from the left buttock, having fractured the neck of the femur. He was conveyed to Washington, and, on May 13th, was admitted into Emory Hospital. A straight splint was adjusted on the outer side of the limb, cold water dressings were continued, laxatives and opiates were administered as required, the diet being restricted. On May 18th, peritonitis set in, and was treated by half-grain doses of calomel, with three grains of Dover's powder every third hour; revulsives being applied meanwhile over the surface of the abdomen. Stimulation by brandy was directed. Death resulted on May 21st, 1864, peritonitis being the immediate cause." The pathological specimen is No. 2309, Sect. I, A. M. M., and shows a complete fracture at the base of the neck, intracapsular, but not extending into the head. The great trochanter is split off. Anteriorly the fissures extend nearly to the anatomical neck. A posterior view of the specimen is shown in the wood-cut. (FIG. 40.) The specimen and history were contributed by Acting Assistant Surgeon T. Walsh.

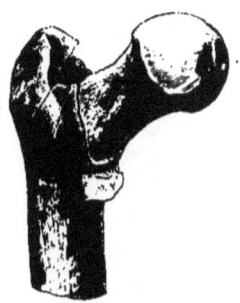

FIG. 40.—Fracture of the neck and trochanter of the left femur by a musket ball. Spec. 2309, Sect. I, A. M. M.

CASE 62.—"Private P. C. Johnson, Co. H, 15th Ohio Volunteers, received at Pickett's Mills, near Dallas, Georgia, about May 27th, 1864, a gunshot fracture of the neck of the femur. He was conveyed to the hospital of the 3d Division, 4th Corps. Death resulted in a few days." Assistant Surgeon W. J. Kelly, 15th Ohio Volunteers, reports the case.

CASE 63.—"Private Hiram Jones, Co. I, 151st Pennsylvania Volunteers, received, at the battle of Gettysburg, Pennsylvania, July 2d, 1863, a gunshot fracture of the right hip-joint. He was admitted into the Camp Letterman Hospital, August 1st. Secondary hæmorrhage occurred. He died on August 5th, 1863." The case is reported by Surgeon H. Janes, U. S. V.

CASE 64.—"Private Michael K———, Co. D, 65th New York Volunteers, aged 36 years, was wounded at the battle of Cedar Creek, Virginia, October 19th, 1864, by a conoidal musket ball, which penetrated at a point in front of the left hip-joint. He also received a simple fracture of the left ulna. He was, within a few days, conveyed by ambulance, a distance of fifteen miles, to Winchester, and thence sent by rail to Baltimore, being admitted, on October 24th, to Jarvis Hospital, under the care of Acting Assistant Surgeon B. B. Miles. The patient was treated only by rest, position, and cold water dressings. Splints were applied to the forearm. Pyæmia supervened, and death occurred on November 1st, 1864. The fractured ulna was ununited." The pathological specimen is No. 3419, Sect. I, A. M. M. It is represented in the accompanying wood-cut. (FIG. 41.) The ball has carried away the upper portion of the middle of the neck. The specimen and history of the case were contributed by Acting Assistant Surgeon B. B. Miles.

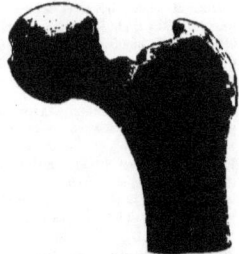

FIG. 41.—Destruction of the upper part of the neck of the left femur by a musket ball.—Spec. 3419, Sect. I, A. M. M.

CASE 65.—"Private Degras A. Kimble, Co. G, 3d Minnesota Volunteers, was wounded in an engagement with Indians at Wood Lake, Minnesota, September 22d, 1862, by a musket ball, which perforated the left thigh very high up, fracturing the neck of the femur. He was conveyed, in a quartermaster's wagon, a distance of forty-five miles, to Fort Ridgely. Pyæmia supervened, and he died on October 14th, 1862." The case is reported by acting Assistant Surgeon Alfred Müller.

CASE 66.—"Private A. J. K———, Co. E, 8th Florida Regiment, aged 20 years, was wounded at the battle of Gettysburg, Pennsylvania, July 2d, 1863, by a conoidal musket ball, which entered at a point in front of the left great trochanter. For several weeks the patient was cared for at the Seminary Hospital, but on the 28th of July he was transferred to Camp Letterman Hospital. It was found that the ball had entered near the base of the neck, and lodged in the cancellated structure of the great trochanter. The patient died on September 27th, 1863. The injured bone represented in the wood-cut (FIG. 42) exhibits a perforation of the anterior wall of the femur, in front of the lesser trochanter, surrounded by fringes of osseous deposit. The ball is impacted near the digital fossa. It has caused but a single fissure along the intertrochanteric line, and this also is bordered with thin layers of callus." The history and specimen were contributed by Surgeon H. Janes, U. S. V.

CASE 67.—"Private Louis P. L———, Co. K, 91st Pennsylvania Volunteers, aged 17 years, was wounded at the battle of Chancellorsville, Virginia, May 3d, 1863, by a musket ball, which entered through the right buttock and emerged in front of the right trochanter major. On the following day the man fell into the hands of the enemy, and was removed from the field to a house near by, where for nine days he was without surgical attention. On May 12th he was paroled and received into a field hospital. The parts were inflamed and swollen. Profuse suppuration ensued, and pus burrowed extensively among the muscles of the thigh. An incision was made and a large quantity of pus was evacuated. Bed sores formed over the sacrum, along the crest of the ilium, and around the trochanters. On June 14th, the patient was conveyed by steamer to Washington, and admitted into the Douglas Hospital, under the care of Acting Assistant Surgeon Carlos Carvallo. He was treated by rest, position, warm water dressings, poultices sprinkled with powdered bark upon the denuded surfaces, egg-nog, brandy toddy, and at night morphia. On the 19th, there was severe pain at intervals in the right knee and ankle-joints, which was relieved by camphor mixture. On June 20th, the patient was moved to a water bed. There was, meanwhile, a copious discharge of bloody fluid from the wound in the buttock. The pulse was at 110 and feeble. On the 23d, he suffered a severe rigor, and again on the 25th. One grain of quinine with one-sixth of a grain of morphia were given meantime every second hour. Stimulants were liberally administered. Death took place on June 26th, 1863. The autopsy revealed only simple congestion of the lungs; the other viscera were nearly normal." The pathological specimen is represented in the adjacent wood-cut, (FIG. 43,) and presents an oblique perforation between the great trochanter and neck of the right femur. The hip-joint was opened; the articular surfaces were roughened by absorption.

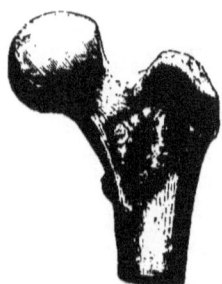

FIG. 42.—Penetration of the base of the neck of the left femur by a conoidal musket ball which has lodged in the great trochanter.—Spec. 1932, Sect. I, A.M.M.

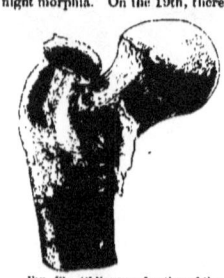

FIG. 43.—Oblique perforation of the neck of the right femur by a musket ball.—Spec. 1343, Sect. I, A. M. M.

CASE 68.—"Private John Laner, Co. F, 39th New Jersey Volunteers, aged 18 years, of temperate habits, was wounded at the battle in front of Petersburg, Virginia, April 2, 1865, by a conoidal musket ball, which entered two inches to the right and a little below the right trochanter major, comminuted the femur, and emerged one and a half inches above and to the right of the anus. He was admitted into field hospital of the 2d Division, 9th Corps, and the next day was conveyed to City Point, whence he was transferred by steamer to Alexandria, and, April 6th, received into the 3d Division Hospital. Nearly four inches of shortening existed. The foot of the wounded limb was everted. He was treated by rest and position. A profuse and offensive sanious discharge issued from the wounds. Stimulants were freely administered, and a nourishing diet was provided. Bed sores formed over the sacrum. On April 20th, there was hectic and intense irritative fever. A tough ropy matter was expectorated. He died April 21st, 1865. At the autopsy, a fracture of the neck of the femur was discovered. Spiculæ of bone were distributed along the course of the missile. A pint of black coagula had collected in the gluteal region. The liver was yellow, the spleen dark, and the kidneys pallid. Peyer's and the solitary glands were enlarged. The pleuræ and pericardium were, in many places, adherent." The case is reported by Surgeon E. Bentley, U. S. V.

CASE 69.—"Captain James M. L———, Co. I, 20th Indiana Volunteers, was twice wounded in an engagement in front of Richmond, Virginia, on June 27th, 1862. The first wound was through the lumbar muscles, and, while lying on the field, he was again struck by a conoidal musket ball, which entered on the outer side of the left thigh, a little below the great trochanter, and, passing upwards and inwards, lodged. He was conveyed to Washington, and on June 29th, was admitted to the Columbia College Hospital. A finger could be readily passed into the perforation of the femur, but the ball could not be reached. There was no shortening or eversion of the limb interfering with the motion of the joint. Three formal attempts to ascertain the position of the ball and accomplish its removal were unsuccessfully made. The patient died from exhaustion August 19th, 1862." The specimen is represented in the adjoining wood-cut (FIG. 44). It shows the upper portion of the left femur perforated between the trochanters on the posterior surface. The track of the ball is carious. The great trochanter has been split off, but is reunited by callus. The space between the trochanters is bridged over by a displaced fragment of bone, attached in its new position by slight osseous deposits. The missile was found resting against the capsular ligament. Assistant Surgeon W. M. Notson, U. S. A., who attended and reported the case, is confident that the ball was external to the joint; but as the grooving of the neck extends upwards nearly to the articular surface of the femur, it is hardly possible that the joint escaped.

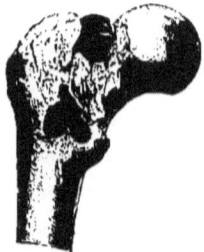

FIG. 44.—Perforation of the neck of the left femur by a conoidal musket ball.—Spec. 565, Sect. I, A.M.M.

CASE 70.—"Private John Lawler, Co. B, 12th New York Cavalry, in a skirmish near Newberne, North-Carolina, August 11th, 1863, was struck by a pistol ball in the thigh near the great trochanter. He was admitted into the Foster Hospital on the same day, and was treated on the expectant plan. The place of lodgement of the ball could not be determined. Pyæmia supervened, and the patient died on September 2d, 1863. The ball was found, post mortem, driven completely into the head of the femur, as evenly as a carpenter could drive a nail into a board. There was a slight stellated fracture of the head, the fissures being not more than three-eighths of an inch in length. The parts about the hip were infiltrated with pus." Surgeon Isaac F. Galloupe, 17th Massachusetts Volunteers, who reported the case, sent the specimen, with a description, to Dr. J. B. S. Jackson, to be exhibited to the Boston Society for Medical Improvement. The specimen was unfortunately mislaid, and, after a diligent search by Drs. Jackson, Dale, Hooper, and Hodges, has not yet been recovered.

CASE 71.—"John McCarthy, Co. E, 76th New York Volunteers, aged 31 years, was wounded at the engagement at Ream's Station, Virginia, October 1st, 1864, by a conoidal musket ball, which entered just behind the right great trochanter, fractured the neck and trochanter, and emerged in front two inches outside of the course of the femoral vessels, and three inches below Poupart's ligament. He was captured, and remained for a few days in the hands of the enemy. Being paroled, he was conveyed to Annapolis, Maryland, where he was admitted to hospital on October 9th. On November 5th, a hæmorrhage of about sixteen ounces took place from ulceration of the profunda artery. A ligature placed upon the femoral artery arrested the hæmorrhage, which recurred, however, November 11th. Digital pressure was kept up until his death, on November 13th, 1864." The case is reported by Surgeon G. A. Palmer, U. S. V.

CASE 72.—"Corporal John G. Mallory, Co. C, 31st Indiana Volunteers, at the battle of Chickamauga, Georgia, September 19th, 1863, was struck in the right hip by a musket ball, and was conveyed to a field hospital, where a fracture of the femur near the hip-joint was recognized. Within a day or two, the wounded man was conveyed by ambulance, twelve miles, to Hospital No. 1, at Chattanooga, Tennessee. Death took place on October 10th, 1863. An intracapsular fracture of the neck of the femur was found to exist." The case is reported by Surgeon Israel Moses, U. S. V.

CASE 73.—"Captain Hezekiah D. M——, Co. K, 79th Illinois Volunteers, was wounded at the battle of Liberty Gap, Tennessee, June 25th, 1863, by a conoidal musket ball, which entered in front of the left trochanter major, and passed backwards and inwards, emerging from the perinæum near the anus. He was conveyed in an ambulance, a distance of about thirty miles, to Murfreesboro', and on June 27th, was admitted into General Hospital No. 1, under the care of Assistant Surgeon W. P. McCulloch, 78th Pennsylvania Volunteers. On June 29th, he had rallied considerably, and an exploration of the wound was made under ether. A perforation of the neck of the femur was detected, with fissures extending to the neck and shaft. The third day after his admission to hospital he began to sink rapidly. He died July 3d, 1863." The specimen is represented in the accompanying wood-cut. (FIG. 45). Posteriorly the fracture extends much higher on the neck. The specimen and history were contributed by Surgeon I. Moses, U. S. V.

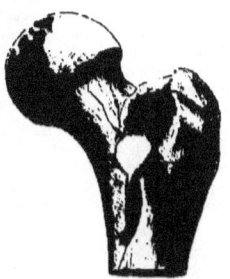

FIG. 45.—Perforation of base of neck of left femur by a musket ball.—Spec. 1717, Sect. I, A. M. M.

CASE 74.—"Private Charles H. M——, Co. G, 3d Maryland Volunteers, aged 16 years, a slender boy, much debilitated by diarrhœa, was wounded in a skirmish, near Frederick, Maryland, July 7th, 1864, by a conoidal musket ball, which is said to have entered the left thigh, two inches below the groin and one inch posteriorly to the femoral artery, and lodged in the head of the femur. He was conveyed to the General Hospital at Frederick. Ineffectual efforts were made to extract the ball. The patient was in such a feeble condition that it was deemed unsafe to cut down on the ball or to attempt excision. Suitable dressings were, therefore, applied to the limb, astringents and tonics were administered, and generous diet provided. On July 15th, the diarrhœa was entirely checked, but bed sores were commencing. The patient was transferred to a water bed, but, at his own request, returned to a hair mattress at the close of the week. The wound, meanwhile, was suppurating freely. At the end of August there was great irritability of the stomach, with bilious vomiting. No solid food was retained. The bed sores extended and sloughed rapidly, and extreme emaciation ensued. On October 15th, he became dull and morose, refusing every kind of nourishment, and died exhausted on October 21st, 1864." The pathological specimen is No. 3923, Sect. I, A. M. M. The head was badly shattered by the ball, and is much absorbed, leaving, however, the shell of the articular surface nearly intact. The lesser trochanter is missing, as though lost by separation at the epiphyseal line. The acetabulum, which was not preserved, is said to have been greatly necrosed. The entire neck is carious, with a coating of partly exfoliated laminated tissue. It is difficult to understand how the ball lodged in its present position, if it entered in front of the trochanter major. The case was reported and specimen contributed by Assistant Surgeon R. F. Weir, U. S. A.

CASE 75.—"Private John Matthews, Co. A, 11th Louisiana Colored Troops, was wounded in the engagement at Milliken's Bend, Louisiana, June 8th, 1863, by a musket ball, which fractured the neck of the right femur. He was treated on the expectant plan. On June 10th, he was placed on board the floating hospital 'Nashville,' near Vicksburg. He died on June 13th, 1863." The case is reported by Assistant Surgeon H. R. Tilton, U. S. A.

CASE 76.—"Corporal Donald McD———, Co. F, 12th New York Cavalry, was wounded July 27th, 1863, in a skirmish, near Plymouth, North Carolina, by a ball, which penetrated to the right of the anus, and lodged in the region of the hip-joint. The wounded man was conveyed to the regimental hospital of the 103d Pennsylvania Volunteers, at Plymouth. There was no shortening or deformity, and neither a fracture nor the position of the projectile was discovered. In a few days, however, Surgeon A. P. Frick, in charge of the hospital, suspected that the bone was injured, and summoned a consultation of surgeons, who decided that operative interference was inadvisable. Nourishing food with stimulants were given, and the patient progressed satisfactorily until the middle of September, when the parts about the wound became inflamed and swollen, and pus burrowed among the muscles toward the knee, though in some measure restrained by the application of a spica bandage. Much pain was obviated throughout the treatment by placing the patient upon a bed, within a quadrangular frame of wood, across which were stretched, at equal distances, strips of canvas, six or eight inches in width, upon which the patient could be lifted with ease into any required position. On September 22d, an incision was made over the trochanter major, through which a large amount of fœtid pus with spiculæ of bone was evacuated. On October 9th, this incision was extended, under ether, for the removal of the missile, which was now discovered upon the inner anterior surface of the neck, and was readily extracted by long bullet forceps On October 26th, pyæmia supervened, and the patient died on November 1, 1863." The specimen is represented in the annexed wood-cut. (FIG. 46). The ball having struck a little above the lesser trochanter, within the intertrochanteric ridge, split off the head, neck, and great trochanter, and lodged in the base of the neck. The detached fragments are all eroded and carious. The specimen and history were furnished by Surgeon A. P. Frick, 103d Pennsylvania Volunteers.

FIG. 46.—Fracture of the neck of the right femur by a conoidal musket ball with consequent caries. *Spec.* 2170, Sect. I, A. M. M.

CASE 77.—"Ebenezer McGee, Secret Service, while conveying despatches from Colonel Dahlgren to General Kilpatrick, in April, 1864, was shot by a pistol ball, which entered just behind the right trochanter major, and, passing upwards, lodged in the posterior surface of the neck of the femur, which it slightly fractured. He was conveyed by ambulance and steamer to Baltimore, and, on April 18th, he was admitted into West's Buildings Hospital. Extensive abscesses formed in the muscles of the thigh and in the gluteal region. The parts ultimately became gangrenous. The patient died, exhausted, on April 27th, 1864." This case is reported by Surgeon A. Chapel, U. S. V., who says: "The ball and head of the femur were preserved, and will be forwarded for the benefit of the U. S. Medical Museum."

CASE 78.—"Second Lieutenant Thomas H. McKinley, Co. B, 29th U. S. Colored Troops, was wounded in the engagement at Deep Bottom, Virginia, September 29th, 1864, by a spherical case shot, which struck the left hip and fractured the neck of the femur. He was conveyed, October 1st, to the Chesapeake Hospital, at Fortress Monroe, and was treated on the expectant plan. He died on January 3d, 1865." The case is reported by Surgeon John F. Stevenson, 29th Connecticut Volunteers, and by Assistant Surgeon Eli McClellan, U. S. A., in the records of the Chesapeake Hospital.

CASE 79.—"Private Charles Miller, Co. A, 9th Illinois Volunteers, aged 26 years, was wounded at the battle of Shiloh, Tennessee, April 6th, 1862, apparently by a small rifle ball, entering at a point midway between the left trochanter major and the tuberosity of the ischium, and fracturing the neck of the left femur. He was conveyed to the General Hospital at Savannah, and a month afterwards was transferred by steamer to Quincy, Illinois, and admitted to hospital on May 7th. An unsuccessful search for the missile was made, and Buck's apparatus was applied. The bone united, and the wound healed. He was discharged from service on October 14th, 1862, having limited motion at the hip-joint. On April 7th, 1864, Dr. John C. Hupp, Pension Examining Surgeon at Wheeling, West Virginia, stated that the limb was shortened by about two inches, a partial luxation of the head of the femur upwards apparently having been produced. Any movement of the thigh created severe pain. The cicatrices were firm, and there were no fistulous orifices." The case is reported by Surgeon R. Nichols, U. S. V.

CASE 80.—"Private William O. M———, Co. C, 24th Iowa Volunteers, aged 21 years, was wounded at the battle of Winchester, Virginia, September 19, 1864, by a conoidal musket ball, which entered behind the right trochanter major and emerged in front, outside of the median line, five inches below the anterior superior spinous process of the ilium. He was admitted, on the 23d of September, into the Sheridan Hospital at Winchester. The limb was shortened; no crepitus, however, could be detected, and no bone spiculæ came away. The limb was placed upon a double inclined plane. The discharges for several weeks contained a free admixture of synovial fluid. It was believed that the margin of the acetabulum had been fractured, and that the head of the femur was luxated. The patient died, exhausted, on February 13th, 1865." The pathological specimen is No. 3702, Sect. 1, A. M. M. The neck is broken off near the trochanteric extremity, and the fracture extends to and involves the head, a portion of which is absorbed, while the whole of its surface is carious. The acetabulum is also roughened by caries. The specimen was contributed and the history recorded by Surgeon L. P. Wagoner, 114th New York Volunteers.

CASE 81.—"Private Joseph W. Moore, Co. H, 6th Pennsylvania Cavalry, aged 27 years, was wounded in the engagement at Old Church, near Cold Harbor, Virginia, May 30th, 1864, by a rifle ball, which fractured the neck of the left femur, and lodged. He was taken to the hospital of the 1st Division, Cavalry Corps, and thence was conveyed, by ambulance and steamer, to Washington, and admitted, on June 4th, into Stanton Hospital. He was treated by rest, position, and ice water dressings. Tonics and stimulants were administered. He died, exhausted, on June 10th, 1864." The case is reported by Surgeon John A. Lidell, U. S. V.

CASE 82.—"Private John M——, Co. F, 63d New York Volunteers, aged 19 years, was wounded at the battle of Antietam, Maryland, September 17th, 1862, by a musket ball, which entered two inches behind and one inch below the left great trochanter, and lodged somewhere about the joint. He was admitted into the 2d Corps Hospital, where the missile was extracted, and cold water dressings were applied. On September 27th, he was conveyed in an ambulance some twenty miles, to Frederick City, and was admitted into Hospital No. 1. No inversion or eversion of the limb existed. The treatment was expectant. On October 21st, an incision was made into the swollen parts at the tuberosity of the ischium, and a large quantity of pus was evacuated, mainly from an abscess in the vicinity of the great trochanter. Tonics and stimulants were administered. On November 4th, the daily discharge was estimated at twelve ounces; there was hectic and profuse sweating. A second incision was made one inch anterior to the trochanter, and a fracture through the neck of the femur was discovered. Death resulted on November 5th, 1862. Upon removing the shaft and opening the capsular ligament, *post mortem*, the head of the femur rolled out as though a billiard ball, the ligamentum teres having been destroyed by absorption. The missile had fractured the neck almost transversely. There was little comminution. Near the fracture the bone was carious and superficially necrosed." The pathological specimen is No. 782, Sect. I, A. M. M., and was contributed with the history of the case by Acting Assistant Surgeon Alfred North.

CASE 83.—"Private John R. Morrill, Co. D, 184th Pennsylvania Volunteers, at the affair at Tolopotomy Creek, Virginia, June 1st, 1864, received a gunshot wound of the left hip. He was taken to the hospital of the 2d Division, 2d Corps, at Cold Harbor. A fracture of the neck of the femur implicating the joint was ascertained. The treatment was expectant. The patient died on June 3d, 1864." Assistant Surgeon R. C. Marshall, 184th Pennsylvania Volunteers, reports the case.

CASE 84.—"Private Peter M——, Co. B, 1st Virginia Rifles, aged 28 years, was wounded at the battle of Williamsburg, Virginia, May 5th, 1862, by a conoidal musket ball, which entered the left thigh just below Poupart's ligament, outside of the femoral vessels, passed in the vicinity of the hip-joint, and emerged through the buttock, directly opposite the point of entrance. He was captured, and, after a few days, was conveyed by steamer to Washington, and on May 17th was admitted into Cliffburne Hospital, under the care of Assistant Surgeon John S. Billings, U. S. A. The limb was shortened one and a quarter inches, and was abducted and rotated outwards. He complained of little pain; the pulse was almost imperceptible, the skin cool and clammy, and diarrhœa existed. The wound of exit was cicatrizing, but from that of entrance sanious pus was discharging. The patient was placed upon a fracture bed, the injured limb was attached by adhesive strips in the usual manner to the foot of the bed, which was then elevated so as to secure moderate extension. Absorbent dressings were applied, stimulants were freely administered, and astringent mixtures given. On May 24th the diarrhœa was arrested, but recurred the next day, and resisted all treatment. The pulse was at 112, weak and irritable. The patient died on the morning of May 26th, 1862. Eight hours after death the autopsy was made, and revealed the existence of a large abscess about the hip-joint, and pus had burrowed among the muscles of the thigh. The liver and spleen were enlarged and congested; the other viscera were normal." The pathological specimen is No. 33, Sect. I, A. M. M. The neck of the femur is comminuted near its junction with the head. The base of the neck is necrosed. Assistant Surgeon Billings, who reports the case, remarks that the patient was first under the care of a Confederate surgeon, who pronounced the injury a flesh wound. When the patient entered Cliffburne, operative interference was out of the question. Dr. Billings adds that the case is especially interesting, as indicating the result of such injuries when left to themselves, and the relative loss of life in comparison with cases of decapitation of the head of the femur.

CASE 85.—"Private John M——, Co. I, 61st New York Volunteers, aged 31 years, was wounded at the battle of Chancellorsville, Virginia, May 3d, 1863, by a round musket ball, which penetrated at a point two inches below the apex of Scarpa's triangle, and lodged in region of the left hip-joint. The patient was conveyed by steamer to Washington, and, on May 8th, was admitted into Mount Pleasant Hospital. The thigh was greatly swollen and very painful, and the countenance expressed anxiety. The pulse was at 120. The limb was supported by pillows, and cold water dressings were applied. Anodynes and stimulants were administered as indicated. A sanious discharge issued from the wound, which, by May 20th, took on erysipelatous inflammation. Tincture of iodine was applied locally, and sesquichloride of iron was administered internally. On June 9th, the erysipelas abated, and diarrhœa set in. On June 16th, the patient referred his pain principally to the hip and knee-joints; the whole limb was œdematous. His appetite was good, however, and he rested well at night. The œdema disappeared under bandaging, but the pain continued, and the pulse was frequent. On July 10th, crepitus was noticed. An incision was made for the escape of serum. On July 29th, troublesome vomiting began, and continued several days. Then involuntary dejections and delirium supervened, and the patient died, exhausted, on August 3d, 1863. At the autopsy, the tissues about the hip-joint were found to be in a gangrenous condition. The acetabulum was uninjured; the head was completely shattered, and partially absorbed. The missile had lodged in it." The preparation is represented in the adjacent wood-cut. (FIG. 47.) It was contributed by Assistant Surgeon C. A. McCall, U. S. A., with the history of the case. Assistant Surgeon Woodhull remarks[1] that this "would have been a fair case for primary excision could the diagnosis have been made."

FIG. 47.—Comminution of the head of the left femur by a musket ball.—*Spec.* 1692, Sect. I, A. M. M.

---

[1] *Catalogue of Surgical Section of Army Medical Museum*, p. 241.

CASE 86.—"Private James B. Mullen, Co. G, 13th Indiana Volunteers, aged 26 years, was wounded at Winchester, Virginia, March 23d, 1862, by a musket ball, which entered at a point just above the right great trochanter, fractured the neck of the femur, and emerged at the lower part of the right buttock. For several weeks the patient was treated in a hospital at Winchester, by position, rest, cold water dressings, and stimulants. The patient suffered from diarrhœa, and abscesses formed in the region of the hip-joint. On May 22d, he was transferred to Frederick City, Maryland, and admitted to Hospital No. 1. He was placed upon a water bed, and twenty ounces of brandy were given daily, with pills of opium and lead for the diarrhœa. No further record of the case appears until the end of July, when it is stated that vomiting of chocolate colored fluid occurred on several successive days. The appetite was poor. The pulse was at 100, and full. Stimulants were still liberally administered. On August 2d, an abscess opened and discharged two pints of unmixed pus. On August 3d, a fragment of bone made its exit at the lower part of the right lumbar region. The stomach and bowels continued irritable. On August 10th, another spicula of bone was eliminated in the discharges. On August 12th, an abscess in the rectum opened and discharged freely. A piece of bone, one fourth of an inch square, was voided at stool. There was complete anorexia; the pulse was 120, and weak. The patient died on August 15th, 1862." The case is reported by Assistant Surgeon R. F. Weir, U. S. A.

CASE 87.—"Lieutenant M. Mullen, Co. G. 69th Pennsylvania Volunteers, at the battle of Gettysburg. Pennsylvania, July 3d, 1863, was struck in the left hip by a musket ball. He was soon conveyed to the Seminary Hospital. An exploration of the wound revealed a fracture of the femur extending into the hip-joint. Cold water dressings were applied. The treatment otherwise consisted of rest, position, and the administration of anodynes and stimulants, with nourishing diet. He died July 7th, 1863." Surgeon F. F. Burmeister, 69th Pennsylvania Volunteers, reports the case.

CASE 88.—"Private Thomas M———, Co. C, 14th Maine Volunteers, aged 31 years, was wounded at the battle of Cedar Creek, Virginia, October 19th, 1864, by a conoidal musket ball, which entered the left buttock near the fold of the nates, and, passing outwards, struck the femur below the great trochanter, and made its exit near the apophysis. After a few days, the patient was conveyed by ambulance to Winchester, and thence by rail to Baltimore, and admitted, on the 26th of October, into the Camden Street Hospital. He was treated on the expectant plan, the limb being supported by pillows. Profuse suppuration took place, and the patient gradually sank, and died, exhausted, on December 5th, 1864. From the appearance of rigors on November 15th, and their recurrence, it was thought that pyæmia had been developed; but the autopsy revealed only congestion of the right lung and adhesion of its pleura to the diaphragm, but no metastatic abscesses in either of them, or in any of the viscera." The specimen is figured in the wood-cut, and shows a long, oblique fracture of the shaft, (FIG. 48,) extending posteriorly to the middle of the neck. The lower part of the great trochanter was crushed by the ball, and a fissure extends to its tip. The neck is carious, and some of the laminated tissue is exfoliating. The specimen and history were contributed by Acting Assistant Surgeon J. G. Keller.

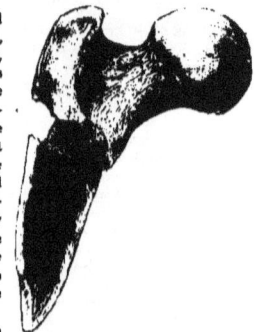

FIG. 48.—Gunshot fracture of the neck and trochanters of the left femur.—Spec. 1729, Sect. I, A. M. M.

CASE 89.—"Sergeant S. W. N———, Co. E, 15th New Jersey Volunteers, aged 22 years, was wounded at the battle of Chancellorsville, Virginia, May 3d, 1863, by a conoidal musket ball, which entered three inches below the anterior superior spinous process of the right ilium, two inches outside of the course of the femoral artery, passed through the hip-joint, and emerged two inches below the tuberosity of the ischium. On May 8th, he was conveyed by steamer to Washington, and admitted into Douglas Hospital. The patient was much prostrated. The limb was placed in an easy position, and cold water dressings were applied; stimulants and concentrated nutriment were administered. On the 15th, he experienced a severe rigor, which was followed by increased constitutional disturbance. The existence of pyæmia was suspected. He failed gradually under the extensive inflammation and profuse suppuration, and died on May 20th, 1863. At the autopsy, no evidence of pyæmia was observed. About one-third of the lower lobe of the left lung was hepatized posteriorly. The viscera, otherwise, were normal, or nearly so. The tissues near the seat of fracture were filled with fragments of bone." The pathological specimen shows a comminuted fracture at the anatomical neck of the femur. It is represented in the adjoining wood-cut, (FIG. 49,) and was contributed, with the history, by Assistant Surgeon Charles C. Lee, U. S. A.

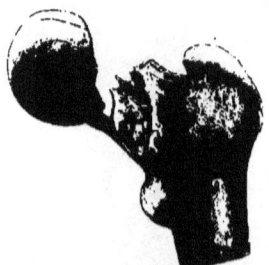

FIG. 49.—Fracture of the head and neck of the right femur by a conoidal musket ball.—Spec. 1253, Sect. I, A. M. M.

CASE 90.—"Sergeant William Norton, Co. 1, 5th Wisconsin Volunteers, aged 30 years, was wounded at the battle of Chancellorsville, Virginia, May 3d, 1863, by a musket ball, which fractured the femur through the neck and trochanter major. He was conveyed by steamer to Washington, and admitted into Stanton Hospital on May 8th. The parts about the hip were swollen and very painful. A long, straight splint was adjusted to the limb, and water dressings were applied. One half grain of morphine was given at night. On May 15th, the wound was suppurating; a small piece of bone was eliminated; pulse 105. On May 17th, considerable febrile action was manifested; the skin was of a yellow hue; the appetite was poor. On May 20th, all the symptoms of pyæmia, with the exception of rigors, began to appear, the icteroid tinge, profuse perspiration, and hebetude. Stimulants were liberally administered, and the limb was placed in Hodgen's splint. On May 23d, the discharge was healthy, and the wound looked well. Five grains of quinine and ten drops of the muriated tincture of iron were directed thrice daily. On June 7th, diarrhœa set in; the pulse was 130 and feeble; rapid emaciation ensued. The diarrhœa was arrested on June 12th, by pills of camphor and opium, but recurred. On June 26th, he had dyspnœa and vomiting, and took but little nourishment. Death took place on June 28th, 1863." At the autopsy, the hip-joint and parts adjacent were found infiltrated with pus. Nearly the entire left lung was in a state of inflammation. The liver and kidneys were fatty. No multiple abscesses were discovered. The case is reported by Acting Assistant Surgeon George A. Mursick.

CASE 91.—"Captain Samuel Oakley, Co. —, 77th New York Volunteers, was wounded near Petersburg, Virginia, March 25th, 1865, by a fragment of shell, which struck the left hip, producing a comminuted fracture of the neck and head of the femur. He was conveyed to the field hospital of the 2d Division, 6th Corps. Death took place within a few hours, on March 25th, 1865." This case is reported by Surgeon S. F. Chapin, 139th Pennsylvania Volunteers.

CASE 92.—"Private William O———, Co. K, 2d United States Cavalry, aged 37 years, was wounded near Old Church, Virginia, June 1st, 1864, by a conoidal musket ball, which struck the left femur on the outside just below the trochanter major, crushed in the laminated structure, and produced a fissure that extended within the capsule, and downwards two inches below the trochanter minor. He was conveyed to Washington, and admitted to Stanton Hospital, on June 4th. Ice was applied to the wound, and stimulants and tonics were administered. Acute irritative fever and profuse suppuration ensued, and death resulted, June 11th, 1864, from exhaustion." At the autopsy, the condition of things represented in the accompanying wood-cut (FIG. 50) were observed. Although the fracture extended within the capsule, little or no alteration had taken place in the articular surface of the femur or in the acetabulum. The specimen was contributed by Surgeon John A. Lidell, U. S. Volunteers.

FIG. 50.—Longitudinal fissuring of the upper extremity of the left femur by a conoidal musket ball.—*Spec.* 2528, Sect. I, A. M. M.

CASE 93.—"Private M. M. Phillips, Co. F, 42d Mississippi Regiment, aged 23 years, was wounded at the battle of Gettysburg, Pennsylvania, July 3d, 1863, by a musket ball, which entered the posterior and outer aspect of the left buttock, fractured the neck of the femur, and emerged in the left groin. He was captured and removed to the hospital for Confederates, and was treated by the expectant plan. On July 22d, he was transferred, by rail, to the De Camp Hospital, in New York Harbor. The wounds of entrance and exit had healed. He was able to walk by the aid of a crutch or cane, and suffered no pain. The limb was shortened one and three quarters inches. By flexing and rotating the limb, true osseous crepitus was obtained. There was no inversion or eversion. The patient, much against his will, was placed upon a fracture bed, where extension was produced for six weeks by a twelve pound weight acting over a pulley. He was then permitted to rise and directed to use passive motion and friction. On September 20th, 1863, he was paroled, being able to walk without assistance of any kind. The limb was shortened one and a quarter inches." The case is reported by Acting Assistant Surgeon George Edwards.

CASE 94.—"Alfred G. R———, Adjutant 134th Pennsylvania Volunteers, aged 24 years, was wounded in the upper part of the left femur by a round ball, which partially fractured the trochanter major, at the battle of Fredericksburg, Virginia, December 13th, 1862. The wound received no attention for some days, and was then dressed with side splints firmly bound by a roller, a plug of lint being tightly inserted in the wound. On the 20th of December, he was admitted to E Street Infirmary, Washington. He stated that for some days he had experienced occasional twitchings in the limb, and had taken large doses of opium. The wound was a little behind the trochanter major. Upon removing the plug of lint, about half a pint of blood and pus was discharged. There was no crepitus upon rotation, nor shortening. Owing to his weakened condition, no extended search was made for the missile. Simple dressings were applied, and half grain doses of sulphate of morphia were given. For the two succeeding days he seemed to improve. The twitchings of the limb occurred every few minutes, with occasional intermissions of a few hours. On the 23d, the spasms became more violent and frequent, and it was deemed advisable to extract the missile. He was etherized, and the wound was enlarged two inches downward and backward. A gum catheter was made to follow the course of the missile behind and beneath the neck of the femur to the body of the pubis, where the ball was found in the scrotum near the spermatic cord. A flattened round musket ball was extracted through an incision at the base of the scrotum. A portion of it had been chipped off. The patient rested well that night, but on the following day the spasms were increased in intensity, commencing in the injured limb and extending over the body. Cloths saturated with chloroform and olive oil were applied to the limb, and an antispasmodic and anodyne mixture was prescribed. He rested quietly until the following morning, when clonic spasms returned and persistently increased. The patient's countenance became pinched, wan, and haggard, and expressive of fright. There was no pain nor trismus, and he partook freely of nourishment. At times there was complete opisthotonos. On December 25th, he took four dozen pills of assafœtida of four grains each, and one half ounce of fluid extract of Cannabis Indica in divided doses, without any benefit. Sulphate of morphia in doses of one grain was then prescribed, to be administered every two hours, and a poultice of powdered opium and cinchona applied to the wound, but, as before, without apparent benefit. The mind, up to this time, continued clear and undisturbed, his pulse moderately full and strong, ranging at about 100. He now became drowsy, and at times lay in a semi-comatose condition. His pulse ran up to 150. Respiration was free, but at times hurried, from 25 to 28 per minute. The skin became

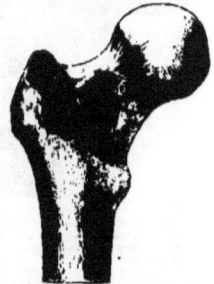

FIG. 51.—Fracture of trochanter and neck of left femur by a round musket ball. A fragment of lead is impacted at the inner end of the groove in the neck.—*Spec.* 545, Sect. I, A. M. M.

bathed in sweat, which exhaled a peculiar pungent odor. The bowels were regular; the urine was scantily secreted and high colored, though voided without difficulty. The discharge from the wound was thin, bloody, and offensive. On December 27th, opisthotonos recurred, and was temporarily relieved by the application of chloroform to the entire extent of the spine. Subsequently, violent epileptiform convulsions set in, and death resulted from exhaustion, on December 28th, 1862." The pathological specimen is figured in the preceding wood-cut, (FIG. 51,) and shows a fracture of the great trochanter of the left femur, and a piece of a leaden ball imbedded in the neck. The specimen and history was contributed by Surgeon C. L. Allen, U. S. V.

CASE 95.—"Private Joseph L. Riley, Co. I, 21st Mississippi Regiment, aged 35 years, was wounded in the engagement at Sailor's Creek, Virginia, April 6th, 1865, by a musket ball, which entered the buttock three inches behind the left trochanter major, perforated the neck of the femur, and made its exit through the groin. He was taken to the field hospital of the 1st Division, 6th Corps, and thence, on April 18th, by steamer, from City Point, to Washington. He was admitted on April 19th, into Lincoln Hospital. He died, exhausted from profuse suppuration, on June 2d, 1865. At the autopsy the neck of the femur was found badly comminuted, and the head of the bone denuded of cartilage." The case is reported by Acting Assistant Surgeon W. A. Finn.

CASE 96.—"Private James R——, Co. C, 69th New York Volunteers, aged 27 years, was wounded at the battle of Petersburg, Virginia, March 25th, 1865, by a conoidal musket ball, which entered the right thigh some three inches below and inside of the anterior superior spinous process of the ilium. He was conveyed by rail to City Point, and thence by steamer to Washington, and was admitted to Lincoln Hospital on March 30th. Suitable support to the limb, simple dressings to the wound, and good food constituted the treatment. Death resulted April 6th, 1865, from exhaustion. At the autopsy, it was found that the missile had passed through the neck of the femur from before backwards and outwards, and was wedged into a fissure near the outer surface of the trochanter major. This fissure is continued upward to the anatomical neck, and extends likewise six inches down the shaft." The specimen is represented in the accompanying wood-cut, (FIG. 52,) and, with the history of the case, was contributed by Acting Assistant Surgeon J. P. Arthur.

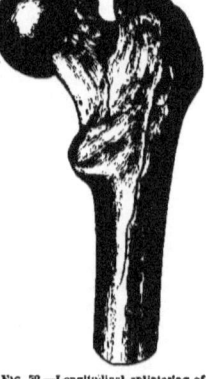

FIG. 52.—Longitudinal splintering of right femur by a conoidal musket ball, which is lodged in the trochanter major. Spec. 88, Sect. I, A. M. M.

CASE 97.—"Corporal John Robinson, Co. I, 3d Maryland Cavalry, was struck, near the Spanish Fort, Mobile Harbor, Alabama, March 26th, 1865, by a fragment from an exploding torpedo, which lacerated the soft parts about the hip-joint and comminuted the upper end of the femur. He was taken to the hospital of the 2d Division, 13th Corps, at Blakely, Alabama. The patient died on April 3d, 1865." The case is reported by Surgeon O. Peabody, 23d Iowa Volunteers.

CASE 98.—"Private Lemuel R——, Co. F, 48th New York Volunteers, aged 18 years, in the assault on Fort Wagner, South Carolina, July 18th, 1863, was wounded in the right thigh by a round ball, which entered posteriorly and fractured the neck of the femur and lodged. He was captured, but soon paroled, and, on July 24th, was received on the steamer Cosmopolitan, and conveyed to New York. He was admitted to the McDougall Hospital, at Fort Schuyler, on July 30th. Death resulted on August 3d, 1863." The pathological specimen contributed with the history by Assistant Surgeon R. Bartholow, U. S. A., is represented in the accompanying wood-cut. (FIG. 53.)

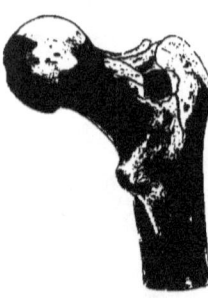

FIG. 53.—Fracture of neck of right femur by a round musket ball, impacted near the base.—Spec. 1601, Sect. I, A. M. M.

CASE 99.—"Private Thomas R——, Co. K, 210th Pennsylvania Volunteers, aged 25 years, was wounded in the engagement on the South Side Railroad, Virginia, March 31st, 1865, by a musket ball, which penetrated at a part just behind the left trochanter. He was conveyed by rail to City Point, and thence by steamer to Washington, and, on April 8th, was admitted into Douglas Hospital. He was much depressed, and the wounded parts were painful, and considerable febrile action was manifested. He lay on his side, with the thighs flexed. An exploration of the wound revealed a fracture of the great trochanter, but it was thought that the joint was not implicated. The limb was placed in a straight position, was supported by sandbags, and moderate extension was made by pulley and weights. Much relief was obtained by this treatment, but he still suffered intense pain in the hip-joint. He failed gradually, and died, exhausted, on April 18th, 1865." The pathological specimen is No. 4168, Sect. I, A. M. M. The great trochanter is comminuted, together with a portion of the neck, which is also broken obliquely off the shaft. The fracture in the neck is intra-capsular. The history and specimen were contributed by Assistant Surgeon W. F. Norris, U. S. A.

CASE 100.—"Sergeant Frank Sallyards, Co. A, 70th Ohio Volunteers, aged 18 years, was wounded at Fort McAllister, Georgia, December 13th, 1864, by a conoidal musket ball, which entered the left thigh in front, and, passing upwards, fractured the neck of the femur. He was conveyed to the field hospital of the 2d Division, of the 15th Corps. Cold water dressings were applied. On December 24th, he was transferred by steamer to Hospital No. 1, at Beaufort, South Carolina. Pyæmia supervened, and he died on January 29th, 1865." The case is reported by Surgeon J. B. Potter, 30th Ohio Volunteers.

CASE 101.—"Private Henry Sault, Co. I, 5th New York Volunteers, was wounded at the second battle of Bull Run, Virginia, August 28th, 1862, by a musket ball, which entered the left hip-joint, fracturing the head of the femur, and lodged. He was conveyed to Washington, and, on September 2d, was admitted to the Ryland Chapel Hospital. He was treated on the expectant plan, and, without the development of any unusual symptoms, he died, exhausted by suppuration, on November 11th, 1862." The case is reported by Assistant Surgeon V. B. Hubbard, U. S. A.

CASE 102.—"Private Charles Saunders, Co. C, 6th Louisiana Cavalry, aged 21 years, was wounded at Fort Blakely, Mobile Harbor, Alabama, April 9th, 1865, by a conoidal musket ball, which fractured the head and neck of the femur. He was conveyed by steamer to New Orleans, Louisiana, and received into the Military Prison Hospital. On April 17th, he was transferred to the St. Louis Hospital. Pyæmia supervened, and the patient died on May 6th, 1865." The case is reported by Surgeon A. McMahon, U. S. V.

CASE 103.—"Captain Edwin F. S——, Co. K, 1st New York Cavalry, aged 23 years, a robust healthy man, was wounded in the action at Sailor's Run, Virginia, April 6th, 1865, by a conoidal musket ball, which entered the buttock, two inches outside the right sacro-iliac symphysis, perforated the neck of the right femur, and lodged about the hip-joint. He was conveyed by ambulance and steamer to the Cavalry Corps Hospital, at City Point, and, for one month, was treated by rest and position. Afterwards, he was transferred by steamer to Washington, and admitted into Armory Square Hospital. The thigh was then swollen and very painful. There were extensive abscesses, and over the lower portion of the sacrum was a small bed sore. An incision was at once made near the trochanter major, which evacuated some two pints of pus. Stimulants were now, for the first time, administered. During part of May 9th and 10th, the patient was somewhat delirious; but subsequently his condition improved, until the 23d, when pyæmic infection, ushered in by a chill, was declared by the peculiar odor of the breath, the rapid circulation and respiration, the prostration, nausea, and anorexia. The suppuration, meanwhile, was healthy and unchanged in quantity. Four days later, a second chill occurred, and was attended with excessive irritability of the stomach, which rejected every sort of food. Porter and lime water, in equal parts, was tried with advantage, and nutriment and stimulants by enema were administered. The salts of quinine, in fifteen grain doses, three times daily, were directed, but apparently did no good. On May 30th, diarrhœa began. Death supervened on June 2d, 1865. The autopsy revealed the lodgment of the missile between the fragments near the trochanter minor. The cartilages of the head of the femur and of the cotyloid cavity had been almost wholly absorbed, and a deposition of callus had so firmly united the head to its socket that it could only be disconnected by the exertion of considerable force. Pus (?) globules were detected in the medulla of the shaft by the aid of the microscope. The viscera were not examined." The pathological specimen is represented in the figure adjoining. (FIG. 54). The neck was perforated from behind, and shattered and separated from the shaft. The head of the femur is eroded and carious. Acting Assistant Surgeon G. K. Smith reported the case.

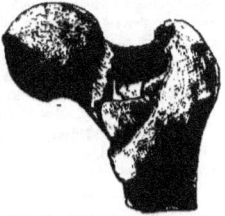
FIG. 54.—Caries of head of right femur from gunshot fracture of the neck.—Spec. 4213, Sect. I, A. M. M.

CASE 104.—"Private Christian S——, Co. F, 22d Michigan Volunteers, aged 21 years, was wounded at the battle of Chickamauga, Georgia, September 20th, 1863, by a conoidal musket ball, which penetrated near the tuberosity of the left ischium. He remained in the hands of the enemy until about the middle of November, when, upon being paroled, he was conveyed to Nashville, Tennessee, and admitted, on November 17th, into Hospital No. 1. The parts in the region of the wound were greatly swollen, and there were extensive abscesses in the left thigh, one of which was opened on the 22d, evacuating a large quantity of pus. Several explorations were made for the missile. Tonics and stimulants were administered, with nourishing food. Great irritability of the system was induced under the excessive suppuration. On February 15th, 1864, a chill with other characteristic symptoms announced the invasion of pyæmia, of which he died on February 18th, 1864. The autopsy revealed ulceration of the capsular ligament, the formation of an abscess in the iliac fossa, and the presence of a crescentic-shaped fragment of bone, eight inches below the great trochanter, beneath the tensor vaginæ femoris muscle. The missile could not be found. It had probably been extracted while the patient was in the hands of the enemy." The pathological specimen is No. 2198, Sect. I, A. M. M., and was contributed, with a history of the case, by Acting Assistant Surgeon H. M. Lilly. The head of the femur is carious, and the acetabulum is also roughened by caries.

CASE 105.—"Captain R. Shaw, Co. K, 56th Pennsylvania Volunteers, in the action at Gravelly Run, Virginia, March 29th, 1865, was struck in the left thigh, near the hip, by a musket ball. He was conveyed to the hospital of the 2d Division, 5th Corps. A fracture of the upper extremity of the femur, involving the hip-joint, was diagnosticated. On the following day he was transferred to the Depot Field Hospital of the 5th Corps. The treatment was on the expectant plan. Death took place on April 11th, 1865." The case is reported by Surgeon A. G. Coe, 147th New York Volunteers, and by Surgeon W. L. Faxon, 32d Massachusetts Volunteers, in charge of Depot Hospital, 5th Army Corps, City Point, Virginia.

CASE 106.—"Private G. H. S——, Co. H, 2d North Carolina Regiment, was wounded at the battle of Gettysburg, Pennsylvania, July 3d, 1863, by a musket ball, which, entering the right hip-joint, shattered the neck of the femur. He was captured and conveyed to the hospital for prisoners. The plan of treatment was expectant. On July 27th, he was transferred to Camp Letterman Hospital, at Gettysburg. He died, exhausted, on October 26th, 1863. The pathological specimen is No. 1067, Sect. I, A. M. M. The fracture involved the head of the femur, and the articular surface is generally eroded. Fragments partly necrosed are attached to the shaft near the tuberosities by new osseous formation." The specimen and history of the case were furnished by Surgeon Henry Janes, U. S. V.

CASE 107.—"Corporal Henry F. Smith, Co. B, 1st Wisconsin Volunteers, aged 25 years, was wounded, near Nashville, Tennessee, March 24th, 1862, at a distance of sixty yards from the enemy, by a musket ball, which entered the left thigh, near the apex of Scarpa's triangle, at the inner edge of the sartorius muscle, passed upwards, perforating the neck of the femur, and emerged half an inch outside of the tuberosity of the left ischium. On March 25th, he was admitted into the University Hospital. He had lost a large quantity of blood, and had suffered much pain, and was still depressed by the shock of the injury. The wound of entrance would admit only the tip of the little finger, while the forefinger could be introduced into that of exit. The position assumed by the limb was that which characterizes a dislocation of the head of the femur upon the dorsum ilii. Anodynes and stimulants were freely administered, and the general condition of the patient seemed much improved after a night's repose. He, however, complained of great pain about the knee. On March 26th, he was placed under the influence of ether. The limb was rotated outwards and shortened three-fourths of an inch. To determine the extent of the injury, the posterior wound was enlarged, when it was ascertained that the projectile had passed through the neck of the femur, producing a fracture above the trochanters. The acetabulum did not seem to be injured. The circulation through the femoral artery was unimpeded, for the pulsation of the dorsal artery of the foot was perfectly natural. The limb was placed upon a double inclined plane, for which a straight apparatus was substituted in a few days. On March 29th, there was a slight oozing of blood from the anterior wound. This was immediately plugged with lint, and the hæmorrhage was arrested. The patient passed a quiet night; pulse about 90. There was scarcely any purulent discharge. Next day, March 30th, at 3 P. M., a torrent of blood gushed suddenly from the anterior wound, and all pulsation ceased at once throughout the lower part of the limb. A ligature was immediately placed, by Surgeon A. H. Thurston, U. S. V., upon the external iliac artery at a point about two inches above Poupart's ligament, and the hæmorrhage was arrested; but the patient had become so far exhausted that he failed to rally, and expired at 11 o'clock P. M., March 30th, 1862." Surgeon E. Swift, U. S. A., reported the case.

CASE 108.—"Private John Spangler, Co. A, 38th Ohio Volunteers, received, in the engagement at Jonesboro', Georgia, September 1st, 1864, a gunshot fracture of the head of the right femur. He was taken to the field hospital of the 3d Division, 14th Corps. Death took place on September 8th, 1864." The case is reported by Surgeon H. J. Herrick, 17th Ohio Volunteers.

CASE 109.—"Corporal J. A. Staunton, Co. H, 1st Florida Volunteers, aged 27 years, was wounded at the battle of Murfreesboro', Tennessee, December 7th, 1861, by a conoidal musket ball, which fractured the neck of the right femur. On the following day he was conveyed to the General Hospital at Murfreesboro'. Death took place on December 20th, 1864." This case is reported by Surgeon Samuel D. Turney, U. S. V.

CASE 110.—"Private John Stewart, Co. B, 26th Ohio Volunteers, received, at the battle of Chickamauga, Georgia, September 19th, 1863, a gunshot fracture of the neck of the right femur. He was conveyed to the hospital of the 1st Division, 21st Corps, near Crawfish Springs, and treated by rest, position, and simple dressings. Death took place on October 10th, 1863." Surgeon I. Moses, U. S. V., reported the case.

CASE 111.—"Sergeant Herman Stutler, Co. D, 53d Pennsylvania Volunteers, was wounded at the battle of Fredericksburg, Virginia, December 13th, 1862, by a conoidal leaden bullet, which entered near the left great trochanter, and produced a comminuted fracture of the head of the femur. He was conveyed to the field hospital of the 1st Division, 2d Corps. Splints were applied, and cold water dressings used. He was thence conveyed to Washington, and, December 26th, was admitted to Stanton Hospital. He died, exhausted, on February 11th, 1863. The missile had not been extracted." Surgeon John A. Lidell, U. S. V., reported the case.

CASE 112.—"Private Joseph S———, Co. B, 43d Ohio Volunteers, aged 23 years, a robust man, was wounded at the battle of Corinth, Mississippi, October 3d, 1862, by a spherical musket ball, which entered anteriorly at the upper part of the right thigh, an inch within the course of the femoral artery, and passed upward and backward, grazing the anatomical neck of the femur, and made its exit at the point of the left buttock. He was conveyed in a hospital transport to St. Louis, and admitted, on October 15th, to the City General Hospital, under the care of Surgeon John T. Hodgen, U. S. V. He died on October 24th, 1862." The pathological specimen is represented in the accompanying wood-cut. (FIG. 55.) At the under surface of the neck, near the head, is a shallow cavity, which marks the point of impact of the missile, which is attached. Around this there is slight caries. Dr. Hodgen contributes this specimen and memorandum.

FIG. 55.—Partial fracture of neck of right femur by a round musket ball.—*Spec.* 463, Sect. I, A. M. M.

CASE 113.—"Private Peter L. Swank, Co. I, 38th Ohio Volunteers, received, in the engagement at Jonesboro', Georgia, September 1st, 1864, a gunshot fracture of the head of the left femur. He was conveyed to the field hospital of the 3d Division, 14th Corps. Death took place on September 5th, 1864." The case is reported by Surgeon H. J. Herrick, 17th Ohio Volunteers.

CASE 114.—"Corporal John M. Thompson, Co. B, 29th Massachusetts Volunteers, aged 27 years, was wounded, near Petersburg, Virginia, June 17th, 1864, by a conoidal musket ball, which entered the left hip, shattering the femur from the head to two inches below the lesser trochanter, and lodged. He was conveyed by steamer from City Point to Annapolis, Maryland, and, June 20th, was received into Hospital No. 1. The soft parts were extensively lacerated. The missile was extracted. Violent inflammation, followed by asthenic fever, ensued; copious suppuration took place. Oakum dressings were applied, and stimulants were administered. He died on June 27th, 1864." This case is reported by Surgeon D. A. Vanderkieft, U. S. V.

CASE 115.—"Sergeant Volney Tidball, Co. H, 122d Ohio Volunteers, aged 27 years, at the battle of Cedar Creek, Virginia, October 19th, 1864, had the head of the left femur fractured by a conoidal musket ball. He was conveyed to the hospital of the 3d Division, 6th Corps, and transferred, November 3d, to the Sheridan Depot Field Hospital, at Winchester. The expectant plan of treatment was pursued. Death took place on November 7th, 1864." The case is reported by Surgeon W. A. Darry, 98th Pennsylvania Volunteers.

CASE 116.—"Private James Vanderbeek, Co. F, 145th New York Volunteers, aged 21 years, was wounded at the battle of Chancellorsville, May 3d, 1863, by a conoidal musket ball, which entered the left thigh above and behind the trochanter major, passed forwards and inwards, fractured the neck of the femur, and made its exit at the groin. He was made a prisoner, and remained in the hands of the enemy eleven days. He was then exchanged, and conveyed to the 12th Corps Hospital, at Aquia Creek. The injured limb was simply placed in a comfortable position, without any attempt at extension. On June 14th, the patient was removed on a hospital transport to Alexandria, and placed in the First Division Hospital. He was in good condition. The suppuration was comparatively slight, and no bone splinters were found loose, and none had come away. Three days subsequently, he was transferred to Philadelphia, and thence, on October 12th, to New York, where he was admitted at the Ladies' Home Hospital. The wounds were closed at this date. The patient was discharged from the hospital and from the service of the United States on November 19th, 1863. At that date he walked with crutches. His limb was shortened two inches, with eversion. He was allowed a pension." On August 2d, 1866, Dr. E. Bradley, examining surgeon of the Pension Bureau, reported that Vanderbeck's general health was good, but that there was much lameness. The fracture was firmly consolidated. Commissioner C. C. Cox reports that Vanderbeck received his pension at the agency in New York city on March 4th, 1868, and that his disability was then rated by the examining surgeon as total.

CASE 117.—"Private Joseph Wagoner, Co. F, 81st Illinois Volunteers, was wounded at the battle of Pleasant Hill, Louisiana, April 8th, 1864, by a musket ball, which entered at a point a little above and behind the left trochanter major, and emerged on the inner side of the thigh, some two inches below the perinæum. An exploration of the wound revealed a fracture of the neck. He was taken to Vicksburg, and thence on the steamer McDougall, and conveyed to Mound City, Illinois, where, on May 5th, he was received into the General Hospital. The parts were very painful. On May 6th, the patient suffered a severe rigor, and increased pain at the hip upon any motion of the joint. He had extensive bed sores, to which applications of myrrh, lead-plaster, and creasote were made. In June he complained much of pain at knee-joint and lower portion of the thigh. Early in July he suffered several days from excessive diarrhœa; the wound began to slough. He died on July 14th, 1864." No doubt was entertained that the hip-joint was implicated, and that ulceration of the cartilages had taken place. Permission to examine the parts post mortem could not be obtained. The case is reported by Acting Assistant Surgeon A. H. Kellogg.

CASE 118.—"Private Richard A. Walker, Co. E, 2d New Hampshire Volunteers, aged 18 years, was wounded at the battle of Williamsburg, Virginia, May 5th, 1862, by a musket ball, which penetrated over the left great trochanter, and lodged about the hip-joint. Shortly after the reception of the wound, the patient was conveyed, by ambulance, some five or six miles to the nearest point on the York river, and thence by steamer to Fortress Monroe, where he was admitted, on May 6th, into the Hygeia Hospital, under the care of Surgeon R. B. Bontecou, U. S. V. For several days no unfavorable symptoms appeared, and permission was granted for the removal of the patient to the house of a lady in the neighborhood, where every attention was afforded him. Profuse suppuration soon supervened, and he died, exhausted, on June 8th, 1862." The autopsy revealed a round musket ball lying loosely in the joint, into which it had passed through a deep furrow in the head and neck of the femur. The articular surfaces were wholly denuded of their covering. Surgeon R. B. Bontecou, U. S. V., reported the case.

CASE 119.—"Private Isaac W. Winans, Co. C, 3d Wisconsin Volunteers, was wounded at the battle of Cedar Mountain, Virginia, August 9th, 1862, by a musket ball, which entered the right thigh one inch below the middle portion of Poupart's ligament, passed obliquely outwards, comminuting the neck of the femur, and lodged. He was conveyed to Alexandria and admitted into St. Paul's Church Hospital, on August 13th. The parts were swollen and very painful. Cold water dressings were applied, and anodynes were administered, but apparently without relief. He died on August 18th, 1862." The case is reported by Acting Assistant Surgeon John A. Cooper.

CASE 120.—"Corporal George W. Wright, of Blount's Virginia Battery, aged 21 years, received, near Petersburg, Virginia, June 17th, 1864, a gunshot fracture of the neck of the femur. He was conveyed to the Washington Street Confederate Hospital in Petersburg. He was treated on the expectant plan, and died, exhausted, on July 7th, 1864." The case is reported by Surgeon W. L. Baylor, C. S. A.

CASE 121.—"Private W. P. Yeargin, Co. E, 22d Georgia Regiment, received at the battle of Gettysburg, Pennsylvania, July 2d, 1863, a gunshot fracture of the neck of the femur, involving the hip-joint. He was captured and conveyed to the hospital for Confederates at Gettysburg. On July 18th, he was transferred, by rail, to the General Hospital at Chester. He died, exhausted, on August 11th, 1863." Surgeon E. Swift, U. S. Army, reported the case.

CASE 122.—"Sergeant David Y——, Co. H, 106th New York Volunteers, aged 29 years, was wounded at the battle of Monocacy Junction, Maryland, July 9th, 1864, by a conoidal musket ball, which entered two inches posterior to and one inch above the right trochanter major, passed forwards and inwards, and lodged in the neck of the femur at its middle portion. He was admitted to the General Hospital at Frederick, Maryland, on the same day, and the wound did well until the 12th of July, when it assumed an unhealthy appearance. A careful examination was made with the finger and by the probe, and the integuments and fascia were divided, giving free exit to sanious and fœtid pus. Large quantities of stimulants and beef tea were given. On July 19th, symptoms of pyæmia made their appearance, such as rigors followed by profuse perspiration and acceleration of the pulse and respiration, dryness of the tongue, and anorexia. Another examination of the wound was made, and the ball was found imbedded in the femur, but, owing to the patient's condition, its removal was deemed inadvisable. On July 20th, another rigor occurred, and gradual aggravation of all the symptoms followed. He died at 3 o'clock P. M., July 22d, 1864." The pathological specimen and history (FIG. 56) were contributed to the Army Medical Museum by Assistant Surgeon R. F. Weir, U. S. A.

FIG. 56.—Fissure of the right femur caused by a conoidal ball lodging in the neck.—*Spec.* 3331, Sect. I, A. M. M.

The foregoing category of one hundred and twenty-two cases comprises the greater number of examples, recorded on the registers of this Office, of gunshot fractures of the upper extremity of the femur, involving the hip-joint, and treated on the expectant plan, which were uncomplicated by lesions of the pelvis or of the great vessels or nerves. In those cases in which the diagnosis was not verified after death, the nature of the injury was explicitly reported. In the one hundred and fourteen fatal cases, the average duration of life, after the reception of the injury, was thirty-one days.

In the next series, those cases of gunshot injury of the hip-joint are included in which the acetabulum was slightly injured as well as the femur. The graver gunshot injuries of the pelvis are, of course, excluded from this category:

CASE 123.—"Corporal Henry Achley, Co. G, 2d New York Heavy Artillery, aged 30 years, was wounded at the engagement at Mechanicsville, Virginia, by a musket ball, which entered the right gluteal region and passed in the neighborhood of the hip-joint. He was conveyed to Washington, and admitted to Emory Hospital, June 4th, 1864, under the care of Surgeon N. R. Mosely, U. S. V. There were symptoms of inflammation of the hip-joint; but no evidence of fracture was discovered. On June 9th, the patient was transferred to McDougall Hospital, at Fort Schuyler, New York. On admission, he was slightly delirious, and wished to put on his trowsers and move about. He seemed to suffer but little pain from motion of the joint. A bed sore formed on the left hip, and profuse suppuration, accompanied by oozing of venous blood, ensued, and the hæmorrhage was so considerable that a ligation of the posterior iliac artery was contemplated; but the patient sank rapidly, and died June 19th, 1864. On the examination of the body, twelve hours after death, it was found that the head of the femur and the rim of the acetabulum had been fractured by the ball. Nearly the whole of the neck of the femur was destroyed, and the trochanters and upper extremity of the shaft were carious. Numerous sinuses, filled with sanious pus, surrounded the articulation, and the tissues in the track of the ball were in a sloughing condition." Acting Assistant Surgeon H. M. Sprague reports the case.

CASE 124.—"Private Francis Baker, Co. I, 3d Vermont Volunteers, aged 21 years, was wounded at the battle of the Wilderness, Virginia, May 5th, 1864, by a conoidal musket ball, which penetrated the left hip. He was taken to the hospital of the 2d Division, 6th Corps. A wound of the soft parts only was recognized. He was conveyed by steamer to Alexandria, and admitted to the 2d Division Hospital on May 24th. A gunshot wound of the left hip, involving the head of the femur, was diagnosticated. He died on June 7th, 1864. The autopsy revealed a comminuted fracture of the head and neck, with a fracture of the acetabulum." Acting Assistant Surgeon E. H. McCartin reports the case.

CASE 125—"Private John H. Brown, Co. A, 19th Massachusetts Volunteers, received, at the battle of North Anna River, Virginia, May 24th, 1864, a gunshot wound of the right hip. He was conveyed to the hospital of the 2d Division, 2d Corps. Simple dressings were applied. It was found that the head of the femur and the acetabulum were fractured. He died May 26th, 1864." Surgeon J. F. Dyer, 19th Massachusetts Volunteers, reports the case.

CASE 126.—"Private C. H. Calhoun, Co. H, 7th North Carolina Regiment, was wounded, at the engagement at Deep Bottom, Virginia, July 28th, 1864, by a conoidal musket ball, which passed through the right hip. He was made a prisoner, and was conveyed to the hospital of the 2d Division, 2d Corps. It was found that the missile had produced a fracture involving the head of the femur and the acetabulum. The patient was treated on the expectant plan, and died on the following day, July 29th, 1864." Surgeon A. N. Dougherty, U. S. V., reports the case.

CASE 127.—"Private John H. Carlon, Co. D, 184th Pennsylvania Volunteers, at the battle of Cold Harbor, Virginia, June 3d, 1864, was struck in the right hip by a musket ball, which fractured the head of the femur and the acetabulum. He was conveyed to the field hospital of the 2d Division, 2d Corps. Simple dressings were applied. Death took place on the same day, June 3d, 1864." Assistant Surgeon R. C. Marshall, 184th Pennsylvania Volunteers, reported the case.

CASE 128.—"Private George W. C———, Co. F, 59th Massachusetts Volunteers, aged 33 years, was wounded, before Petersburg, Virginia, July 30th, 1864, by a conoidal musket ball, which entered one and a half inches posteriorly to the left trochanter major, and escaped two and a half inches below the anterior superior spinous process of the ilium. He was immediately conveyed to the hospital of the 1st Division, 9th Corps, and thence, on August 2d, to Douglas Hospital, Washington. He died, August 5th, 1864, from exhaustion." The pathological specimen is No. 3582, Sect. I, A. M. M. The acetabulum was fractured, and about one-fifth of the head of the femur, at its junction with the superior border of the neck, was carried away. The specimen and report were contributed by Assistant Surgeon William Thomson, U. S. A.

CASE 129.—"Private David Combe, Co. K, 200th Pennsylvania Volunteers, aged 29 years, was wounded at Fort Steadman, near Petersburg, Virginia, March 25th, 1865, by a conoidal musket ball, which, entering the left hip, directly over the articulation, perforated the head of the femur, and carried away the anterior portion of the acetabulum. He was taken, the same day, to the field hospital of the 3d Division, and thence to the general field hospital of the 9th Corps. A fracture of the hip was recognized. Early in April, the patient was conveyed by steamer to Alexandria and admitted, April 6th, to the Slough Hospital. Paralysis of the lower extremities and of the bladder existed, and the catheter was required. Cold water dressings were applied to the wound. He died on, April 8th 1865. At the autopsy, the femur was found splintered three inches from the coxo-femoral articulation; the rim of the acetabulum was fractured; the wound was throughout gangrenous." The case is reported by Surgeon M. F. Bowers, 200th Pennsylvania Volunteers, and by Surgeon Edwin Bentley, U. S. V.

CASE 130.—"Private Charles Cushion, a recruit of the 179th New York Volunteers, was shot in the hip, while attempting to desert, at Fort Porter, Buffalo, New York, on the night of May 5th, 1864. An ounce musket ball, entering through the trochanter major, shattered the femur, and, passing one and a half inches from the femoral artery, emerged near the apex of Scarpa's triangle. He died fifty-five hours after the reception of the injury, on May 7th, 1864. At the autopsy, made eight hours after death, the head of the femur was found severed from the shaft. One buckshot was lodged in the acetabulum, and one in the integuments of the thigh, high up." Acting Assistant Surgeon S. W. Wetmore reports the case.

CASE 131.—"Private H. De Coux, of Depeak's Confederate Battery, was wounded, in front of Nashville, Tennessee, December 16th, 1864, by a musket ball, which perforated the right hip-joint. He was captured and conveyed the following day to the Cumberland Hospital, in Nashville. The limb was supported by pillows, and simple dressings were applied. He died on December 18th, 1864. The head of the femur and acetabulum were fractured." Surgeon B. Cloak, U. S. V., reports the case.

CASE 132.—"Private Samuel N. E———, Co. G, 40th Indiana Volunteers, aged 22 years, was wounded at the battle of Marietta, Georgia, June 18th, 1864, by a rifle ball, which penetrated at a point over the left hip-joint. He was conveyed by ambulance and railcar to Chattanooga, Tennessee, and admitted, on June 25th, into the field hospital. He was treated by rest, position, and the application of cold water dressings. He died June 28th, 1864. The pathological specimen is No. 3390, Sect. I, A. M. M., and shows a deep grooving of the head in a line with the axis of the neck. The neck is fractured obliquely. A small fragment of the acetabulum is also broken off." The specimen and history were furnished by Assistant Surgeon C. C. Byrne, U. S. A.

CASE 133.—"Private John S. Fahus, Co. I, 109th New York Volunteers, aged 24 years, received, in front of Petersburg, Virginia, July 1st, 1864, a fracture of the left hip, by a conoidal musket ball, which entered over the trochanter major, and, passing inward, emerged through the left buttock. He was conveyed to the General Field Hospital of the 18th Corps, at Point of Rocks, and thence, on the following day, by steamer, to Point Lookout, Maryland, where he was admitted, July 4th, into the General Hospital. Pyæmia supervened, the premonitory chill occurring on the 18th July. Tonics and stimulants, with nourishing diet, were liberally given. The patient died on July 24th, 1864." The case is recorded by Surgeon Anthony Heger, U. S. A.

CASE 134.—"Private David F———, Co. G, 111th New York Volunteers, aged 25 years, was wounded at the battle of Gettysburg, Pennsylvania, July 3d, 1863, by a conoidal musket ball, which entered over the left trochanter major, passed inwards and upwards, causing a fracture of the head of the femur, and fissures in the acetabulum. The patient was treated in a field hospital, on the expectant plan, until the 16th of July, when he was sent to the Jarvis Hospital, at Baltimore. He was so completely prostrated that it was thought inadvisable to perform resection, which had been contemplated. Nourishing food was sedulously given, and stimulants were freely administered. The case terminated fatally on the 26th of July, 1863. The pathological specimen is represented in the adjacent woodcut, (FIG. 57). The head of the femur is deeply fissured and irregularly eroded. The ball is impacted in its upper part. The pelvic portion of the specimen was not preserved." The specimen and history were contributed by Assistant Surgeon D. C. Peters, U. S. A.

FIG. 57.—Fracture of the articular surface of the left femur by a musket ball. *Spec.* 1616, Sect. I, A.M.M.

CASE 135.—"Sergeant Samuel Garver, Co. K, 89th Indiana Volunteers, aged 28 years, was wounded, in front of Nashville, Tennessee, December 16th, 1864, by a conoidal musket ball, which perforated the left hip-joint, grooving the head of the femur, and chipping the acetabulum. He was taken to the regimental hospital, and, the following day, to the Cumberland Hospital, in Nashville. He died on December 18th, 1864." Surgeon B. Cloak, U. S. V., reports the case.

CASE 136.—"Private Frederick Geyser, Co. F, 1st Minnesota Volunteers, received, in the engagement at Deep Bottom, Virginia, August 14th, 1864, a gunshot fracture of the left hip, implicating the rim of the cotyloid cavity and the head of the femur. He was taken to the field hospital of the 2d Division, 2d Corps. He died on August 15th, 1864." The case is reported by Surgeon Gilbert Chaddock, 7th Michigan Volunteers.

CASE 137.—"Corporal James T. Glancy, Co. F, 2d Rhode Island Volunteers, received, in front of Petersburg, Virginia, April 2d, 1865, a gunshot fracture of the acetabulum, and of the head and neck of the left femur. He was conveyed to the field hospital of the 1st Division, 6th Corps. He died on April 3d, 1865." The case is reported by Surgeon Bedford Sharp, 15th New Jersey Volunteers.

CASE 138.—"Sergeant M. N. Hinds, Co. D, 152d New York Volunteers, was wounded at the engagement on the North Anna River, Virginia, May 24th, 1864, by a musket ball, which fractured the head of the left femur and the acetabulum. He was conveyed to the field hospital of the 2d Division, 2d Corps, at Hanover Junction. He died within a few hours after the reception of the injury, May 24th, 1864." Assistant Surgeon E. Lyon Corbin, 152d New York Volunteers, reports the case.

CASE 139.—"Private D. M. Johnson, Co. I, 13th Alabama Regiment, received, at the battle of Gettysburg, Pennsylvania, July 3d, 1863, a gunshot wound in the right hip. Left on the field, he was conveyed to the hospital for prisoners, at Gettysburg. On August 1st, he was transferred to the Camp Letterman Hospital. It was ascertained that the head and neck of the femur were fractured, and also the acetabulum. The patient died on September 12th, 1863." The case is reported by Surgeon Henry Janes, U. S. V.

CASE 140.—"Private George L———, Co. D, 5th Louisiana Regiment, aged 21 years, was wounded at the battle of Monocacy, Maryland, July 9th, 1864, by a conoidal ball, in the left hip. He was admitted to the hospital at Frederick, Maryland, on the same day, and, as it was supposed that he was suffering only from a flesh wound with consequent paralysis of the left leg, he was treated by simple dressings and a stimulant and tonic regimen. The wound suppurated freely, and diarrhœa supervening, the patient became very much reduced, having, on August 29th, a large bed sore over the sacrum and hip. He died on the 3d of September. At the post mortem examination, the real extent of the injury was discovered. The ball had entered above the trochanter major, completely fracturing the neck; had chipped the posterior portion of the head of the femur; had fractured the lower portion of the acetabulum, and lodged above the iliac border. The joint was filled with pus, and nothing had been accomplished in the way of repair." The pathological specimen, numbered 3046, Sect. I, A. M. M., was contributed by Assistant Surgeon R. F. Weir, U. S. A.

CASE 141.—"Private William D. Little, Co. G, 100th Indiana Volunteers, was struck, at the battle of Mission Ridge, Tennessee, November 25th, 1863, by a musket ball, which produced a severe wound of the left hip. He was taken to the field hospital of the 4th Division, 15th Corps. He died on December 7th, 1863. It was found that the upper extremity of the femur was shattered and the acetabulum fractured." The case is reported by Surgeon Philander C. Leavitt, 100th Indiana Volunteers.

CASE 142.—"Private L. Lewis Lowe, Co. E, 101st Ohio Volunteers, was wounded in the battle at Stone River, Tennessee, December 31st, 1862, by a conoidal musket ball, which, striking the edge of the ilium, one inch below the anterior inferior spinous process, fractured the anterior margin of the acetabulum and the head of the femur, and produced a nearly transverse fracture through the neck. He was admitted, on January 24th, 1863, into Hospital No. 5, at Murfreesboro'. On February 18th, forty-nine days after the reception of the injury, he was admitted into the No. 3 hospital, at Nashville. The limb was shortened three inches. The patient was very much debilitated from profuse suppuration and diarrhœa. The ball had not been extracted. The limb was supported by pillows, and a nutritious diet with stimulants was directed, with an occasional opiate. No important change took place until the early part of March, when the irritative fever became more intense. The case terminated fatally on March 12th, 1863. At the autopsy, the missile was found one inch from the orifice of the wound. There was a very offensive collection of pus about the hip-joint." The case is reported by Surgeon Alexander Ewing, 13th Michigan Volunteers.

CASE 143.—"Private Peter M———, Co. A, 28th Massachusetts Volunteers, aged 29 years, was wounded at the assault on Petersburg, Virginia, March 25th, 1865, by a conoidal musket ball, which entered over the right pubic bone, passed an inch and a half to the right, and entered the hip-joint, shattering the head and imbedding itself in the neck of the femur. He was conveyed to the field hospital of the 1st Division of the 2d Corps, and the next day taken by steamer from City Point to Washington, and admitted, March 28th, into the Armory Square Hospital. There was no shortening of the limb, and the opinion that the hip-joint was implicated was not then nor subsequently entertained by any surgeon who examined the case before death. No important vessels were touched. He suffered, however, constant pain along the course of the sciatic nerve, and this was thought to be injured. Profuse suppuration ensued. Stimulants, tonics, and anodynes were administered, and nourishing diet was provided. The patient died, exhausted, on June 12th, 1865. A large amount of pus was found between the peritonœum and the walls of the abdomen. The pathological specimen is No. 4227, Sect. I, A. M. M. The upper portion of the neck and fully one-half of the head of the bone are wanting. There is a slight fracture of the pubic border of the acetabulum." The case is reported and the specimen contributed by Acting Assistant Surgeon H. Richings.

CASE 144.—"Private Royal S. N———, Co. A, 26th Massachusetts Volunteers, aged 34 years, was wounded, at the battle of Winchester, Virginia, September 19th, 1864, by a conoidal musket ball, which entered the left groin and passed inwards and upwards, completely disorganizing the hip-joint. On the same day, he was admitted to the hospital of the 2d Division, 19th Army Corps, where he remained until the end of October, when he was sent to the Jarvis Hospital, at Baltimore. He was in a typhoid condition, with diarrhœa. The limb became gangrenous, and the patient died on the 20th of November, 1864." The pathological specimen is No. 3726, Sect. I, A. M. M. The summit of the head of the femur was found grooved, and the superior margin of the acetabulum broken, by the bullet. There was thickening and ulceration of the sigmoid flexure of the colon. The specimen and history were contributed by Acting Assistant Surgeon B. B. Miles.

CASE 145.—"Sergeant Charles G. P———, Co. G, 13th Pennsylvania Cavalry, aged 34 years, was wounded in the engagement near Malvern Hill, Virginia, August 16, 1864, by a conoidal musket ball, which, entering the left hip one inch above and two inches behind the trochanter major, fractured the neck of the femur and glanced upwards. He was taken to the field hospital of the Second Division of the Cavalry Corps, and thence transferred, by steamer, to Philadelphia. He was admitted, August 20th, into Satterlee Hospital. The course of the projectile, as far as could be ascertained by the probe, was transversely forward and slightly downwards. Very little broken bone was found, and the extent of the injury could not be determined. The wound looked well, and the patient's general condition was quite good. Water dressings were applied, and a nutritious regimen was directed. Profuse suppuration took place, and, on September 27th, he had a rigor, followed by fever and profuse perspiration, with a haggard countenance, and delirium. Ten grains of the bi-sulphate of soda in an infusion of quassia, were given every two hours. On September 28th, Acting Assistant Surgeon A. H. Smith having etherized the patient, made an incision in the inner aspect of the upper third of the thigh, and a large amount of unhealthy pus was evacuated. The patient died on October 9th, 1864. At the autopsy, fluid in considerable quantity was found between the membranes of the brain and in the ventricles, which were otherwise healthy. The pia mater was congested. There was an infiltration of pus near the margin of the left lung. The ball was found near the acetabulum." The pathological

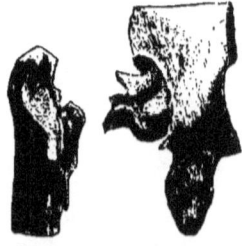

FIG. 58.—Secondary effects of a gunshot fracture of the left hip-joint.—*Spec. 3632*, Sect. I, A. M. M.

specimen is numbered 3632, Section I, A. M. M. A portion of the head of the femur has been absorbed; the margin of the acetabulum is slightly broken, and its surface is eroded. The neck, which was shattered at the trochanter major by the missile, is almost entirely missing. A fragment was split from the shaft in front of the trochanter minor. The cancellated structure of the great trochanter is partially absorbed; a slight deposit of new bone exists in the shaft. The case is reported by Acting Assistant Surgeon A. H. Smith. The specimen is contributed by Surgeon I. I. Hays, U. S. V., and is represented in the adjacent woodcut, (FIG. 52.)

CASE 146.—"Private Christopher Robinson, Co. C, 2d District of Columbia Volunteers, received in the vicinity of Washington, on July 1st, 1862, a gunshot wound of the hip. He was received, on July 2d, into the Ebenezer Church Hospital. He died on July 20th, 1862. The autopsy revealed a comminuted fracture of the coxo-femoral articulation." The case is reported by Assistant Surgeon W. F. Waters, U. S. A.

CASE 147.—"Private C. R———, Co. C, 49th New York Volunteers, aged 23 years, was wounded in General Sedgwick's assault on Fredericksburg Heights, Virginia, May 3d, 1863, by a rifle ball, which entered just outside of the anterior superior spinous process of the left ilium, passed downwards and backwards in the neighborhood of the hip-joint, and emerged from the buttock one inch outside of the tuberosity of the left ischium. After a few days, he was conveyed by steamer to Washington, and, on May 8th, was admitted to Douglas Hospital, under the care of Assistant Surgeon C. C. Lee, U. S. A. The patient's general health was excellent, and the wound caused him little inconvenience during the first week. He sat up with ease for the nurse to prepare his bed. The thigh, however, was constantly flexed and bent across the other, the foot being inverted. On May 16th, considerable febrile action was manifested, and just over the trochanter some fluctuation was detected. The wound was now explored under ether. No crepitus could be distinguished, or other evidence obtained of much injury to the bone, though several small fragments were removed from the wound of exit. Profuse suppuration ensued, attended with hectic and great prostration, insomnia, nausea, anorexia, emesis, and diarrhœa. The therapeutic agents which were administered—opiates, tonics, and stimulants—were only palliative. The abscess near the trochanters rapidly extended, and the whole thigh became greatly swollen. He died, exhausted, May 31st, 1863. At the autopsy, nineteen hours after death, the stomach was found filled with a dark, fœtid liquid, resembling broken-down blood; the other viscera were healthy." The pathological specimen is No. 1247, Sect. I, A. M. M. The head of the femur is slightly chipped by the missile, and is much eroded. The edge of the acetabulum was a little roughened.

CASE 148.—"Private Z. S———, Co. K, 2d Ohio Cavalry, aged 25 years, was wounded near Charlestown, Virginia, on the 22d of August, 1864, by a conoidal musket ball, which entered about midway between the anterior superior spinous process of the left ilium, and the symphysis pubis immediately beneath Poupart's ligament. He was admitted to the Cavalry Corps Hospital, where he remained several days, and thence was sent to Frederick, Maryland, where he was admitted on the 29th of August, 1864. Here Assistant Surgeon R. F. Weir, U. S. A., extracted the ball. Stimulants and nourishing diet were prescribed, and the fracture was treated on the expectant plan. The patient did well until the 4th of September, when he complained of chills and slight tenderness over the abdomen. Opiates were freely administered, and a large poultice was placed over the abdomen. On the following day he felt better, but on the 6th, the chills returned and the tenderness over the abdomen increased, with vomiting and short and quick respiration. He died on the 7th of September. At the *post mortem* examination, it was ascertained that the missile had fractured the pubic portion of the acetabulum, and chipped off a fragment the size of a hickory nut from the inferior segment of the head of the femur. The hip-joint was filled with pus." The pathological specimen is No. 3904, Sect. I, A. M. M., and was contributed, with the history, by Acting Assistant Surgeon J. H. Coover.

CASE 149.—"Private Hiram H. Sturdivant, Co. A, 179th New York Volunteers, was wounded before Petersburg, Virginia, April 2d, 1865, by a fragment of shell, which entered the right buttock. He was admitted into the field hospital of the Ninth Army Corps, at City Point, under the care of Assistant Surgeon W. O. McDonnell, U. S. V. A fracture involving the head of the femur and the rim of the acetabulum was discovered. On April 14th, the patient was placed on the hospital transport 'State of Maine,' to be conveyed to Annapolis, and the diagnosis rendered at the field hospital was corroborated by Acting Assistant Surgeon W. A. Finn." The death of the patient is reported, April 22d, 1865, by Surgeon B. A. Vanderkieft, U. S. V., in charge of the General Hospital at Annapolis.

CASE 150.—"Private William P. T———, Co. D, 121st New York Volunteers, aged 21 years, was wounded, at the battle of Spottsylvania Court House, Virginia, May 10th, 1864, by a conoidal musket ball, which entered the right gluteal region and opened the hip-joint. He was conveyed, on the same day, to the hospital of the 1st Division, 6th Corps, and, on the 14th of May, was sent to Douglas Hospital, at Washington. The discharge from the wound was very profuse and sanious. The usual symptoms of pyæmia—chills, an icteroid hue of the skin, and profuse perspiration—manifested themselves in a few days, and death ensued on May 21st, 1864. At the autopsy, the liver was found softened, and friable. The gall bladder was filled with inspissated bile." The specimen is No. 3525, Sect. I, A. M. M. The ilium is deeply grooved at the upper and posterior margin of the acetabulum, which is chipped. The specimen and history were contributed by Assistant Surgeon W. Thomson, U. S. A.

CASE 151.—"Private George F. Tilton, Co. E, 1st Massachusetts Cavalry, aged 21 years, was wounded at the engagement at Mine Run, Virginia, November 29th, 1863, by a pistol ball in the left hip, the ball entering behind the trochanter major and passed inwards and outwards. He was conveyed to Alexandria, and admitted, on December 4th, into the Third Division Hospital. Any attempt to move the limb caused intense pain. The entrance wound was of a dark, ashy appearance. There was no shortening or deformity. Pyæmia supervened, and death took place December 22d, 1863. At the autopsy, it was found that the ball had entered the neck of the femur, lodged in the head posteriorly, and fractured the posterior edge of the acetabulum. The head of the femur was carious and partially absorbed. The joint was full of pus, and there were abscesses about it. Acting Assistant Surgeon W. G. Elliott reports the case."

CASE 152.—"Corporal James P. White, Co. H, 115th Illinois Volunteers, aged 25 years, was wounded at the battle of Chickamauga, Georgia, September 20th, 1863, by a conoidal musket ball, which, entering behind the right trochanter major, fractured the head of the femur and the upper edge of the acetabulum, and lodged in the cotyloid cavity. He was admitted October 1st, into Hospital No. 2, at Chattanooga, Tennessee. Water dressings were applied, and a supporting regimen enjoined. The discharge from the wound was profuse, and the appetite was poor. He died, exhausted, on October 30th, 1863. At the autopsy, the articular surfaces were eroded, and the parts about the joint were gangrenous." The case is reported by Surgeon Israel Moses, U. S. V.

CASE 153.—"Sergeant William Whitney, Co. K, 174th New York Volunteers, received in front of Petersburg, Virginia, June 20th, 1864, a gunshot wound of the thigh, with an extensive injury involving the upper extremity of the femur and the acetabulum. He was taken to the field hospital of the Third Division, Second Corps, and thence to the base hospital at City Point. He died on June 25th, 1864." Surgeon Frank Ridgeway, 74th New York Volunteers, reports the case.

CASE 154.—"Private Joseph W——, Co. K, 6th Maryland Volunteers, aged 45 years, was wounded, on the 30th of November, 1863, by the accidental discharge of a musket, while on the march. The ball entered two inches to the right of the middle of the os coccygis, and passed outwards towards the right hip-joint. He was conveyed to the Prince Street Hospital, Alexandria, Virginia. Search for the ball was made, but it could not be found. Water dressings were applied to the limb, and cordials and generous diet were provided. The patient had been constipated for four days; a dose of castor oil was administered, and the bowels were moved in a few hours. The wound suppurated freely, and there was considerable pain in the hip and thigh. There was little change until the 12th, when the discharge from the wound became very offensive, and the tongue dry and black. Death took place on the following day. The cavity of the wound, which was large, contained dark, fœtid fluid. The lower margin of the acetabulum was chipped by the ball, which lodged in the summit of the head of the femur, splitting it perpendicularly. The soft parts around the joint were in a sloughing condition, and the ligamentum teres was much softened." The specimen is No. 1908, Sect. I, A. M. M., and was contributed by Acting Assistant Surgeon Jonathan Cass.

CASE 155.—"Private John Wiley, Co. A, 155th New York Volunteers, received, in front of Petersburg, Virginia, June 16th, 1864, a gunshot wound of the left hip. He was conveyed to the field hospital of the 2d Division, 2d Corps. Simple dressings were applied. He died on June 17th, 1864. It was found that the articulation was opened, and that both the head of the femur and the acetabulum had been fractured by the projectile." The case is reported by Surgeon Farand Wylie, 155th New York Volunteers.

CASE 156.—"Private James B. Wilson, Co. E, 20th Massachusetts Volunteers, aged 37 years, was wounded at the battle of Spottsylvania, Virginia, May 12th, 1864, by a conoidal musket ball, which fractured the head of the left femur and the acetabulum. He was taken to the field hospital of the 2d Division, 2d Corps, and was conveyed thence to Washington, and admitted, on May 20th, into Armory Square Hospital. He was treated on the expectant plan. He died on June 1st, 1864." The case is reported by Acting Assistant Surgeon F. H. Woodbury.

CASE 157.—"Private L. Winslow, Co. A, 67th Ohio Volunteers, was wounded at the affair near Kernstown, Virginia, March 23d, 1862, by a fragment of shell, which penetrated to the hip-joint, fracturing the head of the femur and the acetabulum. Death took place on March 25th, 1862." This case is reported by Surgeon W. S. King, U. S. A.

CASE 158.—"Private Albert Wormack, Co. G, 48th North Carolina Regiment, received, at the battle of Antietam, Maryland, September 17th, 1862, a gunshot wound of the hip. He was captured and treated in a field hospital until October 10th, when he was conveyed by ambulance to Hospital No. 5, at Frederick. Death took place on October 14th, 1862. It was found that the neck and head of the femur had been shattered by the projectile, and that the cotyloid cavity was also injured." The case is reported by Surgeon H. S. Hewit, U. S. V.

CASE 159.—"Corporal Moses F. Yoder, Co. G, 51st Ohio Volunteers, aged 21 years, received, near Marietta, Georgia, about June 20th, 1864, a gunshot fracture of the neck of the femur, with injury to the acetabulum. He was conveyed to the field hospital of the 1st Division, 4th Corps, and thence was transferred by rail, June 25th, a distance of about ninety miles, to the General Hospital, Chattanooga, Tennessee. He died on July 2d, 1864." The case is reported by Assistant Surgeon C. C. Byrne, U. S. A.

There are also, on the registers of the Office, a few cases of gunshot wounds of the hip-joint, in which slight injury to the acetabulum occurred, without fracture of the femur:

CASE 160.—"Private James Brandon, Co. F, 119th New York Volunteers, aged 30 years, was wounded at the battle of Chancellorsville, Virginia, May 2d, 1863, by a conoidal ball, which entered the right hip midway between the spine and the trochanter major, passed over the head of the femur, and fractured the upper margin of the acetabulum. He remained a prisoner in the hands of the enemy for about ten days; was, on the 15th of May, admitted to the Third Division, Eleventh Corps, Hospital, and thence sent to the Second Division Hospital, at Alexandria, Virginia, on the 23d of May. He was extremely emaciated, and the wound discharged large quantities of very offensive pus. Tonics and stimulants were freely administered, and warm water dressings were applied to the wound. The patient's appetite became ravenous, and his thirst excessive. Diarrhœa soon supervened. On the 15th of June several pieces of necrosed bone were eliminated. The wound began to look better, and the diarrhœa was partially checked. In the latter part of August the wound was closed, and the patient continued to grow better until the beginning of October, when pneumonia supervened. He died on the 8th day of October. At the post mortem examination, the left cavity of the thorax was found to contain about one quart of sero-purulent effusion; the lungs were much congested. The head of the femur and the acetabulum were carious." The case is reported by Acting Assistant Surgeon A. Walter Tryon.

CASE 161.—"Private Joseph D———, Co. E, 129th Pennsylvania Volunteers, aged 20 years, was wounded at the battle of Chancellorsville, Virginia, May 3d, 1863, by a conoidal ball, which entered immediately over the tuberosity of the left ischium, and made its exit in the left groin a little below the pubis. He was left on the field, taken prisoner, and for several days confined in a shed. He was then paroled, and arrived at the Fifth Corps Hospital on the 13th of May. For four days he walked about with the assistance of a crutch, apparently suffering little inconvenience from the injury sustained. On the fifth day he complained of increased pain in the parts; irritative fever of a typhoid character supervened, which increased gradually until June 3d, 1863, when he died, in a low, muttering delirium. At the *post mortem* examination, the track of the wound was found to communicate with, and lead into, the hip-joint. It was ascertained that the ball had chipped the upper edge of the cotyloid cavity, where there was a slight exfoliation. The cavity of the joint, and the surrounding parts, were full of sanious pus. The head of the femur and the acetabulum were deprived of periosteum, and every trace of cartilage was gone. A section of the symphysis pubis, the whole of the ischium, and part of the ilium, were deprived of periosteum, and showed extensive injury and pyæmia." The pathological specimen is No. 1285, Sect. I, A. M. M. The injury to the bone is very trivial. The specimen and history were contributed by Assistant Surgeon Philip Adolphus, U. S. A.

CASE 162.—"Private Thomas McGowan, Co. H, 121st New York Volunteers, aged 26 years, was wounded at the battle of Chancellorsville, Virginia. May 3d, 1863, by a conoidal musket ball, which entered near the anterior superior spinous process of the left ilium, wounded the capsular ligament near the upper edge of the acetabulum, opening the joint, and lodged at the posterior inferior spinous process, where it was cut down upon and extracted. He was conveyed, by steamer, to Washington, and, on May 8th, was admitted to Finley Hospital. He complained of pain in the hip and thigh, and was unable to move the limb. There was redness and swelling about the wounds, which discharged but little. Some febrile excitement existed. He was placed in as easy a recumbent posture as possible, water dressings were applied, and an opiate given at night. On May 9th, the pain lessened, and the fever abated somewhat. A nourishing diet was ordered. On May 10th, the pulse was at 80, full, and of fair strength; the thirst and heat less. On May 18th, symptoms of depression came on; the patient had a sensation of chilliness; his pulse was accelerated. On May 25th, he suffered less pain in the limb, and his appetite was improved. Towards the end of the month he suffered severe pain at the hip upon the slightest movement. On May 31st, he had a severe rigor. On the evening of June 1st, another severe rigor occurred. He died on June 3d, 1863. At the autopsy, on laying open the track of the ball, several ounces of pus were evacuated, which had burrowed about the joint and the ischiatic notch. The round ligament and the cartilaginous covering of the head of the femur were more or less absorbed. A few days previous to his death, the symptoms were those common to both pyæmia and intermittent fever; but for a long time there had been no miasmatic disease in the hospital, while there had been a number of cases of pyæmia, some of erysipelas, and one of gangrene. No metastatic abscesses were discovered, but the question was asked was not this a case of pyæmia which was not sufficiently advanced for the formation of multiple abscesses?" The case was reported by Acting Assistant Surgeon Lewis Heard.

CASE 163.—"Private Charles Henry Roberts, Co. C, 1st New Jersey Volunteers, aged 19 years, was wounded at the battle of Gaines's Mill, Virginia, June 27th, 1862. by a conoidal musket ball, which entered the left nates and emerged at the root of the penis. While resting his head on both hands, another bullet wounded both hands, passed through the right ear, and upward under the scalp for three inches. The man was taken prisoner, carried to Richmond, exchanged, and sent on a hospital steamer to Philadelphia, where he was admitted into the Master Street Hospital, July 30th, 1862. He rallied, and for some time did well, but on the 15th of September he began to fail gradually, and died on the 20th." The pathological specimen is No. 694, Sect. I, A. M. M. Bordering the acetabulum there is a deposit of callus. The point of impact of the missile appears to have been at the lowest margin of the acetabulum, where a square inch of the ischium is necrosed and nearly separated. The specimen and history were contributed by Surgeon P. B. Goddard, U. S. V.

CASE 164.—"Private Wm. H. W———, Co. G, 4th New York Volunteers, was wounded at the battle of Antietam, Maryland, September 17th, 1862. He was conveyed to Frederick City, and admitted, on September 28th, into Hospital No. 1. It is stated that he had a gunshot wound of the hip. He died on November 2d, 1864." A case-book of the hospital, of this date, contains the following description of a pathological specimen from the person of one W———, who had received a gunshot wound of the hip: "The head of the femur was entirely denuded of articular cartilage, and was much diminished by absorption, being no larger than an English walnut. The base of the acetabulum, the rim of which had been injured, was perforated by ulcerative absorption, and the head of the femur protruded within the pelvis. The partition between the obturator foramen and the acetabulum was much broken down." Assistant Surgeon R. F. Weir, U. S. A., reports the case.

The following cases of gunshot wounds supposed to involve the hip-joint, and uncomplicated by fracture of the articular surfaces, are recorded on the registers of this Office:

CASE 165.—"Corporal William Blair, Co. F, 63d Pennsylvania Volunteers, aged 26 years, received, at the battle of the Wilderness, May 6th, 1864, a wound, by a musket ball, in the right groin, which passed obliquely downwards and backwards, making its exit at the buttock. He was received into the hospital of the 3d Division, 2d Corps, and, after a few days, was transferred, by ambulance and steamer, to Washington, and placed in Lincoln Hospital on May 29th. He died on the following day, and it was found that there was a large abscess about the joint, and that the capsular ligament had been extensively opened." The case was reported by Surgeon Orpheus Evarts, 20th Indiana Volunteers, and by Assistant Surgeon J. C. McKee, U. S. A.

CASE 166.—"Corporal Hiram C. Boyd, Co. H, 39th Illinois Volunteers, aged 20 years, was wounded at a skirmish near Bermuda Hundred, Virginia, June 2d, 1864, by a conoidal musket ball, which entered the right hip, between the tuberosity of the ischium and the trochanter major, and produced an abrasion of the neck of the femur. He was conveyed by steamer to Fortress Monroe, and admitted, June 5th, to the General Hospital. He was transferred thence to New York Harbor, and received, August 7th, into De Camp Hospital. It was there considered doubtful whether there had been any direct lesion of the articu-

COMPARED WITH TEMPORIZATION.   91

lation. He was discharged from service on February 20th, 1865. In February, 1866, Dr. James Neil, Examining Surgeon at Harlem, New York, reported that Mr. Boyd was then unable to walk without crutches, and that it was doubtful if the limb would ever become useful." Dr. James Neil, Examining Surgeon, reports the case.

CASE 167.—"Lieutenant A. D. Bradshaw, Co. A, 3d Kentucky Regiment, received, at the battle of Chickamauga, Georgia, September 19th, 1863, a gunshot wound of the right hip. He was conveyed, September 29th, to the Officers' Hospital, at Lookout Mountain, Tennessee. It was found that the hip-joint was opened. Abscesses formed about the articulation. Death took place from exhaustive suppuration and surgical fever on October 8th, 1863." Assistant Surgeon C. C. Byrne, U. S. A., reported the case.

CASE 168.—"Private Clarence Cornell, Co. F, 149th New York Volunteers, received, at the battle of Resaca, Georgia, May 15th 1864, a gunshot wound of the left hip-joint. He was taken to the field hospital of the 3d Brigade, 2d Division, 20th Corps. It was thought that the joint was opened without injury to the bones forming the articulation. Intense inflammatory reaction occurred, and death resulted May 18th, 1864." Surgeon J. A. Wolfe, 29th Pennsylvania Volunteers, reports the case.

CASE 169.—"Lieutenant Coleman Duncan, Co. B, 18th Indiana Volunteers, received, at Port Gibson, Mississippi, May 1st, 1863, a gunshot wound of the right hip-joint. He was conveyed, by ambulance and steamer, to Memphis, Tennessee, and, on May 18th, was admitted into the Officers' Hospital. Obtaining a leave of absence, he left the hospital on May 27th. He was mustered out at the expiration of his term of service, November 22d, 1864." The case is reported by Acting Assistant Surgeon C. H. Cleveland.

CASE 170.—"Captain John W. Falconer, Co. A, 41st U. S. Colored Troops, was wounded in the engagement at Clover Hill, near Appomattox Court House, Virginia, on April 9th, 1865, by a musket ball, which entered the left nates, passed forward, and opened the hip-joint. He was taken to the field hospital of the 2d Division, 25th Corps. He died on April 24th, 1865." The case is reported by Surgeon E. P. Morong, U. S. V.

CASE 171.—"Private Stephen Finnegan, Co. F, 91st New York Volunteers, was struck, in the engagement at Gravelly Run, Virginia, March 30th, 1865, by a bullet, which produced a wound of the hip-joint. He was conveyed, the following day, to the field hospital of the 3d Division, 5th Corps. He died on March 31st, 1865." Surgeon A. S. Coe, 147th New York Volunteers, reports the case.

CASE 172.—"Private George Green, Co. F, 169th New York Volunteers, aged 24 years, received, at the battle of Cold Harbor, Virginia, June 3d, 1864, a gunshot wound of the left hip-joint. He was conveyed to the field hospital of the 18th Corps, and, June 5th, to White House Landing, and was transferred thence, by steamer, to Washington, where he was admitted, on June 10th, into Emory Hospital. Cold water dressings were applied, and tonics, stimulants, and anodynes were given, with nutritious diet. He died on June 17th, 1864." The case is reported by Acting Assistant Surgeon Samuel Graham.

CASE 173.—"Private F. M. Hate, Co. E, 7th Arkansas Regiment, received, in the engagement at Perryville, Kentucky, October 8th, 1862, a gunshot wound of the hip-joint. He was made a prisoner, and conveyed to the General Hospital, at Perryville. He died, exhausted, on January 18th, 1863." The case is reported by Surgeon James G. Hatchett, U. S. V.

CASE 174.—"Private Joseph Leedy, Co. D, 100th Indiana Volunteers, aged 35 years, received, at the battle of Missionary Ridge, Tennessee, November 25th, 1863, a gunshot wound of the hip-joint. No fracture was detected. He was taken to the regimental hospital. Surgical fever set in, and he died on December 2d, 1863." Surgeon P. C. Leavitt, 100th Indiana Volunteers, reports the case.

CASE 175.—"Private Isaac McMahan, Co. D, 56th Massachusetts Volunteers, aged 18 years, received, at the battle of Spottsylvania, Virginia, May 12th, 1864, a wound of the left hip-joint, by a conoidal musket ball. He was taken to the field hospital of the 1st Division, 9th Corps, and, near the close of the month, was conveyed, in an ambulance, a three days journey, to Fredericksburg, and thence transferred, by steamer, to Washington, and admitted, on May 28th, into Armory Square Hospital. Simple dressings were applied. He died on May 31st, 1864." The case is reported by Surgeon D. W. Bliss, U. S. V.

CASE 176.—"Private George W. Minnick, Co. G, 7th Maryland Volunteers, was wounded before Petersburg, Virginia, July 12th, 1864, by a conoidal musket ball, in the left buttock. Surgeon Thomas M. Flandrau, 140th New York Volunteers, reports that it was believed that the hip joint was involved. The wounded man was transferred to Lovell Hospital, Portsmouth Grove, Rhode Island, and admitted on August 7th, 1864; furloughed on October 12th, 1864; readmitted November 25th, 1864; and transferred to Frederick, Maryland, December 9th, 1864. Surgeon Charles O'Leary, U. S. V., reports that a fracture implicated the left hip. At Frederick, Assistant Surgeon J. H. Helsby, U. S. A., reports the admission of the patient on December 13th, 1864, his furlough on January 12th, 1865, readmission on February 10th, 1865, and discharge from service, in accordance with General Order 77, A. G. O., June 9th, 1865, and regarded the wound, which had firmly healed, as altogether limited to the soft parts." His name is not on the pension list.

CASE 177.—"William M. Moore, Co. B, 3d Ohio Volunteers, received, in the engagement at Perryville, Kentucky, October 8th, 1862, a gunshot wound of the hip. He was conveyed to Hospital No. 7, in Perryville. The report states that the wound involved the coxo-femoral articulation." He was discharged from service on February 17th, 1863. His disability was rated as total. The case is reported by Assistant Surgeon H. S. Wolfe, 81st Indiana Volunteers.

CASE 178.—"Private W. N. Morgan, Co. C, 9th Pennsylvania Reserves, aged 18 years, was wounded at the battle before Richmond, Virginia, June 30th, 1862, by a round musket ball, which entered the left groin, passed near the head of the left femur, and was supposed to have opened the capsule. He was captured and sent to Richmond, where he was confined three weeks.. Being then exchanged, he was conveyed, by steamer, to Philadelphia, and, on July 26th, was admitted into the Fourth and George Streets Hospital. There was anchylosis of the joint, and enlargement. The adhesions were successfully broken up by Professor S. D. Gross. Morgan was discharged from the United States service on October 25th, 1862." The case is reported by Acting Assistant Surgeon J. B. Bowen.

CASE 179.—"Private D. F. Pitman, Co. C, 1st North Carolina Regiment, received, at the battle of Gettysburg, Pennsylvania, July 3d, 1863, a gunshot wound of the left hip. He was captured and conveyed to the hospital for Confederate prisoners, and is reported as having a wound of the joint without fracture. On August 10th, he was transferred to the Camp Letterman Hospital, and died on September 14th, 1863." The case is reported by Surgeon H. Janes, U. S. V.

CASE 180.—"Private W. R. Reeves, 15th Mississippi Regiment, received, at the battle of Gettysburg, Pennsylvania, July 3d, 1863, a gunshot wound of the hip. He was captured and conveyed to the hospital at Hagerstown, Maryland. A wound of the hip-joint without fracture was diagnosticated. There was very grave inflammatory reaction, with profuse suppuration and febrile disturbance; and death took place on July 19th, 1863." The case is reported by Assistant Surgeon J. B. Harrington, 18th Connecticut Volunteers.

CASE 181.—"Private Thomas Smith, Co. C, 82d New York Volunteers, received, at the battle of Antietam, Maryland, September 17th, 1862, a gunshot wound of the hip. He was admitted, October 8th, to the General Hospital at Smoketown. The register states that the wound involved the joint without fracturing the articular surfaces." Smith died in October, and was buried at Antietam, as appears upon the Roll of Honor, No. 15, p. 45, Q. G. O., February 26th, 1868.

CASE 182.—"Private R. Taylor, Co. C, 32d Virginia Regiment, received, at the battle of Cold Harbor, Virginia, June 3d, 1864, a wound of the left hip-joint by a conoidal musket ball. He was conveyed to Richmond, and admitted, on the following day, to the Chimborazo Hospital, and was treated on the expectant plan. He died June 7th, 1864, and at the autopsy it was found that the ball had opened the hip-joint and lodged in the rotator muscles." The case is reported by Surgeon C. M. Seabrook, C. S. A.

CASE 183.—"Private D. J. Wagoner, Co. E, 27th Mississippi Regiment, at the battle of Perryville, Kentucky, October 8th, 1862, received a gunshot wound involving the hip-joint. He was captured and conveyed to the General Hospital at Perryville. He died, exhausted, on January 16th, 1863." The case is reported by Surgeon J. G. Hatchett, U. S. V.

CASE 184.—"Private Joseph Wells, Co. C, 1st Massachusetts Heavy Artillery, aged 46 years, was wounded at the battle of Spottsylvania Court House, Virginia, May 19th, 1864, by a conoidal musket ball, which entered the left hip posteriorly, two inches below the crest of the ilium, and lodged. He was taken to Washington, and was admitted, May 22d, to Emory Hospital. A gunshot wound of the left hip-joint was diagnosticated. On May 24th, the patient was anæsthetized by a mixture of one part of chloroform and two of ether, and the projectile was extracted. Water dressings were applied. On May 26th, the wound was highly inflamed and painful. Poultices of yeast, charcoal, and powdered elm bark were directed. On May 28th, he began to have rigors, which recurred at irregular intervals. He died on May 31st, 1864." The case is reported by Acting Assistant Surgeon J. P. Gilbert.

CASE 185.—"Private Jacob Widman, Co. H, 97th New York Volunteers, aged 28 years, was wounded at the second battle of Bull Run, Virginia, August, 1862, by a musket ball, which perforated the left hip near the articulation, probably opening freely the capsular ligament, for apparent dislocation of the head of the femur was produced. He was conveyed to Washington, and, September 2d, was admitted to the Ascension General Hospital. On January 15th, 1863, he was transferred to the Patent Office Hospital. In February, he was transferred to Stanton Hospital, and in March, to the hospital at Fort Columbus, New York Harbor, where he was discharged from the service on July 8th, 1863." The case was reported by Dr. H. B. Day, Pension Examining Surgeon at Utica, New York, September 5th, 1865, and the disability rated at one-half.

CASE 186.—"Private Henry Witzleben, Co. B, 28th Ohio Volunteers, in the engagement at Carnifex Ferry, West Virginia, September 10th, 1861, was wounded in the hip by a musket ball. He was treated in the regimental hospital, and it is reported that the wound involved the hip-joint. He was discharged from service on May 9th, 1862, having partial anchylosis of the hip-joint." The case is reported by Assistant Surgeon A. Shoenbein, 28th Ohio Volunteers. Witzleben's name is on the Pension List, and his disability is rated at "one-third, and temporary."

The following group of cases may be classed under the rubric of secondary traumatic arthritis, the evidence appearing to indicate that the projectile did not open the coxo-femoral articulation primarily:

CASE 187.—"Private John Delaney, Co. C, 51st New York Volunteers, aged 41 years, was wounded in the battle at Antietam, Maryland, September 17th, 1862, by a conoidal musket ball, which entered one inch external to the anterior spine of the left ilium, and emerged just outside of the lower third of the sacrum. He was admitted to the Frederick Hospital on October 2d, 1862. The patient was greatly exhausted, and the wound discharged profusely unhealthy pus, and his countenance was hectic. Stimulants and nutritious diets were freely given for two weeks without materially altering his condition, when he began slowly to regain strength; the wound, however, continued to suppurate freely. From this time he improved gradually, and the discharge was less. Early in December he walked about the tent, and occasionally outside of it, by the aid of a cane. All along he had suffered more or less pain along the course of the sciatic nerve, but had referred none to the hip-joint. On December 14th, while entering the tent, he fell heavily backwards to a sitting posture. Upon this, all the untoward symptoms returned. The wound discharged more profusely than ever, and the neuralgia became excruciating. Death took place on December 24th, 1862." At the autopsy, an abscess was found burrowing among the muscles of the thigh, and surrounding the hip-joint. The cotyloid cartilage and the round ligament were entirely absorbed, as also the cartilage about the head. The thin portion of the acetabulum was perforated. The case is reported by Assistant Surgeon Wm. M. Notson, U. S. A.

Fig. 59.—Caries of the head of the right femur, from secondary traumatic arthritis.—Spec. 3849, Sect. I, A. M. M.

CASE 188.—"Private Thomas J. D——, Co. E, 18th Mississippi Regiment, was wounded at the battle of Antietam, Maryland, September 17th, 1862, by a conoidal musket ball, which entered the right side on a level with the trochanter major, passed through the thigh, and thence through the upper portion of the middle third of the left thigh. He was admitted to the hospital at Antietam. The wound suppurated copiously; bed sores formed over the sacrum, and on the legs, and paralysis of both lower extremities ensued. On May 3d, 1863, the patient was transferred to Hospital No. 1, at Frederick, Maryland. Tonics and stimulants were administered, and a nourishing diet was provided. On May 20th, erysipelas attacked the wounds of the right thigh, and extended rapidly upon the body as far as the ribs, and down the thigh to the knee. Tincture of iodine was applied along the margin of the inflammation without stopping its progress. Death occurred on June 19th, 1863." The *post mortem* examination revealed no injury to the bones, but secondary involvement of the hip-joint from extensive inflammation. The pathological specimen is represented in the accompanying wood-cut, (FIG. 59.) The head of the femur is eroded. The specimen was contributed, and the history reported, by Assistant Surgeon R. F. Weir, U. S. A.

CASE 189.—"Alexander Hall, Co. M, 1st U. S. Cavalry, aged 23 years, was wounded near Greenville, Louisiana, December 27th, 1865, by a pistol ball, which entered above and behind the left trochanter major and penetrated to the hip-joint. The missile was at once extracted, and the wounded man was conveyed, December 29th, to the Sedgwick Hospital, at Greenville. On January 7th, 1866, severe clonic spasms of the muscles of the left lower extremity and of the lower jaw began. Morphia was administered and chloroform applied locally. The patient died, exhausted by the severity and long continuance of the spasms, on January 12th, 1866. The great sciatic nerve was either injured directly by the missile, or became secondarily involved." This case is reported by A. Hartsuff, Assistant Surgeon, U. S. A.

CASE 190.—"Captain Herbert C. Mason, Co. H, 20th Massachusetts Volunteers, received, at the battle of Gettysburg, Pennsylvania, July 3d, 1863, a wound of the right hip from a musket ball. He is reported among the wounded at the field hospital of the Second Division, of the Second Corps, with an injury of the hip-joint without fracture; but, as his name does not again appear upon the hospital registers, it is probable that he was treated in private quarters. He was mustered out of service March 23d, 1864, on account of disability, and, on September 15th, 1865, the Pension Examining Surgeon at Boston, Dr. George S. Jones, reported that anchylosis of the hip-joint had resulted from inflammation following his wound."

CASE 191.—"Albert McGee, aged about 30 years, was struck, at the first battle of Bull Run, July 21, 1861, by a musket ball, which entered the hip at the level of the trochanter major, and a little posterior to it. He made a complete recovery, with a limb shortened two and a half or three inches, and much everted. Seven years subsequently, he entered the Howard Grove Hospital, Richmond, Virginia, with Bright's disease, and died on November 30th, 1868. An autopsy was made, and the surgeons who examined the injury of the hip were of opinion that the head of the femur had been grooved by the ball at its lower part. The pathological specimen was sent to the Army Medical Museum, and there a vertical section of the epiphysis and upper portion of the femur was made, and it clearly appeared that there had never been a fracture of the head, and that the groove supposed to have been made by the ball was a depression resulting from a former abscess. The appearances closely

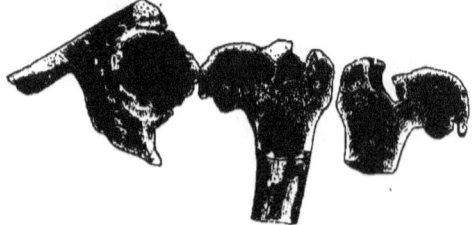

Fig. 60.—Chronic traumatic arthritis of the hip-joint following a gunshot injury. Spec. 5518, Sect. I, A. M. M.

resembled those sometimes observed in cases of chronic rheumatic arthritis of the hip." This very interesting specimen is represented in the adjacent wood-cut, (FIG. 60). It was contributed, with a memorandum of the facts of the case, by Brevet Lieutenant Colonel J. H. Janeway, Assistant Surgeon U. S. A.

CASE 192.—"Corporal Joseph L. Potts, Co. H, 75th Ohio Volunteers, aged 25 years, was wounded at the battle of Gettysburg, Pennsylvania, July 2d, 1863, by a conoidal musket ball, which entered behind the right trochanter major, and, passing horizontally forwards, emerged from the thigh in front. He was taken to the Seminary Hospital, at Gettysburg. The soft structures only were thought to be involved. On July 24th, he was transferred to Camp Letterman Hospital. The report states, August 9th, there was great angular deformity at the seat of fracture, and that partial union had taken place. No splints were applied. The wounds healed, and the patient was transferred, November 17th, to Camp Dennison, in Ohio. He was unable to sustain any weight on the injured limb, but his general health was good. He was furloughed in May, 1864, and, on August 6th, 1864, was discharged from the service of the United States. His disability was rated at three-fourths. The limb was atrophied and shortened." The case is reported by Acting Assistant Surgeon F. A. Koerper.

CASE 193.—"Private Samuel Schafer, Co. K, 81st Ohio Volunteers, aged 21 years, was wounded in the battle at Resaca, Georgia, May 14th, 1864, by a round musket ball, which entered the left hip. He was taken to the regimental hospital. A gunshot wound of the thigh was reported by Surgeon W. Jacobs, 81st Ohio Volunteers. On May 20th, Schafer was conveyed thirty-five miles, by rail, to the General Hospital at Chattanooga, Tennessee, where Surgeon Francis Salter, U. S. V., recorded the case as a gunshot wound of the hip involving the soft parts only. On June 9th, the patient was transferred to the Sherman

Hospital, at Nashville, where Surgeon W. Threlkeld, U. S. V., diagnosticated a gunshot wound of the left hip-joint. Private Schafer was furloughed on August 12th, 1864, and, a month later, was transferred to the Jefferson Hospital, in Indiana. A gunshot fracture of the left femur was here reported by Surgeon M. Goldsmith, U. S. V. On December 10th, the patient was sent to Camp Dennison, in Ohio, and Surgeon C. McDermot reported the case as a gunshot wound of the left hip. Finally transferred, December 21st, to the Seminary Hospital, at Columbus, Ohio, Private Schafer was discharged from the service on February 4th, 1865. There was shortening and distortion of the limb, and considerable stiffness of the hip-joint, and the disability was rated at three-fourths by Assistant Surgeon Gerhard Saal, U. S. V. In November, 1868, Commissioner Cox reports that Schafer's pension was increased, and his disability rated as total and permanent."

CASE 194.—"Sergeant Charles M. Scovil, Co. F, 14th Connecticut Volunteers, aged 29 years, was wounded at the battle of Spottsylvania, Virginia, May 12, 1864, by a conoidal musket ball, which produced a severe wound of the left hip. He was taken to the field hospital of the Second Division, Second Corps. A gunshot flesh wound of the left hip was diagnosticated. He was conveyed thence to Alexandria, and was admitted, May 24th, into the Saint Paul's Church Hospital. Simple dressings were applied. He died on July 14th, 1864, from a secondary involvement of the left hip-joint." The case is reported by Surgeon T. R. Spencer, U. S. V.

CASE 195.—"Lieutenant J. D. Seldon, Co. G, 2d U. S. Cavalry, received, in the battle at Gettysburg, Pennsylvania, July 3d, 1863, a gunshot wound of the left thigh, near the hip. He was taken to the hospital of the Cavalry Corps. The soft parts only were thought to be involved. On the following day he was transferred to Hospital No. 1, at Frederick, Maryland. The hip-joint became involved secondarily. The patient died on September 17th, 1863." The case is reported by Assistant Surgeon R. F. Weir, U. S. A.

CASE 196.—"Private William A. Shingledecker, Co. A, 121st Pennsylvania Volunteers, was wounded at the battle of Gettysburg, Pennsylvania, July 2d, 1863, by a musket ball, which penetrated the right hip. He was taken to the field hospital at Gettysburg, and gunshot wound of the hip and ankle were reported. On October 2d, he was transferred to the Mower Hospital, in Philadelphia. Here a gunshot wound of the right thigh was recorded. On January 22d, 1864, Shingledecker was transferred to the General Hospital at Pittsburg. Partial paralysis of the limb had resulted. He was discharged from the service of the United States on February 20th, 1864, and his disability was rated at three-fourths. Dr. G. McCook, Pension Examining Surgeon at Pittsburg, in his report, received at this Office March 14th, 1866, states that the missile remained lodged in the vicinity of the hip-joint, and that the patient could not rotate the limb or bear any weight upon the articulation." Dr. McCook reports the case.

CASE 197.—"Private Thomas Swartwood, Co. I, 25th Ohio Volunteers, was wounded at the engagement at Bull Pasture Mountain, Virginia, May 8th, 1862, by a musket ball, which entered the right thigh while flexed, one inch below the pubic bone, and near the inner edge of the femoral artery, injured the capsular ligament, and emerged near the great trochanter. He was captured near the close of the month, and remained in the hands of the enemy until the middle of October, when he was exchanged and conveyed to the Camp Chase Hospital at Columbus, Ohio. Slight shortening of the limb existed. He was discharged from service on October 21st, 1862." On April 7th, 1864, C. Hupp, Pension Examining Surgeon at Wheeling, West Virginia, stated that the limb was shortened by about two inches, a partial luxation of the head of the femur upwards apparently having been produced. Any movement of the thigh created severe pain. The cicatrices were firm, and there were no fistulous orifices.

CASE 198.—"First Sergeant Charles B. Wheeler, Co. K, 81st Indiana Volunteers, aged 32 years, received, near Atlanta, Georgia, August 8th, 1864, a gunshot wound of the left hip. He was taken to the field hospital of the First Brigade, First Division, Fourth Corps, and thence, September 8th, to Hospital No. 1, at Chattanooga, Tennessee. Indications that the articulations were secondarily involved now appeared; surgical fever, with profuse suppuration, ensued, and the patient died on October 7th, 1864." Surgeon M. S. Sherman, 9th Indiana Volunteers, reports the case.

The next category comprises cases of gunshot fracture of the trochanteric region, in which it was surmised that fissures extended within the hip-joint, or in which evidence that the lesions were thus complicated was revealed after death:

CASE 199.—Sergeant James B. Bridwell, Co. B, 87th Illinois Mounted Infantry, received, in the engagement at Pleasant Hill, Louisiana, April 8th, 1864, a gunshot fracture of the right femur through the trochanters. It was suspected that fissures extended into the joint. He was conveyed, by ambulance and steamer, to New Orleans, and was admitted to the St. Louis Hospital on April 14th. Simple dressings were applied. He died on April 17th, 1864." The case is reported by Assistant Surgeon Samuel H. Orton, U. S. A.

CASE 200.—"Private William Campbell, Co. A, 23d Missouri Volunteers, was wounded at the second assault on Atlanta, Georgia, July, 1864, by a fragment of shell, which produced a lacerated wound of the right hip, and fractured the femur at the trochanter major. He was conveyed, on July 30th, to the field hospital of the Third Division, of the Fourteenth Corps. It was found that the femur was extensively comminuted at the trochanters, and suspected that the fissures extended upwards within the capsule. The case was treated on the expectant plan. Death took place on July 31st, 1864." Assistant Surgeon H. J. Herrick, U. S. V., reports the case.

CASE 201.—"First Sergeant Peter Casserleigh, Co. A, 38th Illinois Volunteers, aged 20 years, was wounded at the battle of Chickamauga, Georgia, September 19th, 1863, by a conoidal musket ball, which entered behind the right great trochanter, and, passing inwards and forwards, fractured the apophysis, and, it was believed, injured the anterior crural nerve. It was considered possible, though not probable, that the joint was injured. He was taken to the field hospital of the 1st Division, 4th Corps. On December 3d, he was admitted into Hospital No. 1, at Nashville, Tennessee. He was discharged from service on July 11th, 1863. Paralysis of the right lower extremity existed, and his disability was rated as 'total' on his discharge papers. On May 17th, 1865, Surgeon Thomas S. Henning examined Casserleigh as an applicant for a pension. The limb was shortened one and a half inches, was inverted, and considerably atrophied. The motion of the hip-joint was lost." The case is reported by Surgeon G. Perin, U. S. A., and by the Pension Examining Surgeon.

CASE 202.—"Private John E. Clark, Co. E, 23d Pennsylvania Volunteers, at the battle of Cold Harbor, Virginia, June 1st, 1864, was struck in the left thigh by a musket ball, which fractured the femur very high up. He was taken to the field hospital of the First Division, Sixth Corps. It was thought at the hospital that the fracture extended within the capsule. He died on June 10th, 1864." The case is reported by Surgeon W. C. Roller, 23d Pennsylvania Volunteers.

CASE 203.—"Sergeant William G. Davis, Co. H, 19th Ohio Volunteers, aged 25 years, was wounded at the battle of Chickamauga, Georgia, September 20th, 1863, by a conoidal musket ball, which produced a fracture of the great trochanter of the femur, and lodged at the base of the neck. He was taken prisoner, and remained in the hands of the enemy until the end of September, when he was paroled and conveyed to Hospital No. 2, at Chattanooga, Tennessee. He was treated by rest and position; stimulants and a full diet were given. On November 14th, he was transferred to Hospital No. 3, at Murfreesboro. He was doing well, and suppuration was slight. On January 23d, 1864, Davis was transferred to Hospital No. 3, at Nashville. He died on May 15th 1864." The case is reported by Surgeon J. R. Ludlow, U. S. V., and by Assistant Surgeon J. C. Norton, U. S. V.

CASE 204.—"Private John Doody, Co. C, 6th New Hampshire Volunteers, aged 23 years, received, at the second battle of Bull Run, Virginia, August 29th, 1862, a gunshot fracture of the right femur, the missile passing through the hip just below the joint. He was admitted, on December 11th, 1862, into the Fifth and Buttonwood Streets Hospital, in Philadelphia, and, on January 15th, 1863, he was transferred to the Mower Hospital. The wound had, by this time, healed, with about two inches shortening of the limb. He was discharged from the service on March 16th, 1863." Surgeon J. Hopkinson, U. S. V., reports the case. Doody is pensioned at the rate of $15 per month. On September 4th, 1868, his disability was rated as "total and permanent" by the examining surgeon.

CASE 205.—"Private Charles Falk, Co. E, 26th New York Volunteers, received, in the battle of Fredericksburg, Virginia, December 13th, 1862, a gunshot fracture of the femur at the trochanter major. He was conveyed to a field hospital, and died on December 13th, 1862, a few hours after the reception of the injury. The trochanteric region was comminuted, and the fissures extended to the joint." Surgeon Thomas Sim, U. S. V., reports the case.

CASE 206.—"Private W. T. Fostner, Co. F, 23th Virginia Infantry, received, at Fredericksburg, Virginia, December 13th, 1862, a gunshot fracture of the femur through the trochanters. The diagnosis was difficult, but it was apprehended that the bone was splintered into the articulation. He was admitted into the General Hospital No. 3, at Richmond, on December 21st, where death resulted on December 23d, 1862." Surgeon A. Y. P. Garnett, C. S. A., reported the case.

CASE 207.—"Private H. B. Gardner, Co. F, 38th Illinois Volunteers, was wounded at the battle of Chickamauga, Georgia, on September 19th, 1863, by a musket ball, which fractured the femur at the left trochanter major. He was conveyed to Hospital No. 2, at Chattanooga, on October 1st. A sinus extended among the gluteal muscles. Bed sores formed over the sacrum. An effort was made to take off pressure by suspending the hips by a sheet; but so much pain in the joint was caused that this plan was abandoned. The suppuration was very profuse. The treatment was limited to preserving a comfortable position of the limb by means of pillows, and cleanliness by very frequent repetitions of emollient dressings, and to the administration of the salts of iron and quinine, with nourishing diet and stimulants. The patient died from ichorrhæmia, October 15th, 1863. It was found that the fissures from the fracture of the trochanter extended within the capsule." Assistant Surgeon John C. Norton, U. S. V., reported the case.

CASE 208.—"Private Michael Haehl, Co. A, 32d Indiana Volunteers, received, at the battle of Chickamauga, Georgia, September 19th, 1863, a gunshot fracture of the left femur, near the hip, which was supposed to implicate the articulation. He was conveyed to the field hospital of the Second Division, Twentieth Corps, and thence to the General Hospital, at Chattanooga, Tennessee. He died on October 8th, 1863." The case is reported by Surgeon Franklin Irish, 77th Pennsylvania Volunteers.

CASE 209.—"Captain J. D. Irwin, Co. C, 124th Ohio Volunteers, received, at the battle of Dallas, Georgia, May 25th, 1864, a gunshot fracture of the right femur through the trochanters. He was conveyed to Chattanooga, Tennessee, and was admitted, June 16th, to Hospital No. 1. The diagnosis was doubtful, but it was thought that the splintering of the bone primarily involved the articulation. The patient was treated on the expectant plan. Death supervened on June 21st, 1864." Surgeon Francis Salter, U. S. V., reports the case.

CASE 210.—"Private A. Latten, of Morgan's command, received, at Harrisburg, Mississippi, July 15th, 1864, a gunshot comminuted fracture of the femur, near the hip-joint, with an extensive injury of the soft parts. A long splint was adjusted to the limb. He was conveyed, forty miles, by ambulance, and over a hundred by rail-car, to the Lauderdale Hospital. The wound sloughed extensively. Pyæmia supervened, and he died on September 1st, 1864." The case is reported by Surgeon G. M. B. Maughs, P. A. C. S., and is published in the *Confederate States Medical and Surgical Journal*, January, 1865.

CASE 211.—"Thomas Purcell, Co. F, 96th Pennsylvania Volunteers, aged 22 years, was wounded at the battle of Chancellorsville, Virginia, May 3d, 1863, by a conoidal musket ball, which produced a fracture of the left femur, in its upper third, extending, it is stated, into the hip-joint. He was taken, May 11th, to the field hospital of the First Division, of the Sixth Corps. Copious suppuration took place, and an extensive abscess formed in the outer side of the thigh. On June 13th, he was conveyed, by steamer, to Washington, and was admitted into the Armory Square Hospital, where he remained until May 3d, 1864, when he was transferred to the Cuyler Hospital, in Germantown, Pennsylvania. No mention is made in the records of those hospitals that the hip-joint was involved. On May 10th, 1865, Purcell was transferred to the Mower Hospital,

Here it is stated that he had a gunshot wound in the upper third of the left thigh, with loss of four inches of bone. The wound having healed, he was discharged from service on July 8th, 1865." The case is reported by Surgeon Henry Janes, U. S. V. On September 4th, 1868, Purcell received his pension at the agency in Philadelphia, and his disability was rated as "total and permanent" by the examining surgeon.

CASE 212.—"Private Selah Randall, Co. G, 118th New York Volunteers, aged 28 years, was wounded in front of Petersburg, Virginia, July 5th, 1864, by a conoidal musket ball, which entered over the right trochanter major, fractured the femur, and lodged. He was taken to the regimental hospital, where the missile was extracted; thence he was conveyed, by steamer, to Portsmouth, and was admitted, on July 9th, to the Balfour Hospital. The limb was kept in position by means of sand bags, and water dressings were applied. Death took place on July 25th, 1864." The case is reported by Surgeon Edwin Powell, 72d Illinois Volunteers.

CASE 213.—"Private John Sebecta, Co. I, 111th Ohio Volunteers, aged 23 years, was wounded at the battle of Kenesaw Mountain, Georgia, June 27th, 1864, by a conoidal musket ball, which entered the right trochanter major and lodged, it was supposed, in the cervix. He was conveyed to the general field hospital of the Twenty-third Corps, and the missile was extracted. Thence he was transferred, by ambulance and rail, to Knoxville, Tennessee, and was admitted, July 1st, to the Asylum Hospital. He died on July 14th, 1864." The case is reported by Surgeon A. M. Wilder, U. S. V.

CASE 214.—"Private David Schamill, Co. H, 60th Pennsylvania Volunteers, aged 23 years, was wounded in front of Petersburg, Virginia, July 1st, 1864, by a conoidal musket ball, which, entering the right thigh behind, passed upwards, fracturing the lesser trochanter, and lodged in the right buttock. He was conveyed to the field hospital of the Third Division, Ninth Corps. The field surgeon's report stated that only the soft parts were involved. The wounded man was transferred, by steamer, to Washington, and was admitted, July 5th, at Carver Hospital. On July 14th, pyæmia was developed. Stimulants were given, with a nourishing diet. Death took place on July 17th, 1864. The missile had not been extracted. Pus had burrowed among the gluteal muscles. There were metastatic abscesses in the upper lobes of the lungs. The spleen and liver were enlarged and softened. The fissuring of the upper extremity of the bone closely approached the articulation." Surgeon O. A. Judson, U. S. V., reported the case.

CASE 215.—"Corporal Granville Williams, Co. D, 11th Missouri Volunteers, aged 25 years, was wounded in the engagement at Lake Chicot, Louisiana, June 6th, 1864, by a conoidal musket ball, which entered the left thigh in front, fractured the great trochanter, and emerged through the buttock. There was extensive splintering of the bone, and every likelihood that the articulation was involved. He was conveyed, by steamer, to Memphis, Tennessee, and was admitted, on June 11th, into the Gayoso Hospital. An expectant treatment was adopted. He died on June 28th, 1864." His case is reported by Surgeon F. Noel Burke, U. S. V.

In the cases which compose the following series of gunshot fractures of the trochanteric region, it was believed that the hip-joint was secondarily involved, and in many of the cases, evidence that this complication had supervened was obtained after death:

CASE 216.—"Private Fernando Bacon, Co. B, 142d New York Volunteers, aged 23 years, was wounded near City Point, Virginia, May 20th, 1864, by a round ball, which entered the left thigh at the fold below the buttock, and fractured the femur through the trochanters. He was conveyed to a field hospital, and was transferred thence to Point Lookout, Maryland, and admitted, on May 23d, into the General Hospital. Abscesses formed in and about the joint, and the patient died on July 10th, 1864, worn out by the protracted suppuration and irritation." The case is reported by Surgeon Anthony Heger, U. S. A.

CASE 217.—"Private Richard M. B———, 2d South Carolina Cavalry, aged 28 years, was wounded in the streets of Frederick, Maryland, September 14th, 1862, by a round musket ball, which entered in front at the middle third of left thigh, passed upwards, and lodged. The femur was ascertained to have been fractured very obliquely. The missile could not be found. The patient was first treated by a Confederate surgeon, who placed the wounded limb in a Smith's anterior splint, but made no extension, so that the fragments overlapped, and the limb became shortened three inches. On September 23d, the patient was received into Hospital No. 1, at Frederick. Buck's apparatus, with weights and pulleys, was applied, by which a complete reduction was effected. Tonics and stimulants and good diet were administered. On October 8th, pus began to burrow among the muscles of the thigh, but was restrained by the bandage of Scultetus. On October 15th, a free incision was made over the trochanter major, and a large amount of dark pus was evacuated. On the following day the limb was placed in a fracture box filled with bran. The discharge continued copious. His general condition, however, improved. On October 29th, the patient was placed on a water bed. A severe diarrhœa soon afterwards commenced, which was arrested within a day or two; but meantime, the discharge from the wound became unhealthy. Rapid emaciation took place, and the patient became so thin that the pulsations of the aorta were visible. On November 5th, the diarrhœa was renewed. There was an abscess behind the iliacus internus muscle, and slight peritonitis in the vicinity. Death took place November 7th, 1862." The specimen is numbered 709, Section I, A. M. M., and was contributed, with a history of the case, by Acting Assistant Surgeon W. W. Keen, jr. It shows a remarkably oblique fracture, extending from the lesser trochanter seven or eight inches downwards and outwards. Marks of caries and a necrosed piece of the laminated tissue extending beyond the usual line of insertion of the capsular ligament, are seen on the under surface of the neck. The ball lay below and behind the head of the femur, whether within or without the capsule is not stated.

CASE 218.—"Private James B. Crouch, Co. I, 148th New York Volunteers, was struck at the battle of Cold Harbor, Virginia, June 3d, 1864, by a conoidal ball, which broke off the great trochanter. He was taken to the 18th Corps Hospital, and thence conveyed by steamer to Washington, and admitted, on June 10th, to the Douglas Hospital. On the following day,

Assistant Surgeon W. Thomson, U. S. A., explored the wound and extracted the ball; the fracture was found not to enter the capsule. Cold water dressings were applied. An inflammation of the hip-joint ensued, with its usual constitutional effects, and the patient died on July 6th, 1864." The autopsy revealed no evidence of pyæmia, and death seemed due to simple exhaustion.

CASE 219.—"Private Joseph W. Dungan, Co. G, 33d Iowa Volunteers, aged 31 years, was struck in the right hip, in the engagement at Spanish Fort, Mobile, Alabama, March 28th, 1835, by a conoidal musket ball, which fractured the right trochanter major. He was received in the hospital of the Third Division of the Thirteenth Corps, and was conveyed thence, by steamer, to New Orleans, Louisiana, and admitted into the Sedgwick Hospital on April 1st. The treatment was on the expectant plan. Abscesses formed about the hip-joint, and death took place on April 25, 1865." The case is reported by Surgeon Benjamin Durham, U. S. V.

CASE 220.—"Corporal W. H. Harrison, Co. A, 41st Virginia Regiment, received, at the battle of Seven Pines, May 31st, 1862, a comminuted fracture of the femur, the ball passing through the trochanter major. He was received, on June 1st, into the Chimborazo Hospital, at Richmond. The extent of fissuring of the bone was doubtful; but there was evidence of secondary involvement of the joint, and death took place on June 23d, 1862." The case is reported by Surgeon E. H. Smith, C. S. A.

CASE 221.—"Sergeant George P. Morgan, Co. D, 1st Pennsylvania Battery, aged 23 years, was wounded at the battle of Cedar Creek, Virginia, October 19th, 1864, by a conoidal musket ball, which entered the left thigh, in front, fractured the trochanter major, and made its exit posteriorly. He was taken to a field hospital, and transferred, by ambulance, on November 4th, to an hospital at Winchester, and, on December 26th, by railroad, to Camden Street Hospital, in Baltimore. There was partial anchylosis of the hip-joint, and fragments of necrosed bone were eliminated. Early in February, 1865, the patient was transferred to the Satterlee Hospital, in Philadelphia, and was discharged from the service of the United States on June 26th, 1865." The case is reported by Surgeon Z. E. Bliss, U. S. V., and Surgeon I. I. Hays, U. S. V. At the Pension Office, the extent of the disability of this soldier has not yet been definitely ascertained, and his application, No. 130,490, is still pending.

CASE 222.—"Private Samuel McMurran, Co. A, 17th Virginia Regiment, aged 25 years, was wounded in the engagement near Amelia Court House, Virginia, April 4th, 1865, by a fragment of shell, which, striking the left trochanter major, carried away this apophysis, and lacerated the gluteal muscles. He was captured and taken to the field hospital of the Second Division, Ninth Corps. . On April 13th, he was sent to City Point, and thence to Washington, where he was admitted, on April 19th, into Lincoln Hospital. The splintering of the femur had not extended within the capsular ligament, but the articulation became secondarily involved. He died April 25th, 1865." Surgeon J. C. McKee, U. S. A., reports the case.

CASE 223.—"Corporal John R———, Co. C, 27th Indiana Volunteers, aged 22 years, was wounded at the battle of Chancellorsville, Virginia, May 3d, 1863, by a musket ball, which entered the upper and outer aspect of the right thigh, fractured the femur through the trochanter major, and lodged in the lower part of the buttock. He was admitted to the hospital of the Twelfth Corps, at Aquia Creek, on May 15th. The missile was cut down upon and extracted. He was transferred, by steamer, to Washington, and admitted, on June 14th, into Douglas Hospital.' The wounds were suppurating freely, and the patient, though weak, was in tolerably good condition. Diarrhœa set in, June 29th, and lasted several weeks. Astringents, tonics, and stimulants were administered The appetite was good. Several spiculæ of bone were eliminated. On September 12th, an abscess in the right buttock was opened and about a pint of pus evacuated. On October 4th, a profuse diarrhœa set in. Death took place on October 25th, 1863. The ligamentum teres and the cartilage about it on the head of the femur were found almost entirely absorbed." The pathological specimen is numbered 1850, Sect. I, A. M. M. Extensive spongy deposit surrounds the solution of continuity in the bone. About the neck and trochanters there are many traces of caries. Assistant Surgeon William Thomson, U. S. A., contributed the specimen and its history.

FIG. 61.—Result of a gunshot fracture through the trochanters of the right femur.— Spec. 333, Sect. I, A. M. M.

CASE 224.—"Private J. W. Sanford, Co. F, 47th North Carolina Regiment, aged 23 years, was wounded in the battle at Gettysburg, Pennsylvania, July 3d, 1863, by a conoidal musket ball, which entered one inch below hip-joint, perforated the right thigh from within, fractured the femur, and emerged externally. He was captured and conveyed to the hospital for Confederate wounded, and thence, on August 4th, was transferred to the Camp Letterman Hospital. Here it is stated that he had a gunshot flesh wound of the right buttock. On October 13th, he was removed to West's Buildings Hospital, in Baltimore. On January 10th, 1864, he was sent to the General Hospital, at Point Lookout, Maryland. The report states, January 12th, that his general condition was good; that the bone was uniting, and the wound suppurating. On February 17th, there was limited motion of the hip-joint. He was exchanged on March 17th, 1864." The case is reported by Surgeon A. Heger, U. S. A.

CASE 225.—"Private John S———, Co. A, 56th Pennsylvania Volunteers, received, at the battle of Bull Run, Virginia, August 29th, 1862, a gunshot comminuted fracture, involving nearly the upper half of the right femur. He was conveyed to Washington, about thirty miles distant, and was admitted, September 1st, to Armory Square Hospital. Abscesses formed about the joint, and the patient died, exhausted, on November 7th, 1862." The pathological specimen is No. 333, Sect. I, A. M. M. The femur is very badly comminuted below and on the posterior surface of the trochanters. The cancellated tissue is exposed nearly to the usual line of insertion of the capsular ligament. The lines of fracture are bordered by foliaceous callus, which, however, has availed nothing in the way of repair. The broken bone is carious and necrosed. The history and specimen (FIG. 61) are contributed by Surgeon D. W. Bliss, U. S. V.

13

98   EXCISIONS AT THE HIP,

CASE 226.—"Private John S———, Co. B, 5th Wisconsin Volunteers, aged 44 years, was wounded in the battle at Chancellorsville, Virginia, May 3d, 1863, by a musket ball, which, perforating the left thigh from without, three inches below the trochanter major, produced a comminuted fracture of the femur, involving the trochanters. He was conveyed, by steamer, to Washington, and was admitted, May 7th, into the Douglas Hospital. He was very feeble, and there was much constitutional disturbance; the pulse was 90; the tongue coated and dark; the limb was swollen, and the wound suppurating. Stimulants were given, with nourishing diet, and laxatives and opiates were administered as required. On May 12th, there was free suppuration, and excessive pain was created by any movement of the joint. The limb was placed in a fracture box with bran, and the foot was secured by adhesive strips. On May 20th, he had much pain in the vicinity of the hip-joint, and it was believed that the projectile had entered the articulation; the pulse was at 100, the skin hot and moist. On May 26th, a considerable venous hæmorrhage took place, and early in June the average daily discharge from the wound of sanguineo-purulent matter was estimated at three pints. On June 10th, there was subsultus tendinum, hiccough, and rigors. His appetite continued good. He died on June 15th, 1863. At the autopsy, made five hours after death, the muscles of the thigh were found disorganized, and bone fragments floated in a grumous matter." The specimen, No. 1342, Section I, A. M. M., is the upper two-thirds of the femur. Several inches of the shaft are missing, and the fractured extremities are carious and partly absorbed. Caries extends well up on the neck. The case is reported by Acting Assistant Surgeon H. L. Burritt.

CASE 227.—"Private William V———, Co. F, 12th Pennsylvania Cavalry, aged 20 years, was wounded at Winchester, Virginia, July 24th, 1864, by a fragment of shell, which struck the left thigh, in front, high up, crushed in the wall of the femur, in front of the great trochanter, and produced a fissure extending obliquely down the shaft. He was conveyed by rail to Baltimore, and was received, on July 30th, into the Jarvis Hospital. He fell into a typhoid condition, and died on September 29th, 1864." The pathological specimen (FIG. 62) shows a line of demarcation about the crushed lamina, which appears ready to exfoliate. The oblique fissure is occupied in its entire length by a beautiful slender sequestrum, and the borders of the fracture are fringed by callus. The history and specimen were contributed by Acting Assistant Surgeon B. B. Miles, while curator of Jarvis Hospital Museum. The report does not state explicitly that the articulation was opened, but the injury to the neck of the femur extends higher than the usual point of insertion of the capsular ligament.

FIG. 62.—Gunshot fracture of the upper part of the left femur. Spec. 3433, Sect. I, A. M. M.

Among the papers of the medical department of the Confederate Army filed in this Office, are many "certificates of disability to retire invalid soldiers," which include notes of a number of alleged examples of recovery from gunshot injuries of the hip-joint. It would appear that the surgeons who signed these certificates did not insist upon a rigorous application of the rules of diagnosis, but humanely accorded to the disabled patients the benefit of any possible doubt. CASE 32 (*ante* p. 70) would have been included in this series, but for the very positive statement of the board that the fracture involved the articulation.

CASE 228.—Corporal J. J. Atkenson, Co. B, 1st Tennessee (Confederate) Regiment, applied, November 24th, 1864, for "authority to appear before one of the medical examining boards now established by law, to be examined and retired in accordance with an act of Congress, 'to provide an Invalid Corps,' approved February 17th, 1864, and published in General Orders No. 34, A. & I. G. O.," current series, having been "permanently disabled in the service of the Confederate States and in the *line of duty*," by reason of a "gunshot wound of left hand, destroying use of two fingers; also, gunshot wound of left hip, fracturing bone, resulting in shortening and deformity and impairing use of leg; both wounds received at Murfreesboro'." This application having been recommended by the medical director of the Army of the Tennessee, was approved by command of General Hood, and Atkenson appeared at Lauderdale, Mississippi, before Surgeons J. T. Kennedy, D. W. Whimper, and H. Yandell, P. A. C. S., who certified that he was suffering from a "wound of left hip, fracturing bone, resulting in shortening and deformity, and impairing use of leg;" and that there was "permanent disability," though the patient was "able to perform light duty;" and recommended "that he be placed on the retired list of soldiers of the Invalid Corps." He was directed to present himself for examination in six months, in accordance with the provisions of "Paragraph VII, General Orders No. 34, A. & I. G. O."

CASE 229.—Private David J. C. Campbell, Co. E, 3d Virginia Infantry, 36 years of age, alleged that he received "a gunshot wound of the hip, at the battle of Payne's Farm, November 27th, 1863, fracturing the joint bones," and asked to be permitted to appear before the medical examining board of General Gordon's Division, for the purpose of being retired from the service. The board, consisting of Surgeon John H. Stevens, P. A. C. S., Surgeon W. J. Arrington, P. A. C. S., and Surgeon C. C. Henkel, P. A. C. S., reported that they had "carefully examined the said soldier, and found him permanently disabled for duty in any branch of the military service, by reason of gunshot wound of left hip, ball entering near the trochanter major and lodging in the bone in the vicinity of the hip-joint. The wound is still discharging freely, and spiculæ of bone are coming away. The injury was received in the line of duty, and unfits him for any duty."

CASE 230.—Private George A. Crymes, Co. B, 22d Virginia Battery, Walker's Brigade, was struck at the battle of Chancellorsville, Virginia, May 2d, 1863, by a fragment of shell, which produced a fracture of the right femur involving the head of the bone. On November 11th, there was shortening of the limb and partial anchylosis. A furlough which he had received was extended thirty days, and on December 21st, another extension of thirty days was granted. In March, 1864, he reported by order of the enrolling officer. He was very much run down from the injury of the articulation, and recently had had an attack of pneumonia. He was now furloughed for sixty days. The case is recorded in a register containing the "minutes of cases examined by Confederate Surgeons White, Roemer, and Walton, in 1863 and 1864, at the C. S. General Hospital at Farmville, Virginia." Crymes resided in Lunenburg county, Virginia.

CASE 231.—Sergeant F. M. Hunter, Co. E, 24th Tennessee Regiment, Cheatham's Corps, aged 24 years, was wounded at Resaca, Georgia, on May 15th, 1864, and appeared, January 27th, 1865, before a medical examining board at Selma, Alabama. The board, consisting of Surgeons V. B. Bilisoly, A. Hart, and W. Curry, P. A. C. S., certified as follows: "We have carefully examined him, and find him unable to perform the duties of a soldier in the field because of a wound by a musket ball, entering the superior articulation of right thigh, causing partial paralysis. Disability permanent, and received in the service of the Confederate States. This soldier is unable to serve the Government in any capacity. We have therefore retired him," under the Act of the (so-called) Confederate Congress, approved February 17th, 1864.

CASE 232.—"Colonel L. S. Slaughter, 56th Virginia Regiment, received, June 27th, 1862, a gunshot wound of the thigh, with a fracture of the femur, which was believed to involve the hip-joint. Colonel Slaughter appeared before a medical examining board, of which Surgeons Crenshaw, Read, and Peebles were members, and was retired from service, on account of '*vulnus sclopeticum* upper third of thigh, fracturing head of bone, deformity, shortening, and permanent loss of usefulness of limb.'"

CASE 233.—"Private Isaac S. Smith, Co. H, 2d North Carolina Regiment, aged 22 years, was wounded at the battle of Chancellorsville, Virginia, May 3d, 1863. On November 3d, 1864, he applied at Goldsboro, North Carolina, for authority to appear before an examining board to be retired, on account of a 'gunshot wound of right hip-joint fracturing head of femur, resulting in necrosis of the femur.' Surgeon W. A. Holt, and Assistant Surgeon F. W. Henderson, certified to the correctness of the patient's statement. He appeared before a board of examiners of Rodes' Division—Surgeons W. S. Mitchell, L. E. Gott, and J. B. Strachan, P. A. C. S.—who certified as follows: 'We find him permanently disabled, and that he cannot perform duty in any branch of military service in the field on account of the result of two gunshot wounds received at Chancellorsville, Virginia, May 3d, 1863, whilst in the service of the Confederate States, in the line of duty. One a wound of the left wrist, resulting in anchylosis of the joint and partial loss of use of the hand, the other through the right hip, fracturing femur, (wound still open and discharging,) as a result of which we find locomotion both difficult and painful, (requiring the use of two crutches.) We therefore recommend that he be retired.'"

CASE 234.—"Private W. V. Trail, Co. C, 57th Virginia Regiment, aged 27 years, was wounded at the battle of Gettysburg, Pennsylvania, July 3d, 1863, by a musket ball, which entered the left groin and passed out at the buttock. On February 9th, 1865, his application to be retired was endorsed by Assistant Surgeon Edward Pollard, who stated that the injury was complicated with 'fracture of the left thigh bone near the socket.' By order of General Lee, Assistant Adjutant General C. Venable, approved the application, with the following endorsement: 'The attention of the examining board is called to paragraph VII, G. O. No. 171, A. & I. G. O., 1864. If unfit for duty in the field, but capable of performing duty in some department of the service, the board will specify for what position he is best qualified, and if he has been detailed upon any light duty, the board will state how and where employed, and if his services are still desirable in such position.' The board, consisting of Surgeons C. B. Morton, C. E. Lippitt, and James D. Galt, P. A. C. S., reported that the patient was permanently disabled 'on account of gunshot wound of left thigh, the ball passing through and fracturing femur.'"

There are also on the registers, a certain number of cases reported as fractures of the upper extremity of the femur, involving the joint, about which the statements are either indefinite or contradictory, and it is thought best to place them in a separate category. Such are the following:

CASE 235.—"Sergeant James M. Adams, Co. D, 13th Georgia Regiment, aged 37 years, was struck in the action at Monocacy, Maryland, July 9th, 1864, by a round musket ball, which slightly injured the head and neck of the left femur. He was conveyed, the following day, to the General Hospital at Frederick City. A gunshot wound of the left thigh and hip, involving the joint, was diagnosticated. Simple dressings were applied, and tonics and stimulants were given. Extensive abscesses formed in the anterior and posterior parts of the thigh. On September 20th, the patient was conveyed to West's Buildings Hospital, in Baltimore, and was transferred thence, on October 27th, 1864, to Point Lookout, Maryland, for exchange." Surgeon A. Chapel, U. S. V., reports the case.

CASE 236.—"Sergeant Samuel R. Arrison, Co. A, 118th Illinois Volunteers, was wounded at the battle of Champion Hills, Mississippi, May 16, 1863, by a musket ball, which, entering at the lower part of the left gluteal region, passed forwards and upwards, producing, it was believed, an intracapsular fracture of the neck of the femur, and emerged at the groin. On June 4th, he was placed aboard the hospital steamer R. C. Wood, and conveyed to Memphis, Tennessee, and admitted, June 8th, into the Union Hospital. Cold water dressings were applied, and he did tolerable well. On June 27th, an exploration of the wound convinced the attendants that fracture extended within the capsule. Extension was made for one week, after which the patient was required to take daily exercise on crutches. The limb was but little shortened. Early in August he began to fail, and subsequently had frequent attacks of diarrhœa, lasting weeks at a time. Rapid emaciation ensued, and his appetite was poor. In October he had cough and purulent expectoration. He died November 2d, 1863." The case is reported by Acting Assistant Surgeon J. M. Steady.

CASE 237.—"Corporal Daniel Bachler, Co. G, 82d Illinois Volunteers, aged 23 years, was wounded in the battle of Chancellorsville, Virginia, May 2d, 1863, by a round ball, which, entering just above the right trochanter major, fractured the neck of the femur, and emerged beneath Poupart's ligament, about two inches from the symphysis pubis. He stated that he lay thirteen days on the plank road unattended. He was admitted, May 15th, to the field hospital of the Third Division, Eleventh Corps, and was transferred thence, by steamer, to Alexandria, and admitted, May 25th, into the First Division Hospital. The patient had been informed that the wound involved the soft parts only; cold water dressings had been applied. The limb was shortened about three inches. A fracture through the neck was discovered. Partial union had taken place. There was little suppuration, and the pus was healthy. The parts were but little swollen. Smith's anterior splint was applied, and moderate extension made by weights and pulley. He was furloughed on July 14th for sixty days, and went to Chicago, Illinois. On September 10th, Bachler was received into the City General Hospital in Chicago. A gunshot wound of the right hip-joint was recorded. He was transferred to the 2d Battalion of the Veteran Reserve Corps on August 7th, 1864." The case is reported by Surgeon Norman S. Barnes, U. S. V. This man's name is not on the Pension List.

CASE 238.—"Sergeant Albert G. Beebe, Co. A, 85th Illinois Volunteers, received, in the engagement near Perryville, Kentucky, October 8th, 1862, a gunshot wound of the hip, which was believed to involve a fracture of the joint. He was conveyed to Hospital No. 7, in Perryville. He was discharged from service on February 15th, 1863. His disability was rated as total. His name is not upon the Pension List." The case is reported by Assistant Surgeon H. S. Wolfe, 81st Indiana Volunteers.

CASE 239.—"Private David A. Brewer, Co. E, 37th Kentucky Mounted Infantry, in March, 1864, was wounded by the accidental discharge of a musket near Glasgow, Kentucky. The ball entered the right groin and passed backwards and downwards, striking the external rim of the cotyloid cavity, and apparently shivering the head of the femur. It was then deflected, and passed out at the upper part of the nates." The regimental surgeon, J. R. Duncan, reports the case.

CASE 240.—"Private James T. Cone, 7th Virginia Regiment, was wounded at Antietam, Maryland, September 17th, 1862, by a musket ball, which entered the back of the right thigh and penetrated the arch of the femur. He was captured and conveyed to the General Hospital at Boonsboro. On October 11th, the limb was shortened and everted, and there was a very copious discharge of pus from the wound. The ball had not been found. No splints had been applied. The prospect for his recovery was not very good." Surgeon Frank H. Hamilton, U. S. V., reported the case at the meeting of the New York Academy of Medicine, November 28th, 1863. The patient died at Boonsboro, Maryland, but no record of the precise date of death can be found.

CASE 241.—"Private William A. Dibble, Co. C, 106th Pennsylvania Volunteers, aged 21 years, received, at the battle of Antietam, Maryland, September 17th, 1862, a gunshot fracture of the neck of the right femur. According to an endorsement of the Adjutant General of Pennsylvania, upon a letter of inquiry from this Office, he was removed to the house of Captain Henry Neville, Co. C, 106th Pennsylvania Volunteers, and died there on September 19th, 1862. According to the Smoketown Hospital register, he entered that hospital on October 5th, under the care of Surgeon B. A. Vanderkieft, 102d New York Volunteers. The accounts as to the nature and treatment of the case, as well as the records of the date of death, are conflicting."

CASE 242.—"Corporal Morris J. Fitzharris, Co. E, 42d New York Volunteers, aged 22 years, received, at the battle of Spottsylvania, Virginia, May 12th, 1864, a gunshot wound of the left hip and fracture of the femur, probably implicating the joint. He was taken to the field hospital of the Second Division, Second Corps. Death took place on May 14th, 1864." Surgeon J. F. Dyer, 19th Massachusetts Volunteers, reports the case.

CASE 243.—"Private Samuel T. Hook, Co. B, 79th Indiana Volunteers, at the battle of Stone River, Tennessee, December 31st, 1862, was struck by a conoidal musket ball, which was thought to have perforated the neck of the right femur. He was conveyed, the same day, to Hospital No. 1, at Murfreesboro. There the diagnosis was simply 'gunshot wound of the right thigh.' On March 17th, 1863, he was transferred to Hospital No. 14, at Nashville. He was treated by position and rest. Surgeon F. Seymour, U. S. V., in his special report, dated March, 1863, states that the limb was shortened two inches, and that the patient was doing well. On April 4th he was transferred to Hospital No. 11, at Louisville, Kentucky, and thence, on April 13th, to Hospital No. 3, at New Albany, Indiana. The wound in the groin had not yet fully cicatrized. He was discharged from service on May 8th, 1863." Though Surgeon Seymour's statement is precise, the contradictory accounts from the field hospital, and the reports from the Louisville and New Albany hospitals, that the case was simply "gunshot flesh wound of the right thigh," lead one to hesitate to accept the case as an unequivocal instance of recovery from gunshot fracture involving the hip-joint.

CASE 244.—"Private A. Ingraham, Co. B, 25th Illinois Volunteers, was admitted into the Post Hospital at Cairo, Illinois, on the 21st of August, 1862. It is stated that the head of the right femur had been fractured by a musket ball, an injury resulting in anchylosis of the hip-joint. The wound was healed. From the records on file, no previous history of the case can be obtained. The patient was discharged from the service on August 21st, 1862." The case is reported by Acting Assistant Surgeon S. Hamilton.

CASE 245.—"Corporal Herman Koeh, Co. F, 7th Missouri Volunteers, aged 26 years, was struck in the left hip at the battle of Chickamauga, Georgia, September 20th, 1863, by a conoidal musket ball, which splintered the femur above the shaft. He was taken to the field hospital of the Second Brigade of the Third Division, Twentieth Corps. Near the close of October, he was transferred to Nashville, Tennessee, and admitted to Hospital No. 12, and on November 26th he was sent thence to the Marine Hospital at St. Louis, Missouri. He suffered pain on rotating the limb. He was discharged from service on May 16th, 1864. His disability was rated at one-fourth." Surgeon J. H. Peabody, U. S. V., reports the case. He was last heard of by the Pension Office, in March, 1867, when his disability was "rated at five-eighths, and probably not permanent."

CASE 246.—"Private Fred. Kresgor, Co. B, 18th Georgia Regiment, aged 22 years, was struck in the right hip, in the

COMPARED WITH TEMPORIZATION.   101

engagement at Sailor's Creek, Virginia, by a conoidal musket ball. The diagnosis was difficult, but it was reported that the projectile had fractured the head of the right femur. He also received a gunshot wound of the right arm. He was captured and conveyed to the base hospital of the Sixth Corps, at City Point, and was transferred thence, by steamer, to Baltimore, and was admitted, on April 22d, to West's Buildings Hospital. On May 9th, Kreegor was transferred to the hospital at Fort McHenry. He died on May 16th, 1865." The case is reported by Surgeon William Hays, U. S. V.

CASE 247.—"Private James McCabe, Co. A, 12th Massachusetts Volunteers, was wounded on September 17th, 1862, at the battle of Antietam, by a musket ball, which entered just below the right groin, and made its exit at the buttock, fracturing the neck of the femur in its passage. He was conveyed to Hospital No. 5, at Frederick, Maryland, and was treated with the limb in an extended position. On November 9th, he was transferred to Frederick Hospital No. 1. In December, a large metastatic abscess formed about the right shoulder, which was laid open by Assistant Surgeon R. F. Weir, U. S. A., a large quantity of pus escaping. After the healing of the abscess, there was much weakness of the muscles of the acromial and humeral regions, and the patient could not raise his hand above his chin. On June 16th, he was sent, in good condition, to the Jarvis Hospital, at Baltimore, and thence to Point Lookout Hospital, where he remained until July 3d, 1863, when the wound being healed, he was discharged from the service of the United States. On June 10th, 1867, he was examined at the office of Surgeon General Dale, of Massachusetts. The fracture was firmly consolidated. There was but a slight limp in walking. His general health was excellent. He received a pension, and was employed in the "Soldiers' Messenger Corps." He experienced no pain, except on change of weather, or when his walk was extended beyond two miles. He considered his injury but a slight disability in his business, since he had free passes on all the lines of horse cars in Boston."

CASE 248.—"Private Wilborn Miles, Co. A, 60th Illinois Volunteers, aged 18 years, was wounded in the engagement at Bentonville, North Carolina, March 19th, 1865, by a conoidal musket ball, which struck the right hip, involving, as is stated, the neck of the femur. He was conveyed to Newberne, and was admitted, on April 5th, into the Foster Hospital, and was transferred thence, by steamer, to the McDougall Hospital, New York Harbor, arriving on April 13th. On April 30th, he was transferred to the Madison Hospital, in Indiana. He was discharged from the service on June 26th, 1865. His disability is rated at one-third." The case is reported by Surgeon John H. Rauch, U. S. V.

CASE 249.—"Private Bright Page, Co. I, 51st North Carolina Regiment, aged 17 years, received, in the battle at the Weldon Railroad, Virginia, on August 18th, 1864, a gunshot wound of the hip. He was transferred, by steamer, to Washington, and, on August 30th, 1864, was admitted into the Carver Hospital. A gunshot wound of the pelvis was diagnosticated. Death took place on September 3d, 1864. At the autopsy, a wound of the hip-joint and a compound fracture of the neck of the femur was found." The case is reported by Surgeon O. A. Judson, U. S. V.

CASE 250.—"Private Edward Powell, Co. A, 3d U. S. Colored Troops, received, in the engagement at Morris Island, South Carolina, September 3d, 1863, a shell wound of the buttock. He was taken to a field hospital, and it was found that the projectile had comminuted the femur within the articulation. Death took place within a few hours, on September 3d, 1863." The case is reported by Surgeon Horace R. Wirtz, U. S. A.

CASE 251.—"Private Samuel Sellers, Co. I, 108th Ohio Volunteers, aged 28 years, was wounded in the engagement at Cynthiana, Kentucky, June 12th, 1864, by a conoidal musket ball, which penetrated the right hip. He was conveyed to the Seminary Hospital, at Covington, Kentucky, on June 13th. The soft parts only were believed to be involved. Simple dressings were applied. He died on September 27th, 1864. The head of the femur was fractured." The case is reported by Surgeon A. M. Speer, U. S. V.

CASE 252.—"Private Jeremiah Shafer, Co. A, 7th Maryland Volunteers, aged 17 years, received, in front of Petersburg, Virginia, March 31st, 1865, a gunshot wound on the field, which was supposed to involve the left hip-joint. He was taken to the field hospital of the Second Division, of the Fifth Corps. A wound of the left hip, involving the soft parts only, was diagnosticated. On April 2d, he was conveyed to the hospital at City Point, and thence, by steamer, to Washington, and was admitted, on April 7th, to the Campbell Hospital. A gunshot wound of the left hip was entered on the register of that hospital. The patient died on April 20th, 1865. The certificate of the cause of death describes the injury as a 'gunshot fracture of the hip-joint.'" Surgeon A. F. Sheldon, U. S. V., reported the case from Campbell Hospital.

CASE 253.—"Private William C. Watson, Co. A, 4th Michigan Volunteers, was struck in the left hip, at the battle of Malvern Hill, Virginia, July 1st, 1862, by a conoidal musket ball. A fracture through the trochanteric region, with fissures possibly extending to the joint, was reported. He was conveyed to Philadelphia, and admitted into the Episcopal Hospital July 30th, transferred to Master Street Hospital on March 18th, and to the South Street Hospital May 11th. The patient stated that for weeks his life was despaired of, in consequence of the excessive suppuration which took place. Abscesses formed in the upper third of the thigh. Spiculæ of bone escaped at intervals for months from the wound and through the incisions made for the evacuation of the abscesses. On January 1st, 1864, suppuration had nearly ceased. The limb was shortened several inches, and there was much deformity. On March 24th, he was admitted into the Christian Street Hospital, and was discharged from the service on May 5th, 1864. He received his pension in September, 1868, and his disability was rated as total and permanent." Surgeon Paul B. Goddard, U. S. V., reports the case.

CASE 254.—"Private Andrew C. Woodall, Co. G, 55th Pennsylvania Volunteers, aged 17 years, was wounded in the left hip, in the engagement at Drury's Bluff, Virginia, May 16th, 1864, by a conoidal musket ball. It is reported that a fracture involving the articulation was produced. He remained in the hands of the enemy until the middle of August, when he was paroled, and conveyed, by steamer, to Annapolis, Maryland, and admitted into Hospital No. 1. A gunshot wound of the left hip was reported. On September 23d, he was transferred to Camp Parole, and thence, on February 17th, 1865, to Rulison Hospital, at Annapolis Junction. Simple dressings were applied. On April 6th, he was transferred to the Tilton Hospital, in Wilmington, Delaware. Here, it is said, he had a gunshot wound of the left hip and groin. He was mustered out of service on June 8th, 1865." The case is reported by Assistant Surgeon John Bell, U. S. A. Woodall's name appears on the pension list as last paid at the agency in Philadelphia, March 4th, 1866. His disability was rated at one-third, and probably temporary.

CASE 255.—"Corporal George W. Wright, of Blount's Virginia Battery, aged 21 years, received, near Petersburg, Virginia, June 17th, 1864, a gunshot fracture of the neck of the femur. He was conveyed to the Washington Street Confederate Hospital, in Petersburg. He was treated by rest and position. He died, exhausted, on July 7th, 1864." The case is reported by Surgeon W. L. Baylor, C. S. A.

CASE 256.—"Private Jacob Wright, Co. E, 96th Pennsylvania Volunteers, aged 17 years, was wounded at Spottsylvania, Virginia, May 8th, 1864, by a conoidal musket ball, which fractured the head of the right femur. He was captured, and remained in the hands of the enemy until August 14th, when he was paroled and conveyed, by steamer, to Annapolis, Maryland, and admitted to Hospital No. 1. The diagnosis here was wound of the left thigh. On September 21st, he was transferred to Camp Parole, and thence, on February 17th, 1865, to Rulison Hospital, at Annapolis Junction. A fracture of the head of the right femur was diagnosticated. On April 13th, he was transferred to the Satterlee Hospital, at Philadelphia. A gunshot wound in the upper third of the left thigh was recorded. July 16th, he was sent to the McClellan Hospital. A gunshot fracture of the upper third of the left femur was reported. Necrosis of the femur existed. He was discharged from service on September 9th, 1865. His disability was rated at one-third." The case is reported by Assistant Surgeon John Bell, U. S. A. In September, 1868, he received his pension, and his disability was reported as "total and permanent."

There are recorded on the registers the following cases of gunshot injury of the hip-joint treated by the extraction of fragments of bone:

CASE 257.—"Captain Samuel J. Alexander, Co. B, 9th New Hampshire Volunteers, was wounded in the engagement near Jackson, Mississippi, July 13th, 1863, by a ball which entered the left hip-joint, fracturing the acetabulum. He was conveyed to the hospital of the Second Division, Ninth Corps, where, twelve hours after the reception of the injury, Surgeon G. W. Snow, 35th Massachusetts Volunteers, having placed the patient under the influence of chloroform, extracted a portion of the ball, and removed several spiculæ of bone. He died on July 23d, 1863." The case is reported by Surgeon W. C. Otterson, U. S. V.

CASE 258.—"Private Frank Beck, Co. F, 115th Pennsylvania Volunteers, aged 24 years, was wounded in the battle of Chancellorsville, Virginia, May 3d, 1863, by a conoidal musket ball, which entered the right thigh high up in front, and, passing directly backwards, fractured the femur at the great trochanter. He was taken to the regimental hospital near Falmouth, and afterwards to the general field hospital of the Third Corps. Copious suppuration, with diarrhœa, ensued. The patient was conveyed, by steamer, to Alexandria, and was admitted, on June 14th, into the Mansion House Hospital. He was much exhausted. The limb was placed in position, and supported by pillows. He was given eight ounces of sherry wine daily, with tincture of the chloride of iron every eight hours, and opiate enemata as required. On June 25th, the discharge from the wound was dark and offensive, and it was laid open freely, and necrosed fragments of bone were removed. The patient died, exhausted, on July 29th, 1863." The case is reported by Surgeon Robert Reyburn, U. S. V.

CASE 259.—"Private H. T. Elam, Co. A, 11th Virginia Regiment, aged 21 years, was wounded at the battle of Williamsburg, Virginia, May 5th, 1862, by a musket ball, which entered in front of and just below the right trochanter major, and, passing upwards, inwards, and backwards, lodged near the hip-joint. The shock of the injury was severe, and reaction was slow. The patient was taken to a field hospital, and, eight or nine days after the reception of the wound, he was conveyed, in an ambulance, a distance of six miles, to the York river, and placed on board an hospital steamer and conveyed to Washington, where, on May 16th, he was admitted into the Cliffburne Hospital, under the care of Assistant Surgeon John S. Billings, U. S. A. On admission, the injured limb was two inches shortened, and rotated outwards. On May 17th, ether was administered, and Dr. Billings made a curved incision over the trochanter, passing through the entrance wound, and removed several fragments of bone. The neck of the femur was discovered to be fractured. The missile could not be found. The patient was placed upon a fracture bed, the right foot being firmly fastened by means of adhesive plasters to a board screwed to the foot of the bed. This end of the bed was then elevated to produce extension of the limb. The patient reacted slowly, and for four or five hours the pulse was almost imperceptible, and the skin cold and clammy. Brandy and beef essence were freely administered, and he soon began to rally. On May 20th, the wound was suppurating. He had no pain. His tongue was moist and clean, and pulse at 96. He was ordered a liberal amount of stimulants, with generous diet. On May 22d, diarrhœa set in, followed by excessive irritability of the stomach, which rejected every sort of food. The gastric irritability abated under the use of small doses of bi-carbonate of soda and aromatic spirits of ammonia; but the diarrhœa continued, and the wound began to discharge offensive pus. He died on May 29th, 1862." At the *post mortem* examination a round musket ball was found imbedded in the pyriformis muscle. The liver and spleen were congested and much enlarged. The pathological specimen is No. 9, Sect. I, A. M. M. It embraces a portion of the right innominatum, and the upper fourth of the femur, minus the neck and some of the contiguous parts of the head. The articular surface of the acetabulum shows the result of caries. The specimen is figured in the adjoining wood-cut. (FIG. 63.)

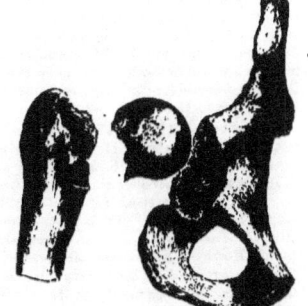

FIG. 63.—Secondary results of a gunshot fracture of the neck of the right femur, treated by extraction of fragments.—*Spec.* 9, Sect. I, A. M. M.

COMPARED WITH TEMPORIZATION.   103

CASE 260.—"Corporal Luke English, Co. E, 2d Wisconsin Volunteers, aged 21 years, was wounded at the battle of Gettysburg, Pennsylvania, July 1st, 1863, by a conoidal musket ball, which entered at a point between the left trochanter major and the tuberosity of the ischium, emerging near the anus, penetrated the right thigh and fractured its trochanters, and lodged. He was admitted the next day into the Seminary Hospital, at Gettysburg. Search for the missile was unsuccessful. Two fragments of bone were extracted. Cold water dressings were applied. On July 17th, he was transferred, by railroad, to the York Hospital. Early in August, the average discharge of pus was about four ounces. The wound of entrance had nearly closed. On November 4th, he could walk a little, the wound, however, discharged considerably. He was discharged from the service of the United States on June 23d, 1864." Surgeon Henry Palmer, U. S. V., reports the case. His attorney, John Hancock, of Oshkosh, Wisconsin, reports that English died in 1867.

CASE 261.—"Corporal Thomas Gallagher, Co. F, 165th New York Volunteers, aged 30 years, was wounded, in the attack on Port Hudson, Louisiana, May 27th, 1863, by a canister ball, which, entering the left thigh, two and a half inches below the great trochanter, passed backwards and upwards, and lodged three inches from its place of entrance. The projectile was cut down upon and extracted sixteen hours after the reception of the injury, and several fragments of bone were removed. The wounded man was admitted, May 29th, to the Convalescent Hospital, at Baton Rouge. He was transferred to New York, and was admitted, October 8th, into the Central Park Hospital. The limb was swollen, and the patient suffered pain along the course of the sciatic and popliteal nerves. There was imperfect rotation of the thigh outwards, and partial anchylosis of the hip-joint. On November 16th, the wound was nearly healed. Water dressings were applied and stimulating lotions. On December 8th, the wound had not entirely healed, and the condition of the limb was not materially improved since admission. He was discharged from service on December 19th, 1863, as unfit for the Veteran Reserve Corps, by reason of a gunshot wound involving the nerves of the left thigh, causing neuralgia, and his disability was rated as one-half." The case is reported by Surgeon B. A. Clements, U. S. A.

CASE 262.—"Sergeant Edward G. Gilliam, Co. C, 11th Virginia Regiment, aged 24 years, was wounded in the engagement at Drury's Bluff, Virginia, May 10th, 1864, by a rifle ball, which, entering the upper portion of the right thigh in front, fractured the neck of the femur and lodged. He was conveyed, the same day, to Richmond, and admitted into Chimborazo Hospital. There was eversion of the limb, but no shortening. On placing the finger upon the seat of fracture, and rotating the thigh, it was observed that the trochanter major moved with the shaft. There appeared to be but little comminution, and there was no escape of synovial fluid. The parts were but little tumefied. The limb was comfortably adjusted on pillows. The patient complained of severe pain in the thigh, especially along its inner aspect, and had frequent twitchings of the muscles. The pulse was 80, the tongue clean, the appetite poor. Half an ounce of whiskey was directed every fourth hour, and a nourishing diet, with one-third of a grain of morphine given at bed time. During the latter part of May, there was severe local pain, with rapid emaciation, and a bed sore formed over the sacrum. On June 15th, the limb was swung in a Smith's anterior splint. The patient's appetite soon began to improve; the pulse about 96. On June 27th, he yet suffered severe pain in the hip and along the inner part of the thigh. On July 1st, synovial fluid in large amount issued from the wound; after which, the pain subsided. His appetite continued to improve, and he now slept well at night. One or two small spiculæ of bone were eliminated in the discharges. On July 10th, he was stronger, and had no pain; the bed sore was healing; the discharge from the wound was small in quantity, and the œdema was disappearing. For several days, about July 20th, he suffered pain at the knee, and the splint was removed. On July 30th, the bed sore was healed, and his health was greatly improved; but any movement of the hip-joint continued to create severe pain. By August 10th, he had become stout and fleshy, and could suffer the limb to be moved; and August 22d, it could be moved freely without producing pain. At this date, a small spicula of bone was removed. There was only a slight discharge from the wound. On September 5th, he was able to raise the limb a short distance from the bed, and September 14th, sat up in a chair. There was limited motion of the hip-joint. He was furloughed for sixty days on September 20th, 1864." The case is reported by Surgeon E. M. Seabrook, C. S. A.

CASE 263.—"Private Theron Hayward, Co. F, 5th Pennsylvania Reserves, aged 23 years, was wounded at the battle of Fredericksburg, Virginia, December 13th, 1862, by a conoidal musket ball, which entered in front of and a little below the right hip-joint, and shattered the trochanters, besides producing a complete fracture through the neck, and an oblique one down the shaft. He also received a gunshot wound of the right arm. He was conveyed to the field hospital of the 1st Corps. The ball was extracted, and the wounds were treated on the expectant plan. On December 23d, the patient was conveyed, by steamer, to Washington, and received into Lincoln Hospital. An excision was attempted by Surgeon H. Bryant, U. S. V., the sharp end of the inner portion of the femur being sawn off one-fourth of an inch below the trochanter minor. Owing to the extent of the injury, the operation was abandoned. The patient died January 2d, 1863. The specimen is represented in the adjoining wood-cut. (FIG. 64).

CASE 264.—"Private Samuel Hensel, Co. H, 114th Pennsylvania Volunteers, aged 31 years, was wounded at the battle of Chancellorsville, Virginia, May 3d, 1863, by a musket ball, which entered in the central portion of the left buttock. He was conveyed by steamer to Washington, and, on May 8th, was received into Armory Square Hospital. The missile, at the end of a fortnight, presented itself at the point of entrance, and was extracted by the patient. He states that it was a round ball, flattened on one side, with a cleft filled with osseous matter. The treatment was expectant. On June 16th, Hensel was transferred to McKim's Mansion Hospital, in Baltimore. About the middle of July, the nurse, while dressing the wound, discovered in it a

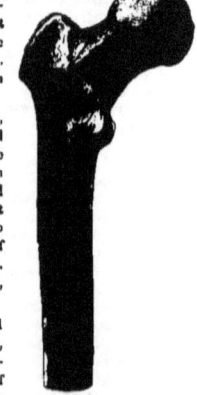

FIG. 64.—Partial excision in a gunshot fracture of the neck of the right femur.—*Spec. 202, Sect. I. A. M. M.*

foreign substance, which, upon removal, proved to be a portion of the haversack. The patient was transferred to Mower Hospital, in Philadelphia. Early in September, a splinter of bone, nearly two inches in length by one in width in its central portion, was extracted from the wound. Subsequently, ten other pieces, varying in size from one-fourth to one inch in length, were eliminated. Soon the wound closed, and the patient was able to use crutches. On September 24th, he was transferred to the McClellan Hospital, where, on May 13th, 1864, he was discharged from the service of the United States, with a disability rated at one-half. On November 1st, 1866, a so-called "resection apparatus" was fitted to the disabled limb by Gemrig, of Philadelphia. In a communication to this Office, dated Philadelphia, February 13th, 1868, Mr. Hensel states that he is unable to walk without the use of the apparatus and a cane, but that, by the aid of these, he can walk from one-half to three-quarters of a mile, being obliged, however, to rest for two or three minutes once or twice by the way, on account of pain and weakness in the limb.

CASE 265.—"Corporal William Hernka, Co. D, 1st Maryland Cavalry, aged 20 years, was wounded in the engagement at Deep Bottom, Virginia, July 26th, 1864, by a rifle ball, which, passing from the left great trochanter to the symphysis pubis, shattered the apophysis and fractured the neck of the femur. He was conveyed to Portsmouth, and was admitted, July 29th, to Balfour Hospital. Water dressings were applied. On August 11th, Acting Assistant Surgeon Thomas Welsh removed fragments of bone. The granulations were pale and flabby, and the patient was much debilitated. Tonics and stimulants with nourishing diet were given. He died, exhausted, on September 16th, 1864." Assistant-Surgeon J. H. Frantz, U. S. A., reported the case.

CASE 266.—"Private Joseph Leehart, Co. C, 16th Mississippi Regiment, was wounded at the battle of Antietam, Maryland, September 17th, 1862, by a conoidal musket ball, which shattered the left trochanter major, and emerged in the inguinal region. He was captured, and was admitted, November 25th, to the Hospital No. 5, at Frederick City. On December 5th, Surgeon Henry S. Hewit, U. S. A., made an exploratory incision and removed several large fragments of bone. On December 6th, the patient was comfortable. Abscesses formed about the hip-joint and extended into the middle third of the thigh. He died, exhausted, on December 15th, 1862." Acting Assistant Surgeon Redfern Davies and Surgeon Henry S. Hewit, U. S. V., report the case.

CASE 267.—"Private T. L. Lemax, Co. K, 30th Virginia Regiment, was wounded at the battle of Antietam, Maryland, September 17th, 1862, by a conoidal musket ball, which fractured the neck of the left femur. He remained in a field hospital for several weeks, and, on October 20th, was conveyed, by ambulance, to Frederick City, and admitted into General Hospital No. 5. Acting Assistant Surgeon A. V. Cherbonnier, who had charge of the case at the latter hospital, states, in a letter dated November 11th, 1868, that 'young Lomax recovered after operative interference involving the neck of the femur.' No further details in regard to the operation and the treatment are recorded. Lomax was transferred on October 25th, 1862, probably for exchange."

CASE 268.—"Private Ernest Longyear, Co. D, 72d Pennsylvania Volunteers, aged 21 years, was wounded at the battle of Antietam, Maryland, September 17th, 1862, by a musket ball, which entered in front, near the groin, and penetrated the neck of the femur, reaching the acetabulum. He was taken by ambulance to Frederick, and was admitted into Hospital No. 1 a week after his injury. On October 2d, fragments of the neck were removed. October 3d, he complained of pain in the left side, and his features were haggard, and on the following day he failed rapidly and was delirious. He died October 4th, 1862." The case is reported by Assistant Surgeon R. F. Weir, U. S. A.

CASE 269.—"Private L. N. P. Rodenbough, Co. D, 55th Illinois Volunteers, was wounded at the battle of Shiloh, Tennessee, April 6, 1862, by a musket ball, which, entering the left hip, impinged against the anterior portion of the neck of the femur, and lodged. The missile was extracted, and the patient was conveyed to the General Hospital at Savannah. Early in May he was transferred, by steamer, to Quincy, Illinois, and was admitted, May 7th, to the General Hospital. The wound in the soft parts healed, and for a while he was about on crutches. Subsequently the wound reopened, when necrosis of the neck was detected, and fragments of bone were removed. An extensive abscess formed in the muscles in the inner part of the thigh. He died, exhausted, on November 16, 1863. At the autopsy the neck of the femur was found injured for an inch and a half on its anterior surface, and within the capsular ligament. The head of the bone was wholly carious. Osseous matter had been abundantly deposited above and below the place of impact." The case is reported by Surgeon R. Niccolls, U. S. V.

CASE 270.—"Captain Henry A. Sand, Co. D, 103d New York Volunteers, was wounded at the battle of Antietam, Maryland, September 17th, 1862, by a musket ball, which entered the left thigh, at a point one and a half inches below Poupart's ligament, and at an equal distance outside of the course of the femoral artery, fractured the neck of the femur, and emerged at the digital fossa of the great trochanter. He was conveyed to the German Reformed Church Hospital at Sharpsburg. The fracture of the femur was believed to be partial. Splints, however, were adjusted to the limb, a long, straight one outside, and a short inside one, and counter extension was made by a perinæal strap. On the night of September 29th, the wounds bled considerably, and, September 30th, he was pale and almost bloodless; his pulse was at 108; yet he was cheerful and hopeful, and comparatively easy; his appetite was good, and the secretions were normal. On October 2d, the patient's countenance expressed anxiety; his pulse was 130; the least movement of the limb caused acute pain in the neighborhood of the hip-joint. Unnatural mobility and distinct crepitus were discovered on a renewed careful exploration of the injury. The finger detected much comminution of the neck, and an oblique fracture of it, extending into the great trochanter; the wounds had begun to suppurate. Buck's apparatus was applied; one half grain of morphia was directed. That night he obtained some sleep, having previously been very wakeful. October 3d, the pulse was 100; he spoke cheerfully. October 6th, he had two painful bed sores over the sacrum; the wounds were suppurating freely. Excruciating pain, which was referred to the hip-joint principally, was created by the slightest movement of his person; his pulse was at 110, his appetite was fair, and he spoke hopefully. October 12th, an ounce of castor oil, to be assisted by mild enematic, was ordered, and acted well. October 13th, the appetite was improved; the pulse 112, and feeble. Emaciation was advancing; he slept but little, on account of pain. Morphine was administered every night. On October 15th, he had a slight rigor. Four grains of quinine, in wine, were given thrice daily, and concentrated nutriment was provided. October 16th, the patient was placed upon a Burgess fracture bed. The daily discharge from the wound was estimated at a pint and a half. On October 20th, the thigh was swollen and œdematous; the apparatus was removed and the limb bandaged, and the next day he rested easier; pulse 125, and thready. October 26th, a long strip of disorganized tissue was removed from the wound, through which, afterwards, a free incision was made for the evacuation of pus, by Assistant Surgeon Gooley, U. S. A., who then removed several splinters of bone. The limb was now bandaged and the apparatus readjusted. Diarrhœa and vomiting took place October 29th, and the patient, rational and able to speak to within an hour of his death, succumbed at ten and a half o'clock P. M., October 30th, 1862. The pathological specimen was forwarded to the A. M. M. On February 26th, 1863, Dr. McDonnell, in respect to the above case, and others similar where there had been such destruction of bone, says that Buck's extension apparatus, and all others, produced such a hiatus between the fractured ends as leaves, by the absence of periosteum and nutrient vessels, no chance for repair, and is destructive of any hope of a good result. He would, in all cases of badly compound comminuted fractures in the upper third of the femur, dispense with *all* extension apparatus, and simply place the limb in an easy position, regardless alike of deformity and shortening, (amputation being out of the question.) This, Dr. McDonnell is persuaded, is the treatment to save most lives. Burgess's fracture bed does not suit such cases."

CASE 271.—" Private Louis Schmidt, Co. H, 8th Kansas Volunteers, aged 35 years, was wounded in the battle of Chickamauga, Georgia, September 19th, 1863, by a conoidal musket ball, which fractured the neck of the right femur. He was conveyed to the General Field Hospital at Chattanooga, Tennessee. The field hospital report states that an excision had probably been performed. The limb was supported upon a double inclined plane of very slight elevation at the knee. Suppuration was copious until near the 20th of November, and there were no indications of osseous union, but there was now a fine appearance of fresh granulations. Pus burrowed among the gluteal muscles, and was evacuated at the anterior border of the gluteus maximus. His general condition continued good. On November 26th, the discharge was less in quantity, and consistent. The limb was straight and shortened one and a half inches. He was transferred, December 6th, and admitted, December 20th, to Hospital No. 3, at Murfreesboro. Here a gunshot fracture of the upper third of the femur was reported. On November 30th, 1864, he was transferred to Hospital No. 2, at Nashville, thence to Louisville, Kentucky, and subsequently to Jefferson, Indiana. Here a simple 'flesh wound of the right thigh' was recorded. Simple dressings were applied. On August 16th, 1865, he was transferred to Saint Louis, Missouri, and received, August 30th, into the Marine Hospital. A gunshot fracture of the right femur and hip was diagnosticated. He died from chronic diarrhœa, on January 3d, 1866." Surgeon Joseph A. Stillwell, 22d Indiana Volunteers, reports the case.

CASE 272.—Lieutenant Colonel James C. Strong, 38th New York Volunteers, was wounded at the battle of Williamsburg, Virginia, May 5th, 1862, by a conoidal musket ball. The ball entered the right thigh in front, a little below the groin, and made its exit through the buttock, over the lower right hand border of the sacrum. Surgeon A. J. Berry, 38th New York Volunteers, examined the wound, and found that the ball had deeply grooved the head of the femur, and had fractured the upper portion of the rim of the acetabulum. A detached fragment of the acetabulum, an inch or more in length, a part of it covered with articular cartilage, together with portions of clothing, were extracted from the wound. On the 8th of May, the patient was transferred, by a steamer from Queen's Creek Landing, to the Hygeia Hospital, at Fort Monroe. Here he remained until the 13th, when, after a painful journey of five days, on a litter, by steamer and rail, he reached his home, in Buffalo, New York. The injured limb was semi-flexed and rotated inwards, the head of the femur being dislocated upon the dorsum of the ilium. Attempts to place the limb in position produced such acute suffering that such efforts were abandoned. For ten weeks there was profuse suppuration, with burrowing of pus in the thigh, and intense pain, with chills, profuse perspiration, and great prostration; after which, a very gradual amendment took place. On December 12th, 1862, the patient was removed to Philadelphia, and entered at the Officers' Hospital, at Cammack's Woods, where he was able to bear treatment by Buck's method of extension by weights. Here a number of spiculæ of bone were extracted or washed from the wound. On January 6th, 1863, the patient was discharged from the hospital. On June 1st, the wounds were nearly closed, and he rejoined his regiment on crutches, and was mustered out with his regiment on June 22d, 1863. On September 29th, he was appointed colonel in the Veteran Reserve Corps. He was subsequently breveted brigadier general. In July, 1866, he visited the Army Medical Museum, and his limb was examined and his photograph was taken. He was then in good health. His limb was shortened nearly five inches, but, by the flexibility acquired by the lumbar vertebræ, the inclination of the pelvis, and extension of the toes, he was enabled to walk with surprising ease and activity, with or without a cane. The head of the femur was firmly anchylosed on the dorsum of the ilium. The cicatrices appeared sound. Early in 1869, General Strong was in Washington, at Willard's Hotel, and sprained his ankle in falling in a dark corridor. The compiler of this report was summoned to see him, and found that he had but little trouble with his anchylosed hip-joint, and that his general health was excellent, seven years after the reception of so grave an injury.

CASE 273.—" J. T. Tindell, Co. K, 18th Mississippi Cavalry, aged 19 years, was wounded at Harrisburg, Mississippi, July 15th, 1864, by a musket ball, which extensively comminuted the femur at the upper third, and passed up to the joint. Great injury was done to the soft parts. He was conveyed, by ambulance, some thirty or forty miles to the rail-car, by which he was taken over a hundred miles to the Lauderdale Hospital. Surgeon C. M. B. Maughs, P. A. C. S., who reports this case, among twenty or more others of gunshot fracture of the femur treated conservatively, states that the wounds were carefully examined, and all foreign bodies, including spiculæ of bone, in the track of the ball, were removed. The limb was placed in Smith's anterior splint, or simply a straight splint, or what was preferred, placed in position and retained there by soft cushions or pillows, and all unnecessary probing or handling carefully avoided. He died on September 18th, 1864."

CASE 274.—"Private Philo Wilbur, Co. I, 165th New York Volunteers, aged 19 years, was wounded in the action at the Southside Railroad, Virginia, March 25th, 1865, by a conoidal musket ball, which, entering over the left trochanter major, produced a comminuted fracture of the process, and emerged at the lower and inner border of the left buttock, one and a half inches above the fold of the nates. He was taken, March 29th, to the field hospital of the First Division of the Fifth Corps, and thence to the base hospital at City Point, whence he was sent, by steamer, to Washington, and admitted, on April 2d, into the Mount Pleasant Hospital. The patient's general condition was good. On April 7th, some febrile action was manifested, and there was anorexia. On April 9th, Assistant Surgeon H. Allen, U. S. A., having placed the patient under the influence of a mixture of equal parts of chloroform and ether, cut a flap three by four inches over the trochanter major, and removed several detached fragments of bone. At the time of the operation the foot was turned outwards, the superficial veins of the thigh were enlarged, and the integuments of the middle portion of it were of a yellow and purple color. Subsequent to the operation, he had irritative fever and wakefulness; anorexia continued. He complained of much pain in the limb, and the parts about the wound were inflamed. Cold water dressings were applied, and antiseptic and stimulating injections several times daily. Tonics, stimulants, and anodynes were administered. On April 14th, he had a rigor, and the existence of pyæmia in the system was apparent, by the sweetish odor of the breath and the icteroid tinge of the skin; he had, also, vomiting and diarrhœa. Quinine was given, and camphor and opium pills, as required. On April 16th, the symptoms were typhoidal. An emulsion of turpentine was ordered. On April 18th, a secondary hæmorrhage of from six to eight ounces occurred from branches of the external circumflex artery, and was arrested by temporary compression of the femoral, and by lowering the limb which had been kept raised in order to facilitate the discharge from the wound. He died on April 20th, 1865." The pathological specimen (FIG. 65.) exhibits the upper third of the femur. The great trochanter has been carried away by the projectile, the fracture entering the capsule. There is some erosion of the inferior portion of the head as it joins the neck. Necrosis of the shaft borders the fracture. The history and specimen were contributed by Assistant Surgeon H. Allen, U. S. A.

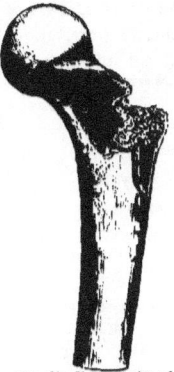

FIG. 65.—Upper portion of femur from which fragments have been removed after a gunshot fracture of the neck.—*Spec.* 3143, Sect. I, A. M. M.

Among the one hundred and twenty-two cases of the first category: cases of alleged gunshot injury of the hip-joint with fracture of the head or neck of the femur, there were eight recoveries. The thirty-seven cases of the second category, complicated by injury to the acetabulum, all proved fatal. The five cases of slight injury to the acetabulum, in which the joint was opened but the femur uninjured, terminated fatally. In the series of twenty-two alleged examples of gunshot wounds of the hip-joint without direct lesion of the osseous articular surfaces, there were seven recoveries. In the category of twelve cases of secondary traumatic arthritis, five patients survived. Among the seventeen cases of gunshot fractures of the trochanters with possible primary injury to the joint, three recoveries are reported. Of the twelve cases of secondary involvement of the hip-joint after fractures in the trochanters, two recovered. A series of seven alleged examples of gunshot injury of the hip-joint all resulted favorably. In a category of twenty-two cases of supposed gunshot injury to the hip-joint, classified apart as not well authenticated, there were eleven recoveries. In Case 239, in which the termination is not recorded, it should have been stated that the patient died, March 2d, 1864. Finally, in a series of eighteen cases of alleged gunshot fracture involving the articulation, and treated by the extraction of fragments, there were six recoveries. The aggregate of two hundred and seventy-four cases furnishes forty-nine examples of recovery. The validity of these results will be discussed further on, and it will appear that, in the great majority of the instances of recovery, the evidence of direct injury to the hip-joint is altogether inadequate. Exclusive of a certain number of cases treated by amputation and excision at the hip, or complicated by grave lesions of the pelvis, and of two cases recorded further on, in a communication from Dr. William Thomson, the foregoing categories comprise all the instances, recorded on the registers of this Office, of gunshot wounds, in which any reasonable evidence is adduced of the existence of injuries implicating the hip-joint.

Coxo-femoral Amputations.—In the surgical report published in *Circular* No. 7, S. G. O., 1867, the histories of fifty-three amputations at the hip-joint performed during the late war were recorded, and notes of one hundred and eight other examples of coxo-femoral amputations for gunshot injury were collected from various sources. This total of one hundred and sixty-one operations was divided as follows:

COXO-FEMORAL AMPUTATIONS FOR GUNSHOT INJURY.	CASES.	DIED.	RECOVERED.	DOUBTFUL.
Primary	72	68	1*	3
Intermediate	62	58	4	
Secondary	19	12	7	
Reamputations	8	4	4	
TOTAL	161	142	16	3

The results recorded as doubtful, refer to the operations by Baron Larrey, Dr. Gilmore, and Dr. Compton, in which the patients are known to have survived three months, two months, and five months, respectively, after which all traces of them are lost.†

Since the publication of that report, the particulars of two additional primary amputations at the hip, performed in the Army of the Potomac, have been ascertained. The first is the much contested case of Private Dadds,‡ which had been variously described as an excision of the head of the femur, an amputation of the thigh, and a comminution of the trochanter submitted to expectant treatment. The original history of the case, by the operator, Surgeon Enos G. Chase, 104th New York Volunteers, has finally been discovered among some detained papers of the Army of the Potomac. The case is of the greatest interest, inasmuch as it supplies an indication for amputation at the hip-joint which has hitherto escaped the attention of authors, viz: comminution of the neck and trochanters of the femur, complicated by a severe compound fracture of the same limb lower down. With the general introduction of arms of precision, the infliction of several wounds upon the same individual or in the same limb has become a much more frequent occurrence than formerly, and is likely to prove, hereafter, one of the complications which may demand the extreme remedy of coxo-femoral disarticulation:

CASE.—"Private John W. Dadds, Co. B, 4th Maryland Volunteers, was wounded on the morning of May 12th, 1864, at the battle of Spottsylvania, by two musket balls, one of which shattered his left tibia and fibula, while the other passed through the left thigh, comminuting the femur at the trochanters. He was carried to the field hospital of the Second Division of the Fifth Army Corps, when a consultation was held by the senior surgeons present. After a careful exploration, under chloroform, it was determined that the thigh wound involved the hip-joint. It was decided that an excision of the head of the femur, or a resort to expectant treatment, were alike forbidden by the grave compound fracture of both bones of the leg, which complicated the thigh fracture, and that there was no alternative but coxo-femoral amputation. Accordingly, stimulants were administered, and every means were employed to bring about reaction, and, three hours after the reception of the injury, Surgeon Enos G.

---

* The noted case of Private James E. Kelly, 56th Pennsylvania Volunteers, amputated by Surgeon E. Shippen, U. S. V., April 29th, 1863, seven hours after the reception of a gunshot fracture of the upper part of the left femur, *Circular* No. 7, S. G. O., 1867, p. 20. Kelly was in good health in the winter of 1868.

† Larrey's patient, a subaltern of dragoons, wounded at the battle of Borodino or Mozaisk, was seen at Orcha, in Russia, three months after the operation, cured; but he did not live to regain his home. (*Mém. de Chir. Mil.* T. IV, p. 50.) Dr. Gilmore's patient, Private Williamson, 13th Mississippi regiment, wounded at Seven Pines, June 4th, 1862, was sent from Richmond to his home, well, towards the end of July; and Dr. Compton's patient, Private Robinson, of a Louisiana regiment, wounded at Battery Pemberton, March 13th, 1863, left the Yazoo City Hospital, well, on April 20th, 1863, and was reported to Dr. J. M. Green, six months subsequently, as in fine health. Persevering inquiries have failed to discover anything of the ulterior histories of Williamson and Robinson. (See *Circular* No. 7, S. G. O., 1867, pp. 24, 26.)

‡ See *Circular* No. 7, S. G. O., 1867, p. 21.

Chase, 104th New York Volunteers, proceeded to remove the limb. He performed the double flap operation, sometimes described as Béclard's, transfixing the thigh and forming a large flap in front, disarticulating and then cutting from within outwards to make a posterior flap. The vessels were then rapidly secured. The patient survived the operation twelve hours."

The particulars of the following case have been obtained from the morning report of October 2, 1864, of the Field Hospital of the Second Division, Second Corps, and from a Second Army Corps register:

CASE.—Private Isaac C. Fulton, Co. I, 4th New York Heavy Artillery, was wounded, on October 1st, 1864, in the trenches before Petersburg, by a fragment of shell, which shattered the upper extremity of his left femur, and lacerated the soft tissues on the outside of the thigh, without, however, implicating any important vessels or nerves. He was immediately carried to the Second Corps field hospital, under the charge of Surgeon F. F. Burmeister, 69th Pennsylvania Volunteers, and a consultation was held, at which it was determined that an amputation at the hip-joint was the only resource that could possibly preserve life. The wounded man was, therefore, immediately placed under the influence of chloroform, and Surgeon J. W. Wishart, 140th Pennsylvania Volunteers, did the operation. The ordinary method by antero-posterior flaps formed by transfixion was employed. The operation was rapidly performed, and but a trifling quantity of blood was lost. The patient survived the shock of the injury and operation but a few hours, and died at City Point, October 1st, 1864.

Brevet Lieutenant Colonel J. H. Janeway, Assistant Surgeon U. S. A., has learned that Dr. Hunter McGuire, Surgeon-in-chief of Ewell's Division, performed a primary amputation at the hip after the engagement at Ball's Bluff (?), upon a Union prisoner of war, with a frightful injury of the upper part of the femur, caused by a large projectile, and that the patient succumbed to the combined shock of the injury and operation, soon after the completion of the latter. It is also known that Dr. B. D. Lay, of Paducah, Kentucky, a surgeon in the Confederate service, performed two unsuccessful intermediate amputations for gunshot injury, besides the primary case he has reported in detail.[1] These operations were performed on Confederate soldiers in one of the Western campaigns.

Dr. B. Rohrer, of Germantown, Pennsylvania, late Surgeon 10th Pennsylvania Reserves, records[2] the following case of intermediate amputation at the hip. It is not mentioned in the returns of the Fifth Army Corps Hospital transmitted to this Office:

CASE.—Frank G———, a private in a Texan regiment, was wounded at the battle of Gettysburg. The wound was in the left thigh, caused by a grapeshot, which entered two inches below the trochanter major, shattered the bone up into the neck, and lacerated the soft parts terribly between the place of entrance and the knee. He remained on the battle-field from the 2d until the 4th day of July, with very little attention, until he was brought to the hospital of the Fifth Army Corps. After a consultation with a number of surgeons, and the conclusion being in favor of amputation at the hip, the patient was placed upon the table, and, when fully under the influence of chloroform, I performed the antero-posterior operation, assisted by Joseph A. Philips, Surgeon General of Pennsylvania, and Henry Grimm, Surgeon 12th Pennsylvania Reserve Volunteer Corps. Surgeon Philips controlled the femoral artery, and not over three ounces of blood were lost. Death followed in thirty-six hours.

In July, 1867, when a report on the hip-joint amputations of the late war was printed by the War Department, about one-third of the sixty thousand monthly regimental reports of sick and wounded accumulated in this Office had received only a cursory preliminary examination. A further search among these reports has discovered memoranda of three more coxo-femoral amputations hitherto unpublished. Two of these were primary cases, in which the operations were, so to speak, compulsory:

CASE.—Private Elisha Wayland, Co. E, 34th Iowa Infantry, had his right thigh almost completely torn off at the hip-joint, at the siege of Vicksburg, July 3d, 1863, by an unexploded shrapnel shell. He was hurriedly conveyed to a field hospital, but bled very profusely during the short transit, and is represented as almost exsanguinous upon reaching the hospital. His regimental surgeon, Charles W. Davis, 34th Iowa Volunteers, having mastered the hæmorrhage, had the patient placed under chloroform, ligated the femoral artery, divided the lacerated soft parts which still connected the mutilated limb with the trunk, and then formed a single large anterior flap, and exarticulated the femur. The patient survived the operation two hours; and died July 3d, 1863.

The next case is one of the very few recorded illustrations of amputations at the hip for gunshot fracture complicated with a wound of the femoral artery:

---

[1] See *Circular* No. 7, S. G. O, 1867, pp. 27, 78.
[2] *American Journal of the Medical Sciences*, January, 1869, N. S., Vol. LVII, p. 285.

CASE.—Corporal Wad Brookins, Co. C, 49th United States Colored Troops,* was accidentally shot, in the regimental camp, near Transylvania Landing, Louisiana, on September 3d, 1863. He was struck in the upper third of the left thigh, at close range, by a musket ball, which divided the femoral artery and badly comminuted the thigh bone. Assistance was immediately rendered, and the copious hæmorrhage was controlled by compression. Prompt preparations having been made, the thigh was amputated at the hip by Surgeon Sylvester Lanning, 49th U. S. C. T. The patient died from the hæmorrhage and shock, four hours after the completion of the operation, September 3d, 1863.

The next case refers to an intermediate operation necessitated by a sudden hæmorrhage:

CASE.—Private C. Hamilton, Co. II, 3d Regiment, U. S. Colored Troops, while employed as a stretcher bearer, in carrying a wounded man from the field, during the assault on Port Hudson, Louisiana, June 14th, 1863, was struck by a musket ball, which passed through the upper part of his left thigh. The missile entered behind, near the gluteal fold, and, having fractured the upper part of the femur badly, passed out in front, in close proximity to the track of the femoral artery. He was taken to his regimental hospital. The limb was shortened and rotated inwards, and great swelling and inflammatory mischief speedily supervened. It was ascertained that fragments of bone were detached and driven into the soft tissues, and there was excessive pain in the limb. It was resolved to make an exploratory incision, and to remove the displaced splinters. Chloroform was administered for this purpose, and the patient was then removed to an operating table. While being moved from his bed to the table, arterial hæmorrhage of an alarming character took place, and was so copious as to threaten a speedy dissolution. The bleeding was controlled by digital compression of the femoral at the crural arch, and a hasty consultation was held by Surgeon E. P. Gray, 3d U. S. C. T., Surgeon Pierce, U. S. C. T., and Assistant Surgeon George P. Percival, 3d U. S. C. T., and it was determined that amputation at the hip-joint afforded the only chance of preserving the man's life. The operation was immediately performed by Surgeon Gray, assisted by his colleagues. He disarticulated the thigh by the ordinary antero-posterior flap method, the operation being quickly accomplished without much hæmorrhage. Although the patient was reduced by the irritation and pain caused by his wound, and prostrated by the sudden profuse hæmorrhage, his constitution was robust, and he bore the operation remarkably well. The wound being approximated by sutures and adhesive strips, he was put to bed and carefully nourished and watched. He expressed great gratitude for the operation, declaring that it had entirely relieved him of his excessive suffering. He rallied and appeared, for forty-eight hours after the operation, to be in a very hopeful condition. Then the vital powers seemed to flag. Thenceforward he sank gradually, and died from exhaustion, June 29th, 1863, four days after the operation.

These nine operations increase the aggregate hip-joint amputations in the American War to sixty-two, of which twenty-four were primary, twenty-two intermediate, and nine secondary operations, and seven reamputations. Forty of the operations were performed in the United States service, and twenty-two in the Confederate service.†

In the report in *Circular* No. 7, S. G. O., 1867, besides the account given of fifty-three hip-joint amputations in the American armies, one hundred and eight such operations are collected from former annals of military surgery; fifty-two of these are reported in sufficient detail to be tabulated.‡ No errors have been discovered in this enumeration;

---

* The regiment was styled at the period "14th Louisiana Regiment of African Descent." Its designation was changed, March 11th, 1864, to "49th United States Colored Troops."

† In the *Cincinnati Lancet and Observer*, Vol. XI, No. 5, p. 257, May, 1868, Dr. John Wright, of Clinton, Illinois, records the history of Private John W. Spradling, Co. A, 33d Illinois Volunteers, who received a gunshot flesh wound in the loins, May 17th, 1863, and subsequently had a bad bed sore over the left trochanter. The femur became diseased, the patient was bedridden for years, and false anchylosis formed in the hip and knee. On February 20th, 1867, Dr. Wright, with the sanction and assistance of Dr. Goodbrake, Dr. Adams, and three Drs. Edmiston, amputated at the hip by Guthrie's oval method. In February, 1868, the patient was reported to be still alive, with fistulous sinuses in the stump. He was still bedridden, the right lower extremity being powerless. I have not cited this case among the secondary coxo-femoral amputations for gunshot injury, because the original wound in the lumbar region appears to have had only a remote connection with the disease for which such an extraordinary remedy was adopted. I was informed, October 18th, 1867, by Dr. Abram Love, of Albany, Georgia, that Dr. Willis Westmoreland, of Atlanta, Georgia, and Dr. S. E. Chaille, of New Orleans, Louisiana, had each performed, during the late war, the operation of amputation at the hip-joint on account of gunshot-injury, and I find in a Confederate Hospital register, that: "Private Eli Loftus, Co. F, 23d North Carolina Infantry, was admitted, May 1st, 1864, to Chimborazo Hospital No. 3, Richmond, Virginia. Diagnosis, amputated hip. Furloughed June 14th, 1864, for 60 days." I have been unable to obtain any particulars to authenticate these cases. The last, in all probability, refers to a recovery from a high amputation in the continuity.

‡ Particulars of a few of the non-tabulated cases have since been collected. After the Alma, the British Director General, Thomas Alexander, amputated at the hip in three cases. Two are included in the tables of *Circular* 7, (p. 56). The third was the case of Peter Cleary, 23d Fusileers, amputated September 21st, 1854, who died on the passage to Scutari, (*Guthrie's Commentaries*, 6th ed., p. 620). The nine amputations at the hip in the Italian War of 1859, (*Circular* 7, p. 15), were: one primary, performed by Bertherand; three intermediate and fatal, by Gherini, at San Fillippo Hospital, Brescia, and Scotti, at the Ospedale Maggiore, at Milan, and by Tassani, at the Ospedale Maggiore; and five successful secondary cases, by Isnard, J. Roux, Arnaud, and Neudörfer, which are particularized in the report in *Circular* 7. (ROCCO GRITTI, *Delle Fratture del Femore*, etc., p. 80).

but a few omissions have been detected.[1] Dr. Stephen Smith has kindly called my attention to one of these, as having, strangely enough, also escaped his own notice in compiling the important statistical article which he published in 1852. He refers to the operation by Dr. D. Quarrier, of H. B. M. ship Leander, who, during an action before Algiers, in August, 1816, amputated primarily at the left-hip joint, in the case of Timothy Sullivan, seaman, whose thigh was cruelly lacerated by a cannon ball, the femur being fractured to its head, and the nerves and blood vessels torn asunder. There was a wound in the breast also, and the right arm was fractured. The patient earnestly besought that an operation might be performed. He expired immediately after its completion.[2] It is stated in *Circular* No. 7, p. 13, on the authority of Drs. Tripler and Porter, that no hip-joint amputations were performed in the Mexican war of 1846; but I have since been informed by several medical officers who served in that campaign, that they were present when Dr. Steiner undertook such an operation for a very grave gunshot injury of the upper extremity of the femur. It was a primary amputation, and the patient died of shock before the dressings were completed. After the insurrection in Lombardy in April, 1848, three intermediate hip-joint amputations were performed at Milan, one by Trezzi, on March 14th, 1848, one by Baroffio, and one by Restelli.[3] The last operation was a successful one. Two fatal intermediate amputations were also performed in Algeria by Gilgencrantz and Bertherand.[4] The first, in the case of a soldier of the 60th Infantry, performed July 14th, 1854, for a comminution of the neck of the femur inflicted June 20th, was fatal in a few hours. The second, of a soldier of the same regiment, performed May 31st, 1857, for a similar injury received May 24th, also resulted fatally from shock soon after the completion of the operation. Adding these two primary and five intermediate operations to the one hundred and eight enumerated in the "Historical Summary" in *Circular* No. 7, 1867, there results the aggregate of one hundred and fifteen operations, of which fifty-five were primary, forty-nine intermediate, ten secondary, and one a reamputation.

Since the close of the American War, a few additional cases of hip-joint amputation for gunshot injury have been recorded. One is reported by Dr. J. Fayrer,[5] of the Bengal Army. It was an intermediate amputation performed thirteen days after the reception of a gunshot fracture of the head and neck of the left femur. The patient died in three hours. He had pyæmic foci in the lungs. A second is recorded by Dr. T. D. Johnson,[6] of San José, California. A patient of Dr. T. H. McDougall, of St. Juan, Monterey county, received, in July, 1862, a wound from a large pistol ball, which entered at the great trochanter and passed down the medullary canal of the femur nearly to the knee. Two days subsequently, he was amputated at the hip. He made a rapid recovery, and was living in 1868, at the New Almaden mine, in Santa Clara county, California.

[1] Dr. J. J. Chisolm correctly remarks (*Richmond Medical Journal*, November, 1867, Vol. IV, p. 452) that the statement at page 22 of *Circular* 7, S. G. O., 1867, that Surgeons Michel and Eve had amputated at the hip for gunshot injury was not made by him. It should have been accredited to Surgeon O. A. White, P. A. C. S.

[2] *Medico-Chirurgical Transactions*, 2d ed., 1820, Vol. VIII, p. 3.

[3] BAROFFIO, *Delle Ferite d'Arma da Fuoco*, p. 286, and GRITTI, *Delle Fratture del Femore*, etc., p. 80.

[4] BERTHERAND, *Campagnes de Kabylie, Histoire Médico-Chirurgicale des Expéditions de* 1854, 1856, 1857. Paris, 1862, pp. 145, 238.

[5] *British Army Medical Department Report for* 1866, London, 1868, Vol. VIII, p. 514, and *Edinburgh Medical Journal*, March, 1868. Dr. Fayrer has had much experience with this operation, having reported three cases, in one of which the operation was performed for gunshot injury, in his *Clinical Surgery in India*. London, 1866, p. 631, *et seq.*

[6] *Pacific Medical and Surgical Journal*, December, 1868. Dr. Henry Gibbons, jr., of San Francisco, informs the compiler that there can be no doubt of the authenticity of this case.

In the Seven Weeks' war in Germany, it is probable that this operation was occasionally performed; but few cases have been published. Dr. J. Heyfelder mentions three.[1] Two of these were early intermediate operations, performed at the Château of Hradeck, on July 4th or 5th, 1866, after the battle of Sadowa. Both terminated fatally within forty-eight hours. The third operation was also an early intermediate one, performed at the hospital at Negolisch or Nederlischt, a Bohemian town, by the surgeon in charge, Dr. Wilde. The operation was practised two days after the battle of Sadowa, or Königgrätz, and was promptly fatal. I am informed by Professor P. Pelechin, of the Medico-Chirurgical Academy of St. Petersburg, who visited the Prussian and Austrian hospitals during the Seven Weeks' war, that he witnessed this operation on several occasions, but I have been unable to obtain reports of these cases.

In January, 1868, Dr. John Ashhurst, jr., performed, at the Episcopal Hospital in Philadelphia, an unsuccessful secondary coxo-femoral amputation for gunshot injury.[2] The case was complicated by a fracture of the pelvis, a lesion which should forbid this operation. The following is an abridgment of Dr. Ashhurst's report:

CASE.—E―― B――, a woman, 22 years of age, was wounded at Tacony, Pennsylvania, by a conoidal musket ball, at short range, in the right hip, August, 1867. The ball entered at the fold of the right buttock, when the woman was in a stooping posture, and shattered the acetabulum and head and neck of the femur, passing out at the groin. The patient was treated on the expectant plan by Dr. W. S. Hendrie until the following December. In January, 1868, large abscesses existing about the hip, discharging through many apertures, the limb being immovable and the seat of severe pain, and the sacrum being occupied by a bed sore, the patient being, moreover, subject to the constant recurrence of an exhausting diarrhœa, and the pulse-rate averaging 150, a consultation of the surgeons of the Episcopal Hospital, at Philadelphia, determined unanimously to recommend ablation of the thigh. On January 14th, 1868, the patient was brought completely under the influence of ether, the circulation was controlled by an aortic tourniquet, and the limb was amputated at the hip by Dr. Ashhurst, jr., by forming, with a short knife, anterior and posterior flaps from without inwards. There was little hæmorrhage, no blood whatever being lost from the general circulation, and the patient reacted satisfactorily; but soon after began to sink, and died three hours and a quarter after the completion of the operation, January 14th, 1868. Every step of the operation had been most carefully planned and executed, and the unsuccessful issue was much deplored. The preparation of the injured femur, the head and greater part of the neck being disintegrated, was placed in the Museum of the Episcopal Hospital. It is represented in the adjacent wood-cut. (FIG. 66.)

FIG. 66.—Upper part of right femur, the epiphysis nearly destroyed from the effects of gunshot fracture.—[After Ashhurst.]

Twenty-two cases of coxo-femoral amputation are thus added to those cited in the report published by this Office in 1867. Nine of these pertain to the surgery of the American War, and increase that list of operations from fifty-three to sixty-two; seven were cases of earlier date, to be added to the one hundred and eight already cited, and six have been recently reported. But two of these operations, those by Restelli and McDougall, resulted successfully. This gives a mortality rate of 90.9, corresponding very closely with the results of cases of amputation at the hip for injury heretofore published, and compels the admission that further experience of this operation only confirms the generally received opinion of its formidable fatality.

Grouping the one hundred and fifteen amputations at the hip for gunshot injury reported from various campaigns prior to 1862, the sixty-two cases of the American War, and the six cases more recently recorded, the results set forth in the following

---

[1] *Rapport sur le Service Sanitaire de l'Armée Prussienne pendant la Guerre de 1866 contre les Saxo-Autrichiens.* Gazette Médical de Paris, 1867, p. 541.

[2] *The American Journal of the Medical Sciences,* January, 1869, N. S., Vol. LVII, p. 94.

table are obtained, and this may be regarded as a summary of the authentic published cases at the present date:

Coxo-Femoral Amputations for Gunshot Injury, January 1, 1869.	Cases.	Primary.	Intermediate.	Secondary.	Reamputations.
In various campaigns	115	55	49	10	1
In the American War	62	24	22	9	7
Later cases	6		5	1	
	183	79	76	20	8

The one hundred and eighty-three operations resulted as follows:

	Cases.	Died.	Recovered.	Doubtful.	Death Rate.
Primary	79	75	1	3	98.68
Intermediate	76	70	6		92.10
Secondary	20	13	7		65.
Reamputations	8	4	4		50.
	183	162	18	3	*90.00

Coxo-femoral amputation for gunshot injury, first performed by Larrey in 1793, has now been practised for three-quarters of a century. The introduction of anæsthetics, the use of the aortic tourniquet, and the familiarity of surgeons with the operative procedure, have possibly slightly diminished the fatality of the operation, but it must ever remain a grave and precarious resource, to which conscientious surgeons will reluctantly resort. These observations apply especially to those cases in which alone the question of direct injury to the hip-joint arises, that is, to the primary and early intermediate amputations. The secondary operations and reamputations are performed for consecutive disease of the upper extremity of the femur or of the joint, are analogous to what Malgaigne terms "pathological" amputations, and have had a fair measure of success.

Excision at the hip-joint for gunshot injury, though first practised by Oppenheim in 1829, could not be regarded as an established surgical resource until the Crimean war of 1855. The published examples of this operation are less than half as numerous as the coxo-femoral amputations; yet the list of successes already comprises, three primary, three intermediate, and one secondary excision. The surgical reader need not be reminded that as late as 1861, MM. Larrey and Legouest were able to affirm that it was impossible to cite a solitary authentic instance of primary coxo-femoral amputation from the annals of military surgery.

The expectant treatment of gunshot injuries involving the hip-joint is, of course, of ancient date; but the authentic instances of recovery under this plan are few indeed. At the beginning of this chapter, I have shown that most of those cited by authors are unre-

---
* The three doubtful cases are omitted in computing the percentages.

liable. I propose to show that the forty-nine instances of recovery from such injuries recorded on the registers of this Office are generally illustrations of mistaken diagnosis or clerical carelessness. Before discussing these, however, I shall introduce a paper on gun-shot injuries of the hip by a surgeon who has investigated the subject with thoughtful and discriminating care, and with a wide field of observation, for, as acting medical inspector of the Army of the Potomac, and surgeon in charge of Douglas Hospital, the author of the paper, Dr. William Thomson, late Brevet Major and Assistant Surgeon U. S. A., was enabled to observe personally, in a large number of cases, the primary or remote results of this class of injuries when treated either by amputation, or excision, or temporization:

"1607 Locust Street, Philadelphia, October 26, 1868.

"Dear Doctor: Your letter of October 14th, requesting me to give you my opinion as to the value of excision at the hip-joint, affords me a coveted opportunity to give my testimony in favor of an operation which I should have done more frequently, had I possessed the experience at the commencement of the war which I obtained at its termination.

"As I recall the cases of gunshot wounds of the hip which I have seen, they may be classed as follows:

"1st. Into those attended by injuries of other vital parts so serious as to prohibit any operation.

"2d. Cases where the joint was clearly wounded without such complications.

"3d. Cases where the diagnosis was uncertain until, under conservative treatment, the exhaustion caused by profuse suppuration rendered any operation hopeless:

"4th. Cases where the joint became secondarily inflamed, and death ensued from the extensive suppuration.

"All of these cases terminated fatally. Two men, of class second, were treated by excision at the Douglas Hospital, one by Dr. Norris, and one by myself; in both the operation was made in the most unfavorable period, about eight days after injury, and upon men prostrated by long transportation and surgical fever; both died. My note-book contains the records of many cases of class third that terminated fatally, some of whom might have been saved by an operation; and this unvarying mortality, under an expectant method of treatment, should indicate the necessity for the most careful examination of wounds near this joint, and their prompt treatment by operative procedure when the diagnosis can be certainly made.

"The brief histories of a few cases will illustrate the difficulties of diagnosis, as well as the natural causes that lead to the fatal result under conservative treatment:

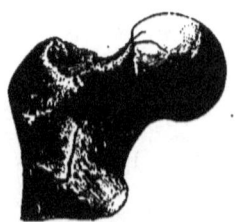

Fig. 67.—Fracture of the neck of the left femur by a conoidal musket ball.—Spec. 4941, Sect. 1, A. M. M.

"Case 275.—'Private Aslac Swanson, Co. H, 3d Michigan Volunteers, was shot at Fair Oaks, Virginia, on May 31st, 1862, and admitted to hospital at Portsmouth, Virginia, on June 4th. The bullet entered above his left trochanter, escaped at the gluteal fold, entered his right foot, and was removed from the sole. He was kneeling when shot, and his hip wound was never explored, nor was I aware of its danger until the period for operation had passed. He had great constitutional disturbance, with marked irritability and peevishness of mind, and death ensued from exhaustion on June 3d, 1862. At the autopsy, abscesses were found in the gluteal region; the joint was disorganized, and a slight abrasion, caused by the bullet, at the neck of the bone, with a fissure extending towards the head of the femur, revealed the primary injury of the joint.' The specimen, represented in the accompanying wood-cut, (Fig. 67), was sent to the Army Medical Museum.

"Case.—'Corporal Gilbert Greenwood, Co. D, 13th Massachusetts Volunteers, was struck on May 3d, 1863, at the battle of Chancellorsville, by a bullet, which entered near the right trochanter. He was admitted to Douglas Hospital on May 8th, when an examination, under an anæsthetic, revealed an abrasion of the upper part of the femur, but did not lead to the discovery of the ball. On May 19th a deep-seated and extensive abscess existed in the gluteal region, attended by great constitutional disturbance, mental irritability, and pain on pressure or motion of the hip-joint. On May 16th, death took place; and at the *post mortem* there was found a comminution of the head of the femur, with penetration of the acetabulum by a bullet, which was removed from the pelvic cavity. Extensive erysipelatous inflammation and diffuse suppuration were found in the gluteal region. The excision of this joint, with its free incision at the primary stage, might have given to this case a different termination.'

"The history of the following case illustrates well the difficulty of a diagnosis in many instances, teaches the symptoms, local and constitutional, of secondary inflammations of the joint, and points to the certainty of fatal results under an expectant treatment, consequent upon the extensive dissecting abscesses in the neighboring soft parts, which are so constantly observed:

"CASE 276.—'Sergeant Edmund Scott, Co. A, 1st New Jersey Cavalry, aged 26 years, on January 15th, 1864, whilst mounted on horseback, was shot accidentally at close range by a Colt's pistol, the ball from which entered the upper part of his right thigh anteriorly. He entered Douglas Hospital on February 1st. He stated that his wound had been pronounced a flesh one, and that he did well until January 25th, when he became feverish, and lost his appetite. There was slight discharge from the wound of entrance, which was not explored, as the wound through the integument did not correspond with the track through the muscles as he lay in bed. He complained of pain in the right hip, but also of pain in the back and left hip. His constitutional disturbance continued to increase, marked by peevishness and depression of mind, and on February 25th, his leg was placed in the position in which it was when wounded, and, after a prolonged examination, a probe was passed to near the small trochanter, where an injury to the femur was detected. The bullet was not found, and the diagnosis still remained obscure. His medical officer reported that he had grown worse; had had a profuse discharge from the wound, and complained of pain in the knee, and on March 13th, an examination revealed distension of the hip-joint with pus, and crepitus on rotation. He was now too much prostrated by his prolonged sufferings and profuse suppuration to bear an excision of the joint, but a free incision was made to evacuate the retained pus. He now grew rapidly weaker, had several severe chills, and died on March 18th, 1864. At the *post mortem* examination, a large abscess occupied the iliac fossa, and a second extensive collection of pus was found, following the course of the large vessels in the thigh. The head of the femur and the acetabulum were denuded of their cartilages, and had an eroded appearance. An opening was found in the femur near the small trochanter, and when the bone was divided by a saw, the bullet was found lodged in the cancellated structure of the neck of the femur.' (See FIG. 68.)

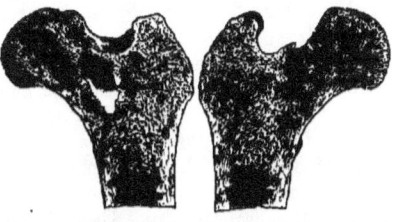

FIG. 68.—Longitudinal section of upper part of right femur, exhibiting a pistol ball impacted in the neck.—*Spec.* 3530, Sect. I, A. M. M.

"From a study of these cases, and others similar in character, it would appear that the fatal terminations, under expectant treatment, are due to the following causes: the joint becomes inflamed primarily or secondarily; the capsule becomes distended by the products of inflammation, gives way, and the contents escape into the neighboring parts, and give rise to those extensive dissecting abscesses which are found at the autopsies, and which account so entirely for the fatal results. If these views should be accepted, a full and free incision into the joint, at an early period, would be the proper surgical procedure, and this is accomplished by its excision. The removal of the head of the bone severs, to that extent, the connection between the body and the lower extremity, prevents that constant disturbance at the joint that follows every motion of the body, and thus places the seat of injury at comparative rest. The division of the bone through its cancellated structure may increase the risk of osteomyelitis with its purulent infection, and experience may yet demonstrate that a full and free incision alone, in the primary stage, may be the best resource of surgery. I must thank you for showing to me one case of recovery without operation, in the person of an officer of rank, whom I saw at the Museum.* His right hip-joint had been traversed by a ball antero-posteriorly, which, at its exit, had caused an extensive laceration of the soft parts. It would seem that this was, in effect, a primary excision, accomplished by the missile, and its good result should encourage surgeons to accept it as a precedent for an early operation.

"With much respect,
"Your obedient servant,

"WM. THOMSON, *M. D.*

"DR. G. A. OTIS, U. S. A."

---

* See *ante* p. 105, CASE 272—Gen. James C. Strong.

The unfavorable impression of the results of temporization in this class of injuries acquired by Dr. Thomson, must become general among surgeons as the evidence on the subject is more closely examined. The results of an analysis of the cases recorded on the registers of this Office are conclusive on this point. Among one hundred and twenty-two cases treated by temporization, and recorded from p. 65 to p. 84 of this report, cases in which the testimony is direct and positive that gunshot wounds of the hip-joint with fractures of the head or neck of the femur had been detected by competent observers, eight examples of recovery are recorded. This gives a mortality of 93.4, a more deplorable result than either amputation or excision present. And yet, notwithstanding the precise statements of the surgeons who communicated the particulars of these cases, one cannot read the histories of the eight successes without entertaining doubts of the accuracy of the diagnoses. In CASES 11 and 16, all the evidence adduced is perfectly compatible with the hypothesis of extra-capsular fracture. As much may be said of CASE 32. In CASE 43, an injury involving the hip-joint was suspected by the Pension Examining Surgeon only, who saw the patient two years after the reception of the wound, and probably gave the patient the benefit of any doubt that he might entertain. In CASE 59, the evidence consists of the affidavit of an unknown surgeon, presented to an "Association for the Relief of Maimed Soldiers," through an executive committee composed of Surgeon General S. P. Moore, C. S. A., Dr. W. A. Carrington, and Dr. J. B. McCaw. It is probable that these gentlemen, supervising a work of charity, received the allegations brought before them in anything but a critical spirit. CASES 79, 93, and 116, were probably examples of fractures of the trochanteric region or base of the neck without the capsule, since no evidence of traumatic arthritis is adduced, and recovery took place without anchylosis. The cases complicated with injury to the acetabulum, from 123 to 164 inclusive, were all fatal. But in the next category of twenty-two cases of supposed gunshot wounds of the joint without lesion of the articular surfaces, seven recoveries are reported: (CASES 166, 169, 176, 177, 178, 185, 186.) In none of these was the escape of synovial fluid, or other unequivocal sign of lesion of the joint, observed. However positive the statements of the reporters, none of these cases can be seriously considered as gunshot injuries directly involving the hip-joint. CASE 178, which resulted in anchylosis, whereupon "the adhesions were successfully broken up," is obviously a pleasant laudation of his professor by a clinical clerk. These seven cases, and the five instances of recovery (CASES 190, 192, 193, 196, 197) in the next series of twelve cases of secondary traumatic arthritis, were probably all illustrations of injuries of the soft parts in the immediate vicinity of the hip-joint, followed by grave inflammatory mischief, by abscesses about the joint, by false anchylosis in some instances, and much loss of motion in all. There is no absolute proof, in any of these twelve recoveries, that the joint was immediately implicated. The specimen from one of these cases was sent to the compiler of this report as conclusive evidence of recovery after the grooving of the head of the femur by a musket ball. A vertical section made through the upper extremity of the femur promptly rebutted this hypothesis. (See FIG. 60, p. 93.) In the category of twenty-nine cases of fracture of the trochanteric region, with primary or secondary involvement of the joint, five cases recovered (CASES 201, 204, 211, 221, 224.) In CASE 201, the field and hospital surgeons disbelieved in any injury of the articulation. It was reserved for the pension examiner to detect this lesion. If CASE 204 were accepted, it would be necessary to believe that a compound intracapsular fracture of the neck of the femur could firmly unite in six months. In CASES 211, 221, and 224, the reports are contradictory, and in the last every sign of fracture was wanting. None of the five will bear examination for a moment as unequivocal examples of recovery from gunshot injuries of the hip-joint. The cases from 228 to 234, inclusive, are of historical interest simply. They illustrate the method of invaliding disabled soldiers in the Confederate service, and the stringency of the conscription regulations enforced in the latter years of the war, but cannot be regarded as examples of recovery from gunshot injuries of the hip-joint. In the series of twenty-two supposed fractures involving the joint, (CASES 235 to 256, inclusive,) there are eleven recoveries. In CASE 235, the patient is represented as completely cured, in three months, of a compound fracture of the head and neck of the femur! If the history of Corporal Bachler, of the Eleventh Corps, (CASE 237,) is accepted, it is necessary to believe that after the battle of Chancellorsville, May 3d, 1863, he lay thirteen days on the plank road unattended, and that notwithstanding this protracted abstinence and privation, a compound intracapsular fracture of the neck of the femur had nearly united on May 25th, and that four months afterwards, August 7th, 1863, the Corporal was in a condition to re-enlist. This is surely an unwarrantable demand upon human credulity. The patient whose history is recorded as CASE 243, was examined by numerous surgeons, and all, with the exception of Surgeon Seymour, U. S. V., regarded his injury as a flesh wound of the hip. This patient recovered in three months. CASES 244, 245, 247, and 248, are clearly instances of mistaken diagnosis. In CASE 253, there is some room for argument, but the weight of evidence is adverse to the supposition that the hip-joint was implicated. CASE 254, in all probability, refers to a flesh wound; and CASE 256 to a fracture of the upper third of the thigh, the histories being misrepresented through clerical inadvertency. The eleven recoveries recorded in this category are interesting in connection with the question of the treatment of gunshot fractures of the upper third of the femur; but must be set aside in considering gunshot injuries of the hip joint.

The series of eighteen instances of gunshot injury of the hip-joint treated by the extraction of fragments presents six successful results: (CASES 260, 261, 262, 264, 267, 272.) After a careful examination of the evidence in these cases, I am convinced that English, Gallagher, Gilliam, and Henzel, (CASES 260, 261, 262, 264,) recovered from fractures of the trochanters without injury to the articulation, and that the free incisions and removal of detached fragments prevented any serious secondary inflammation of the joint. Gallagher had subacute traumatic arthritis, with partial anchylosis. His disabilities were regarded as permanent by the Pension Examiner, in September, 1863. In the case of Gilliam, the breaking of an abscess of the hip was probably mistaken for the escape of synovia, otherwise it would be incredible that the motion of the joint should have been retained. In the case of Lomax, (CASE 267,) it may well be doubted if the lesions were intracapsular.* CASE 272, that of General Strong, is the only one of the forty-nine

---

* Evidently the operator, Dr. Cherbonnier, M. S. K., is doubtful on this point. He writes me as follows: "When Lomax was admitted in the hospital, it was supposed that he was suffering with a fracture of the thigh high up. He was removed to a private residence and his wound was explored, and the extent of the injury was ascertained. It was found to be serious, involving the upper portion of the trochanter, and, if I remember correctly, the neck. In the presence and with the assistance of Surgeon Hewit, U. S. V., and Assistant Surgeon Bill, U. S. A., I removed all the loose fragments of bone, and with the gouging forceps, a considerable portion of rough and dead bone. The operation was repeated a fortnight afterwards. The whole of the trochanter major was removed, going deep into the shaft, and leaving but a shell of bone. When I left Frederick, I had the satisfaction of seeing young Lomax walking on crutches."

recoveries from injuries supposed to implicate the hip-joint, in which there is really a great probability that the articulation was directly injured. In this case, a competent surgeon, Dr. A. J. Berry, declared "that the ball had deeply grooved the head of the femur, and had fractured the upper portion of the rim of the acetabulum." Dr. Berry died several years ago; but General Strong is positive that he made this statement, and, furthermore, exhibited to the bystanders "a detached fragment of the acetabulum an inch or more in length, a part of it covered with articular cartilage." It is possible that these pieces of bone may have been broken from the sacrum; but the fact that the head of the femur was luxated on the dorsum of the ilium eight days after the reception of the wound, renders it highly probable that the acetabulum was fractured. I have frequently examined this officer, and from the close resemblance of the condition of his limb to the results in some forms of morbus coxarius, was long disposed to regard the case as an example of secondary traumatic arthritis. But it is more probable that the early luxation was due to a fracture of the margin of the cotyloid cavity, and that the favorable result achieved after innumerable perils and intense suffering, was due to the free counter-opening made by the ball, and the free incisions employed during the treatment.

I am not satisfied by the evidence in any of the alleged examples of recovery, without operative interference, from gunshot wounds involving the hip-joint, reported by authors or recorded on the registers of this Office. I continue to share the convictions of Guthrie and the elder authors as to the uniform fatality of such injuries when abandoned to the resources of nature. I am fortified in this opinion by the fact that, with the exception of Dr. Winne, (see CASE 11, p. 67) none of the officers of the regular staff have ever observed an unequivocal instance of recovery under such circumstances; by the testimony also of a majority of the surgeons of the volunteer staff, and the concurrent testimony of a large number of those regimental surgeons who, as chiefs of division or of field hospitals, contributed so largely to the credit of American military surgery. Had it been impracticable to offer the conclusive evidence on the subject recorded in the preceding pages, I would have willingly opposed to the assertions of Drs. Pirogoff, Demme, and Gross, the convictions of this great body of experienced men, who have enjoyed unprecedented opportunities of observation, and who, with a solitary exception, aver that they have never seen a recovery from an undoubted gunshot injury of the hip-joint treated on the expectant plan.

DR. J. R. GIBSON'S SUCCESSFUL EXCISION OF THE HEAD OF THE FEMUR.

The preceding portion of this report was already in print, when the following account of a successful secondary excision at the hip for gunshot injury was received from Brevet Lieut. Colonel J. R. Gibson, Assistant Surgeon, U. S. A.:

CASE.—"Private Charles F. Read, Co. I, 37th Infantry, was shot on June 6th, 1863, at three o'clock in the morning, at the picket post of Missouri Bottom, sixty-one miles from Fort Stanton, New Mexico. From the patient's narrative of the affair, and that of one of the non-commissioned officers on duty at the station, it appears that Read left his bed, and, whilst in a sitting posture in the bushes, or in the act of rising, was challenged by a sentinel, twenty-five or thirty yards distant. The challenge not being responded to, after the third repetition, the sentinel fired, the ball striking about the middle of the posterior aspect of the left thigh, and lodged. An express messenger was immediately dispatched to Fort Stanton for surgical assistance. Assistant Surgeon Joseph R. Gibson, U. S. A., reached the picket post about noon on the following day, and found the patient in a tolerably comfortable condition, having suffered much, however, the previous night. On examining the wound, but one orifice was found, the ball supposed to be lodged in the thigh, the patient expressing the belief that he could feel it in the groin. Search was made, and the track of the wound was followed up through the posterior muscles of the thigh for a distance of eight or nine inches. No clue to the location of the ball could be discovered. The bone seemed to have escaped injury entirely; hæmorrhage had been very slight; no pain was complained of other than that in the inguinal region; the limb could be moved, and its appearance and position were natural. In view of the uncertain situation of the ball, and doubt as to the diagnosis, it was decided to transport the patient to Fort Stanton, as his condition seemed to warrant the journey, and there to make a more thorough examination. It was impracticable for Dr. Gibson to be longer absent from Fort Stanton. The patient reached the fort about noon on the 9th, much fatigued, having made the journey without excessive suffering. The wound was again examined, the orifice enlarged, and diligent search was made for the ball. The track of the wound passed through the deep muscles of the thigh, and close up to the pubic bones, and its direction could not be further traced. The opinion formed at the previous examination was confirmed, that is, that the shaft of the femur had escaped injury entirely, the ball apparently taking its course parallel with and approaching to the femoral vessels. It seems beyond doubt that a missile fired at the distance mentioned, and entering the limb direct without impinging first upon another body in its flight, would pass entirely through the limb, or make a wound of exit somewhere, unless the femur obstructed its passage, or the pubic and ischial bones served as a lodgement for it. The first hypothesis being excluded, it might be possible that the ball had lodged in the pubic bones. On June 10th, the patient was tolerably comfortable, but had slept very little through the night. The limb was placed on a double inclined plane, which seemed to afford ease and comfort. Water dressings were applied to the wound. On June 11th, Dr. Gibson found that the patient had passed a restless night, complaining of much soreness in his back and limbs, probably from his constrained position in the wagon in which he was transported, and also from the jolting. There was a tendency to diarrhœa; the pulse was quiet; no fever nor special pain in the region of the wound, which was beginning to slough. The thigh was considerably swollen, and slight discoloration of the skin with fœtor from the wound were observed. Dressings of lint dipped in cold water and a lotion of chloride of zinc were applied, and pills of opium and ipecac were given to check too frequent alvine discharges. On June 12th, the patient was visited, at 7 A. M. He had passed a comparatively comfortable night, but complained of fatigue and soreness. His pulse was about 80 per minute, tongue clean, wound sloughing, with slight inflammation about the edges. On June 13th, at 7 A. M., the patient reported a comfortable night; the wound was suppurating; the discharge was sanious; the edges were slightly inflamed and covered with a grey slough; and there were excoriations of the skin of the buttocks from pressure and drippings from dressings. On June 14th, the patient passed a very comfortable night, pulse rate 75, skin cool and moist, tongue clean, appetite fair. He was ordered tea, toast, and eggs. The wound, which looked angry and inflamed, and was filled with a dark grey slough, was dressed with a solution of carbolic acid reduced one half with water. On June 15th, the patient passed a restless night, could assign no other reason than fatigue and soreness, not complaining of wound, which discharged a thin sanious pus, its edge and surface having a healthier look than on the previous day. Pulse about 70, appetite fair, tongue clean, skin moist, no fever. The dressings with carbolic acid in water, to be renewed by clean pieces of lint three or four times daily, were continued. On June 16th, the diary records that the slough is separating; surface of wound granulating healthily beneath slough; patient in good condition; had slept well the previous night; tongue clean and moist; pulse 70; surface moist; no pain; appetite good. The diet prescribed consisted of eggs, a small piece of steak, and tea and toast. Simple water dressings applied to wound. On June 17th, the patient is reported as restless and sleepless through the night, without any specific reason; wound looks clean, with healthy granulations springing up; no burrowing or bagging of pus. The same dressings and diet were continued. On June 18th, the wound was granulating finely, the discharge was much diminished and healthy, pulse quiet and soft, tongue clean, appetite fair; but the patient was restless and feverish, without equivalent constitutional symptoms. He had passed a sleepless night, and could give no reason. His position was frequently changed in bed. He had no pain, but some soreness in the limb. There was no tendency to burrowing of pus in the thigh, and no clue to the location of the ball. On June 19th, the patient was restless, and complained bitterly of pain, soreness, and inability to move. He had passed a sleepless night, and his nervousness was extreme. The wound was discharging freely, granulating, and cicatrizing. The pain was not referred to the region of the hip-joint, on which great pressure could be borne. The region of the joint was natural in appearance, and motion was full while the patient was under the influence of an anæsthetic. He had a bed sore over the sacrum, but did not complain of it. The appetite was poor, pulse irritable, and the expression of his face was anxious. Half grain doses of sulphate of morphia were ordered to be administered in the night. The patient slept well after the administration of three doses. On June 20th, he reported that he had slept from midnight the night previous until daylight undisturbed. His nervousness was so extreme that anæsthetics were administered in order to dress the wound and change his position in bed. The wound was probed with a Nélaton probe, and its track followed for a distance of eight or nine inches, and there lost. There was no pointing or burrowing of matter to indicate the position of an abscess. The patient was dressed with adhesive straps to the thigh, and water dressings to the wound. June 21st the report states that the patient had passed a quiet and comfortable night under the influence of half a grain of morphine. He complained much of stiffness and soreness. The least approach to the wounded limb, or the slightest handling, produced bitter complaint, partly due to great nervous agitation. Ether was administered and position in bed changed to facilitate free discharge of pus. Expression of countenance anxious and haggard, pulse irritable, disinclination for food. The entry for this day continues: Prognosis grave; still continue expectant treatment; no clue existing to the location of the ball, and appearances not warranting it, further surgical interference was considered improper, gangrenous inflammation having followed incisions when the ball was first sought for. Diet, milk, bread, etc. The diary in the case-book continues as follows: June 22d: Patient more comfortable than the day previous; can be turned in bed with far more ease, and can assist himself; wound rapidly cicatrizing; free discharge from it when he turns on his right side, to favor which he is turned several times through the day; still no indications pointing to the lodgement of the ball. June 23d: Much the same; appetite slightly improved; fed on fresh milk, *ad libitum*, eggs once or twice daily, and occasionally a beefsteak. June 24th: Continues the same as previous day; complains of fatigue; a third of a grain of morphia administered at tattoo, as he complains of wakefulness; has been taking, for a day or two past, six grains of sulphate of quinia in solution with aromatic sulphuric acid, as there is great tendency to prostration, with night sweats. June 25th: Patient comfortable this morning; looks well, and symptoms favorable; visited about 2 P. M.; becoming impatient and nervous; complains of soreness every way. June 26th: Better; appetite improving.

From this time on, the patient gradually gaining strength, bed sore healing, discharge from wound decreasing. July 24th: The patient has been rapidly gaining strength; appetite excellent; no night sweats; the ball still undiscovered, and, at times, severe pain in the thigh and hip complained of; has no power with the wounded limb, no suffering from it when motionless. The patient is lifted into a chair and allowed to sit up for three or four hours daily, appetite good, diet full, with wine occasionally, sleeps well, digestion good. August 1st, general condition good, pulse fair, tongue clean, secretions natural with the exception of a slight diarrhœa, which, though not excessive or prostrating, is readily excited, and from which the patient was suffering at the time of the injury. August 2d: Complained piteously and urgently this afternoon of violent pain in the wounded thigh and region of the hip-joint. He was put under the influence of an anæsthetic, the wound extensively enlarged and dilated, and explored with a Nélaton probe. After much difficulty the probe passed upwards about ten inches, and was then felt to impinge upon a rough surface; it was now feared that this was the head of the bone. A further exploration was not persisted in, on account of the great depression of the patient from the administration of the anæsthetic, of which he had inhaled considerable before the desired impression could be induced. The anæsthetic used was a mixture of chloroform and sulphuric ether in the proportions, by measure, of one to two. August 4th: Patient far more comfortable than yesterday; had passed a good night after 12 P. M. This morning uncomplaining; the limb is in a semi-flexed position, on a double inclined plane, formed of pillows and cushions, a position which was found to be most comfortable to him, some fullness about the thigh, but apparently no deformity or mal-position of the limb. At 2 P. M., the patient was again placed under the influence of anæsthetic for a further examination of the wound. I deemed it best to continue the exploration through the back of the wound, which had been already much dilated, and, inasmuch as the diagnosis was involved in so much doubt, an incision over the great trochanter, which, to place the joint within the reach of exploration, would involve the division of the important muscles and ligaments forming the complement of the joint. I felt confident that the shaft of the bone was not injured, and was disinclined to believe that the joint was completely opened, thinking it quite possible that the ball might have struck on the body of ischium, below the rim of the acetabulum, and partially opened the joint. Certainly the constitutional symptoms never were sufficiently marked to confirm the diagnosis of fracture of the head of the bone. The wound was greatly enlarged, with much trouble from hæmorrhage of minute blood vessels, and explored with the finger; a fracture was felt and the finger passed into an opening in bone, evidently the track of the bullet. This, without doubt, is the head of the bone. No fragments of bone were felt. On motion of the limb at the joint, the bone under the finger followed apparently the complete motion of the femur. No bullet was discovered either by the finger or the Nélaton probe. At this stage of the examination the patient was very much depressed, both from the administration of the anæsthetic, which, as usual, had been inhaled in large quantities before any effect was obtained; also from hæmorrhage, which, though not alarming or excessive, told on the pulse, the patient being rather anæmic. The wound was packed with sponge wet with cold water, and patient allowed to come out from the influence of the anæsthetic; wine was administered him, and in an half hour sulphuric ether was inhaled, an effort was made to find and detach any fragments that might be in the wound, but, with the exception of one or two slivers of bone, none were found. The limb was found to be very little, if any, disproportionate to its fellow. No crepitus was felt on motion. Under the circumstances, I determined to dress the limb as usual, and wait the development of stronger indications for more serious surgical interference. Of course no professional assistance was at hand for consultation, my steward assisting me, and even that assistance I had been without until within a few days. The patient expressed himself averse to any grave operative interference, expressing the opinion that he would recover. A long sponge tent was introduced into the most dependent part of the wound to facilitate full drainage, and the wound dressed with oakum wet with a solution of carbolic acid. Sloughing was feared, as it had invaded the wound after the first exploratory incisions. The bed sore over the sacrum suppurates freely and cicatrizes slowly; another is threatened over the right buttock. These complicate the treatment, and make dressing of the wound tedious and difficult. Air cushions are used as much as possible to relieve pressure. August 7th: The patient has been very comfortable since the examination; suffers but little pain; there is full exit for matter; considerable fœtor from the wound. Dressed this morning with the patient under the influence of anæsthetic, as it was necessary to change his bedding and to examine and dress more thoroughly the bed sore. August 13th: Wound discharging freely; much pain complained of in the hip-joint; continued same treatment; bed sore cicatrizing. August 14th: Complains excessively of the wound and pain; expresses willingness to submit to any operation which will give relief from the intense pain; no appetite at all this morning; has been freely and generously fed, also stimulated with milk punch and wine. At noon he was placed under the influence of anæsthetics, a mixture of chloroform and ether, with the purpose of again freely examining the parts, and such operation performed as considered necessary under the circumstances. The wound was explored with the probe and finger, and the passage of the ball into the head of the bone unmistakeably recognized. A T-shaped incision was made over the joint, the head of the bone turned out of the acetabulum and exscinded through the neck just within the great trochanter. The ball was found lodged in the head of the femur; no fracture through the neck. The incisions closed with one or two metallic sutures, the limb temporarily placed between splints, with a pillow under the knee. A Smith's anterior splint was not on hand, nor could the material, at the post, be obtained for making one. A long external splint, made in two parts, connected by iron braces, was devised and put in course of construction. That the ball did not fracture the neck of the bone is remarkable, especially in view of the short distance at which the missile was fired, 75 or 100 feet, as stated by the sentinel. Exsection was deemed the best under the circumstances, giving the patient a better chance of recovery than amputation at the joint. August 14th: Had passed a wakeful but quiet night. Disinclination for food, with much irritability of stomach, the effects of etherization the day previous; the pulse sharp, quick, and irritable; much thirst, milk the only nourishment that he could retain or had any desire for. August 15th: Appearances much brighter; complains of weariness and soreness in the muscles; fatigued by his long confinement on his back; much discharge from original wound, which is dressed with great difficulty, as also bed sore. Partook rather freely of milk in forenoon, and ate with much relish in the afternoon some chicken, tea, and toast. August 16th: Doing well. August 17th: Patient comfortable, free from pain, but annoyed by an irritable diarrhœa; from attacks of which he has suffered since the receipt of wound. This involved frequent disturbance of the limb and moving of the body. Through the night had frequent discharges from the bowels. August 18th: Patient weak, with irritable pulse; slept but little through the night; wound dressed with carbolic acid lotion. August 19th: Patient weak, diarrhœa somewhat checked. The frequent movings and changes in bed render it impossible to keep the limb fixed and immovable; the long external splint was found impracticable and uncomfortable; it was removed—pulley and weight substituted. Much distortion of the limb is feared. It is impossible to restrain the patient from shifting the position of his pelvis, to relieve his back and bed sore. The original wound is foul, and suppurating freely; the dressings are changed frequently, and wet with carbolic acid lotion. A perforated stretcher is placed between the body and the mattress, upon which the man is raised from the bed, to use the night pan, also to dress lower wound and bed sore. From the constant motion, the limb is frequently displaced, sand bags are applied the length of the limb, and the foot fixed by a bandage around the ankle and tied to the foot of the bed. September 3d: The patient continued slowly gaining strength from the time of last notes; occasional attacks of diarrhœa occurred, which prostrated him for a few days; notwithstanding this, his general condition is much improved. Appetite excellent, sleeps well, and is very cheerful. Carefully dieted. Beefsteak, egg, chicken, and mutton alternately, constituted the nourishing part of the diet, with farina and rice pudding, preserved peaches, cocoa and tea the lighter part. A sulphate of iron pill, with aromatic sulphuric acid, thrice daily, is the only medication. September 11th: Continues to improve in strength, wound healing rapidly, but a tendency to burrowing of pus. Openings enlarged, and upper part of thigh strapped with broad pieces of adhesive plaster, to prevent burrowing, and evacuate pent up pus. October 1st: Patient gaining strength rapidly; secretions good, appetite excellent, wounds and bed sore thoroughly cicatrized; has some motion at hip-joint, exercised through his own efforts; pulley and weights still attached, but are removed daily for an hour or two. October 4th: Weights entirely removed. November 13th: Allowed to get up from bed with limb suspended with bandage slung around the neck; walks around the ward on crutches. November 20th: Permitted to walk outside about the hospital buildings, doing finely, good motion at hip-joint, some anchylosis at knee, which is gradually yielding to forced flexion and extension. December 12th: Patient doing finely up to date; last evening taken with a violent headache; to-day high fever temperature 103.°8. fever lasted until 2 or 3 P. M., and followed by profuse sweating, skin surrounding line of cicatrix of the operation inflamed and pungent to the touch. Erysipelas or pyæmia apprehended. December 14th: Headache passed away, temperature decreased, redness of the skin fading, complains of a slight pain deep in the thigh about the femoral region; this condition continued, with fluctuations of temperature from 99 to 103 for several days, part of the time with cool skin and no headache; at others, high fever, violent headache, and occasional sweating, until the 19th, when a swelling made its appearance between the groin and buttock, rather tense, and with but little fluctuation. No pain in the limb, skin in the vicinity of cicatrix has resumed its normal color; poultices of flaxseed applied to the swelling; patient kept quietly in bed, limb extended. December 21st: Skin over tumor or swelling ruptured, and gave vent to a free discharge of pus, probably about ten ounces; opening is in line of cicatrix of original wound. December 22d: Patient quiet and comfortable, tongue clean, temperature 95° for several days; has been put upon treatment by Quin. Sulph. and Sodæ Hyposulphas gr. viii three times daily. December 24th: Slight discharge from abscess, patient comfortable and in good condition constitutionally; tongue clean, secretions natural. Complains of pain or soreness in the limb when handled; dressing of linseed oil and carbolic acid applied to wound; treatment by sulphate of quinia and hyposulphate of soda continued constitutionally." From this time forward the patient steadily regained health and strength, and his limb acquired power and usefulness. The stiffness of the knee was overcome, and the shortening did not exceed an inch and three quarters. Early in 1869, the patient was discharged. The specimen, presented by Assistant Surgeon Gibson to the Army Medical Museum. is represented by the wood-cut. (FIG. 60.) The patient still used crutches; but, with a shoe made to compensate for the shortening, Dr. Gibson believed that he could walk very well with a cane.

FIG. 60.—Excised carious head of left femur with an impacted musket ball.—*Spec. 5576*, Sect. 1, A. M. M.

A memorandum of yet another case of excision at the hip for gunshot injury has been discovered since this report was in press. It was a primary operation, and would have been included, on account of the insufficiency of the details, among the unclassified cases on pp. 54, 58 *ante*, had the paper been received in season. The memorandum was found on the monthly report for September, 1864, of the Jackson C. S. A. General Hospital, at Richmond. It states that: "Private J. W. Epton, Co. I, 5th South Carolina regiment, having received a gunshot wound of the hip, was admitted to the third division of Jackson Hospital, underwent a primary excision of the head and neck of the femur, and died September 2, 1864." This report is signed by Surgeon J. G. Cabell. On the register it is stated that the wound was inflicted August 16 [probably at Deep Bottom, Virginia], by a conoidal musket ball, upon the right hip.

CONCLUDING OBSERVATIONS.

In the preceding sections of this report, a large amount of evidence is set forth in regard to the three plans of treatment of gunshot injuries of the hip-joint. If the question as to the most eligible treatment was susceptible of a purely arithmetical solution, it might readily be computed that in eighty-five cases of excision, (pp. 61 and 117,) the mortality was 90.6 per cent; that in one hundred and eighty-three amputations, it was 90 per cent., while one hundred and twenty-two cases, treated on the expectant plan, gave a fatality of 93.4 per cent., and concluded dogmatically that operative interference was always indicated, and that amputation was preferable to excision. But the variety of the conditions under which these patients were placed, the diversity in the extent of their injuries, and the inevitable imperfection of all surgical records, forbid any such rigorous comparison. In order to attain just conclusions, it is necessary to analyze the different categories of injuries, to weigh the opinions of careful and candid observers, and to avoid an undue reverence for naked numerical returns.

Of the coxo-femoral amputations, it is highly probable that all of the successful cases have been recorded, while it is almost certain that some fatal cases remain unpublished. Hence the results of amputation, in a statistical point of view, are too favorably represented. Yet it is to be borne in mind that the worst cases were subjected to amputation, and that this category includes many instances in which the extent of the injuries demanded instant interference, and precluded the consideration even of excision or temporization. Such were the conditions on the occasion of the felicitous operation by Dr. Compton, in several of Baron Larrey's cases, and in those of Drs. Jewett, Ladd, Chase, Lanning, and Yandell. There is an obvious unfairness in comparing the secondary hip-joint amputations and reamputations, in which the original injury was inflicted low down in the thigh, with cases of direct lesion of the hip-joint treated by excision or by the expectant plan. It should be remembered, also, that a certain number of the amputations were performed simply to mitigate suffering or as the readiest mode of arresting hæmorrhage in cases in which there was no hope of a successful issue under any treatment.

Among the excisions, also, there were many cases in which a fatal issue was inevitable on account of the nature of the original injury. Not less than nine of the sixty-three American cases were complicated with such lesions of the pelvic walls or viscera as rendered any operative interference useless. In other instances, excision at the hip was performed when the femur was fractured below the trochanters, and the mortality-rate of the operation was thus augmented by cases which might have recovered under expectant treatment. And, above all, a large proportion of the subjects of excision were denied, in the after treatment, the rest and immobility of the injured limb so essential to the success of the operation.

The statistics of cases treated by the expectant method, bear the burthen, in the first place, of doubt regarding the diagnosis; secondly, of including the fatal instances in which discriminating and courageous surgeons would possibly have secured fortunate results by operative interference; and lastly, of embracing some complicated cases necessarily fatal because of injuries in some other parts of the body.

Ample evidence has been presented, in the preceding chapter of this report, to prove that there is scarcely a recorded example of recovery under expectant treatment from a gunshot wound involving the hip-joint, where legitimate doubts of the accuracy of the diagnosis may not be entertained. In the great majority of the thirty-six instances cited by authors, and of the forty-nine quoted from the registers of this Office, of successful results under such circumstances, it is found, where materials for analysis are available, that the injuries were, in reality, extra-capsular. With the possible exception of CASE 272, there is perhaps not an instance in which the evidence of direct injury to the joint is absolutely conclusive. CASE 191 was so regarded, but an examination of the specimen completely disproved this supposition. When the old discussions on the union of simple intra-capsular fractures of the neck of the femur are remembered, it is hardly exacting too much of the advocates of expectant treatment in compound intra-capsular fractures to require the production of an anatomical demonstration of repair. The researches of Dr. G. K. Smith,* on the great variability in the line of insertion of the capsular ligament, especially on the posterior surface of the neck, explain the liability to error in this class of cases, even after the most careful anatomical exploration. Practically, the differential diagnosis between a gunshot fracture involving the trochanter or base of the neck, and one complicated by fissures extending within the joint is frequently out of the question, unless a free exploratory incision is made.

Assuming that the successes claimed for the expectant treatment of gunshot injuries involving the hip-joint are illusory, and that such accidents are generally fatal under any treatment, it remains to inquire what experience teaches of the measures whereby the largest proportion of patients may be saved, and of the conduct to be pursued in cases in which the diagnosis is doubtful.

A sufficient number of instances have been collected to permit precise rules on this subject to be established:

Amputation at the hip-joint for gunshot injury, notwithstanding its great fatality, cannot be altogether discarded, and should be performed under the following circumstances: 1. When the thigh is torn off, or the upper extremity of the femur comminuted with great laceration of the soft parts, in such proximity to the trunk that amputation in the continuity is impracticable; 2. When a fracture of the head, neck, or trochanters of the femur is complicated with a wound of the femoral vessels; 3. When a gunshot fracture involving the hip-joint is complicated by a severe compound fracture of the limb lower down, or by a wound of the knee-joint.

There are two other possible contingencies under which primary or early intermediate coxo-femoral amputations for injury may be admissible: 1. When, without fracture, a ball divides the femoral artery and vein near the crural arch; 2. When a gunshot fracture in the trochanteric region is complicated by such extensive longitudinal fissuring as to preclude excision. Experience has yet determined nothing on these points. Secondary

---

* *The Relation of the Insertion of the Capsule of the Hip-Joint to Intra-capsular Fractures of the Neck of the Femur.* American Medical Times, Vol. III, p. 389, and *Med. and Surg. Reporter*, Vol. VII, pp. 244, 270, 381, 416, 508, 536, 577, 605.

amputations and reamputations at the hip in military surgery should be performed, when, from caries, or necrosis, or chronic osteomyelitis following gunshot wounds or amputations in the continuity, the patient's life is in jeopardy.

Restricted to the classes of cases above enumerated, coxo-femoral amputation will occasionally save lives that would otherwise be inevitably lost.

Primary excisions of the head or upper extremity of the femur should be performed in all uncomplicated cases of gunshot fracture of the head or neck. Intermediate excisions are indicated in similar cases where the diagnosis is not made out till late, and also in cases of gunshot fracture of the trochanters with consecutive arthritis. Secondary excisions are demanded by caries of the head of the femur or secondary involvement of the joint, resulting from fractures in the trochanteric region or wounds of the soft parts in the immediate vicinity of the joint.

Expectant treatment is to be condemned in all cases in which the diagnosis of direct injury to the articulation can be clearly established.

Although the great majority of cases complicated by lesions of the pelvis terminate fatally, the successful operation by Dr. Schönborn (*ante* p. 60) proves that a slight injury of the margin of the acetabulum does not contra-indicate the operation of excision.

Experience teaches that considerable portions of the shaft may be with propriety removed with the head, neck, and trochanters, in cases in which splintering extends below the trochanter minor. I have shown that the limitations respecting excisions of the humerus laid down by the older writers on military surgery were unfounded.* Their rules regarding excisions of the upper part of the femur are equally untenable.

In cases of gunshot fractures of the trochanters, or wounds in the vicinity of the hip-joint, where the diagnosis is uncertain, the surgeon should make such incisions as will permit a satisfactory exploration. If the head or neck of the femur are found to be injured, he should resort to excison; if the articulation is not involved, the incisions will be useful in providing for the escape of the products of inflammation, or in permitting the extraction of foreign bodies or bone-fragments, and thus tending to arrest secondary traumatic arthritis. The expediency of such practice is illustrated by the eighteen cases recorded on pp. 102, 106, (CASES 257 to 274 inclusive,) one-third of which resulted successfully. The necessity of accurate diagnosis cannot be insisted upon too strongly, since there can be but little doubt that gunshot fractures of the upper third of the femur not involving the articulation, are best treated by conservative measures.†

The utility of excision in gunshot injuries of the hip in diminishing suffering, even when unsuccessful in preserving or prolonging life, is insisted on by many who have practiced the operation, and is forcibly illustrated in the following communication from Dr. Billings:

"WASHINGTON, 1869.

"DEAR DOCTOR: In accordance with your request, I offer the following memorandum of my opinions on the subject of excision of the hip-joint:

---

* See *Surgical Report* in *Circular* No. 6, S. G. O., 1865, p. 56.
† Not that the expectant plan is so successful, but because amputation high up is so fatal. It is probable, however, that expectant treatment succeeds better in gunshot fractures of the upper than in those of the middle third. In the large series of photographs I have collected of patients who have recovered from gunshot fractures of the femur, the representations of united fractures of the upper third are next to those of the lower third in number, the consolidations in the middle third being the least numerous.

"The operation is an easy one, can be performed, if necessary, without skilled assistants, and should, I think, be preferred to amputation in all cases in which the injury is confined to the head and neck of the femur. Operating, as I did, upon men whose vital force had been diminished by scorbutus and malaria, and exhausted by transfer from a distance, I had little hope of successful results. But in what I did expect to accomplish I was not disappointed, namely: relieving my patients from much suffering, and that without shortening life. On the contrary, in two cases, I think, actually prolonging it. This consideration of the comfort of the patient should have its weight in cases in which it is doubtful whether the ball has entered the pelvis, a question which it is sometimes impossible to determine previous to operating. I have seen one case of gunshot injury to the hip in which there was little or no pain, the man being in that peculiar state of shock described by McClellan. In this case I did not think it best to operate. I do not think there would be much difference of opinion by surgeons in any given case as to the propriety of excision of the hip-joint, or as to the method of performing it. It is in regard to the mode of after treatment to be adopted that doubt would arise, and it is precisely here that the greatest skill and judgment are needed, and that there is room for improvement. I am not satisfied with the methods which I tried in my own cases, and had I to repeat the operation, I would try Dr. Buck's weight-extension apparatus with a rubber tube perineal band. Or I would try an Amesbury's splint with elastic extension in the direction of the axis of the thigh. And I would use my opiates hypodermically instead of by the mouth, in order to disturb the alimentary canal as little as possible.

"Very respectfully yours,

"(Signed) J. S. BILLINGS,
"*Assistant Surgeon, U. S. A.*

"Dr. GEORGE A. OTIS, U. S. A."

In the fatal cases detailed in this report of alleged injury of the hip-joint treated on the expectant plan, the average duration of life was a little over thirty days. In the fifty-eight fatal cases of excision during our late war, the average duration of life was twenty-three days from the reception of the injury, and less than thirteen days from the date of the operation. Making every allowance for the mistaken diagnoses in the first category, and the promptly fatal complicated cases among the excisions, it must be admitted, according to the evidence at present accessible, that the operation of excision, if unsuccessful, while it mitigates suffering, will abbreviate the duration of life, although its shock is incomparably less than that of amputation.* The large experience acquired of excisions for coxalgia proves that decapitation of the femur is not, in itself, and in patients inured to suffering, a very dangerous operation. But in patients laboring under the reaction consequent upon a severe wound and the concussion and shock both moral and physical incident to being stricken down in battle, the operation is unquestionably often followed by great depression of the vital powers, and by speedy death.

It has been intimated, at page 20, that the small measure of success obtained in the primary excisions was less discouraging than a simple numerical summary would indicate. Of seven operations prior to the American War, Mr. O'Leary's case completely recovered; Oppenheim's patient, who was doing well until the seventeenth day, died, it was believed, in consequence of the invasion of the hospital by the Plague; Seutin's patient died with gangrene of the entire limb, the comminution of the femur and laceration of the soft parts having been so great that the majority of the surgeons in attendance considered amputa-

---

* The assertion of Dr. S. W. Gross, (*Loc. cit.*, p. 450,) that: "the shock of either operation scarcely deserves mention, provided chloroform be used," is so opposed to common observation as to require no refutation.

tion the appropriate treatment; Mr. Blenkins's patient did well for five weeks, when suppuration of the knee-joint supervened. Dr. Crerar ascribed to debility, induced by insufficient rations, the fatal termination, after fifteen days, of his case. The operation apparently hastened the issue in but one of these fatal cases, that of Dr. Baum. Among the thirty fatal cases of the thirty-two primary excisions of the American War, there were not less than six instances (CASES III, IV, V, XI, XXVII, XXVIII) of such grave injury to the pelvis that the operations would not have been undertaken had the extent of the lesions been previously ascertained. In a seventh case, Dr. Blackman's, (CASE I,) the patient was exhausted by hæmorrhage prior to the operation. Surgeon Ladd's patient (CASE XXII) survived sixty days, and is believed to have succumbed for want of suitable nourishment. Of Dr. Cullen's case, (CASE XII,) it is reported that there was profuse hæmorrhage in consequence of the division of the gluteal artery during the operation. After the operations by Surgeons Reamer (CASE XXVI) and Fuller, (CASE XXXII,) the patients having survived fifteen and twenty-eight days respectively, the happiest prospects of success were dispelled by the frequent removals of the patients, necessitated by the military regulations in force. In Surgeon McGuire's second operation, (CASE XV,) the patient survived seventeen days in spite of removal under unfavorable circumstances. Dr. Carey's patient (CASE VIII) was depressed by primary hæmorrhage, and was transported hundreds of miles, yet lived ten days. Dr. Hewit was convinced that his patient (CASE X) would have recovered under less unfavorable circumstances. The removal of the patients to a distance immediately after the operation, was an important element in the fatal results in the cases of Drs. Grant and Clark, (XXIII and XXXI,) and, in Dr. Gibson's two cases, (XVII and XIX,) the prolonged transportation to which the patients were subjected prior to the operation had a very depressing effect. Both Union and Confederate surgeons were ill supplied with immovable apparatus for patients with fractures of the lower extremities who were to be subjected to transportation to a distance. Dressings with plaster, or starch, or dextrine, were probably too much neglected, and none of the various modifications of straight, angular, or suspensory splints secured complete immobility during the transportation of patients over the rough country they were compelled to traverse. In fractures or excisions of the upper extremity of the femur, no apparatus or dressing yet devised ensures entire rest under such circumstances. Although Dr. Dement's patient (CASE XIV) survived a ride of twenty-five miles in an ambulance wagon, and a subsequent journey of seventy miles by rail, and ultimately recovered, many others, who enjoyed the fairest prospects of recovery, succumbed under such treatment. These examples teach that rest and immobility of the limb are indispensable after primary excisions at the hip. The hope may be entertained that in future wars, through international conventions, such safeguards of field hospitals may be established as to preclude the necessity of their hasty evacuation, and to ensure their inmates undisturbed quiet and appropriate after-treatment. Of the thirty-nine subjects of primary excision, but eighteen were placed under favorable conditions for the operation, and there were three recoveries—a result far from discouraging.

The results of the intermediate excisions also are more encouraging than would appear from a numerical estimate. Thirty-three of these operations are recorded. Three prior to 1861, performed by Schwartz, Macleod, and Coombe resulted fatally. In the first, the ischium was badly fractured, and purulent deposits formed in the ankle and shoulder joints. Dr. Macleod's patient progressed favorably till the fifth day, when the patient was suddenly seized with cholera, which was prevalent in the camp, and died the following day. Dr. Coombe's patient died exhausted by suppuration in a fortnight. Of the twenty-two intermediate excisions during the American war, two of the patients recovered, and recovered with really useful limbs. In CASE XLI, the patient survived two and a half months, and the wound had nearly healed, but in the intense heat of midsummer in New Orleans, his strength gave way. In CASE XXXVI also, the wound had nearly healed, and everything was progressing favorably, when, from some error in diet, a colliquative diarrhœa set in, and death followed on the twentieth day. CASES XXXVII, L, and LI were complicated by fractures of the pelvis, and afforded little hope of a favorable issue, and in the remaining fifteen cases the patients were subjected, in every instance, to distant transportation, either prior or subsequently to the operation. In the eight intermediate excisions reported since the American war, there was one recovery, the more remarkable because there was a slight injury of the acetabulum. Dr. Perin's patient was an habitual drunkard, and Dr. Stromeyer's a very debilitated subject. When the unfavorable constitutional condition of the patients, their privations and distant transportation are considered, and the fatal complications in three instances, the result of three successes and two partial successes in thirty-three operations, is anything but discouraging.

The thirteen secondary excisions include two performed prior to 1861, nine performed during the American war, and two of later date, with two recoveries in the thirteen cases. In the two earlier cases, disease of the acetabulum had supervened, and the prospect of recovery would have been better had not the operations been deferred so long. Three of the nine secondary excisions of the American war were delayed until caries of the cotyloid cavity had attained a fatal progress. In CASE LVIII, and Dr. von Langenbeck's case, the acetabulum was badly fractured.

When all the facts adduced are taken into consideration, it must be admitted that the eight recoveries in eighty-five excisions at the hip after gunshot injury constitute an encouraging result, and are examples of the saving of life in an almost hopeless class of cases.

The evidence on the subject has been so fully presented, that it is hardly necessary to quote the opinions of authors. It has been already shown that the teachings of Guthrie, Ballingall, Syme, Textor, and Jæger have been vindicated by the practical experience of modern military surgeons who have enjoyed the largest opportunities for observation. In Great Britain, Mr. Blenkins and Dr. Macleod; in France, MM. Larrey and Legouest; in Italy, Drs. Porta and Gritti; in Germany, Drs. Stromeyer, Langenbeck, Klopsch and Heyfelder; in this country, Drs. Hewit, Clements, Thomson, Billings, and McGuire[*] advocate excision in those cases of injury of the femur involving the hip-joint, in which amputation is not imperatively demanded by the extent of the lesions.

It may seem superfluous to accumulate testimony; but the following notes of European military surgeons are not devoid of interest:

It is well known that, in the Crimea, all the hip-joint amputations terminated fatally; that one of six excisions of the head of the femur was successful; that all cases treated on the expectant plan resulted fatally, unless the vague statements of Pirogoff are accepted.

---

[*] A host of other authorities might be cited, if necessary, to controvert the assertion of Dr. H Demme that: "the leading military surgeons of the present day prefer expectant treatment in gunshot fractures of the neck of the femur."

In the war in Schleswig Holstein in 1848, 1849, and 1850, Dr. L. Stromeyer reports thirteen examples of gunshot fracture of the hip. "In seven, amputation at the hip was performed, with one recovery (Langenbeck). Excision was performed without success in one case (Schwartz) and an expectant treatment was adopted in five cases, all of which terminated fatally."*

After Langensalza, in 1866, Dr. L. Stromeyer observed four fractures implicating the hip, in one of which excision was unsuccessfully performed, while three were submitted to expectant treatment with fatal results.†

Dr. Bernhard Beck records as follows his observations of gunshot injuries involving the hip-joint during the Seven Weeks' War:‡

"I have received the very interesting preparations from two cases, in which death occurred from pyæmia, of intracapsular fracture of the neck with fissure of the head of the femur, where the balls had entered the hip and passed inwards, and where all appearances of an intracapsular fracture being wanting, and fracture of either the head or neck of the bone was not suspected. In one case the ball had, without injuring the edge of the acetabulum, entered the neck of the femur and split off the head transversely. The joint was filled with fluid, the patient having rapidly succumbed to septicæmia. I happened to be present at the dissection and was so fortunate as to obtain the specimen. In the other case the report of the post mortem was as follows: 'Comminution of the neck of the left femur and the acetabulum, a thin bridge of bone remaining at the upper half of the neck still connecting it with the trochanter major. Great burrowing of pus in the hip-joint and vicinity. The muscles extensively infiltrated. Equal infiltration at the ileo-psoas up to its origin, without involving the peritonæum. The missile was imbedded in the posterior part of the thigh. Pyæmic abscesses of the size of a pea in the left lung and about three pounds of bloody serous exudation in the corresponding pleural sac. Death supervened four weeks after the injury.' These two were the only cases in which I could assure myself of the existence of the local injury of the hip-joint. In all other cases the main injury was extra-capsular, although involving the base of the neck and the great trochanter, and the joint became only involved secondarily. In six cases the projectile penetrated directly in the direction of the great trochanter, once passing from the anterior superior spinous process of the ilium into the upper portion of the great trochanter, comminuting the neck of the femur and the trochanters. In four cases pyæmia soon resulted fatally; in the fifth, I performed intermediate resection of the hip-joint (the patient having been placed under my care at a late period) unsuccessfully; in the sixth, disarticulation was resorted to, soon resulting fatally; in the seventh case, I recommended secondary resection six weeks after the injury, there being every appearance of traumatic phthisis, and no union of the displaced bony fragments being expected. Whether the operation was performed after my departure or not, I am unable to state.

"According to my opinion, too much reliance had been placed upon conservative treatment in these severe cases, and neither on the field nor in the hospital had the extent of the injuries been comprehended at first. My view, expressed sixteen years ago, I have found again confirmed, viz.: that conservative treatment results fatally, rare cases excepted; that success can only be expected from a timely operation, and that early resection is proper in these cases."

In his comments on the surgery of the Prusso-Austrian War, Dr. B. von Langenbeck earnestly protests against the abandonment of amputations and resections at the hip, and,

---

* STROMEYER. *Maximen der Kriegsheilkunst.* Hannover, 1861.
† STROMEYER. *Erfahrungen über Schusswunden im Jahre 1866, als Nachtrag zu den Maximen der Kriegsheilkunst.* Hannover, 1867.
‡ BERNHARD BECK, *Kriegs-chirurgische Erfahrungen während des Feldzuges 1866 in Sud-deutschland.* Freiburg, 1866.

in referring to the statistics published in the surgical report in *Circular* No. 6, S. G. O., 1865—incomplete returns, representing these operations in a less favorable light than that in which they now appear—he expresses himself as follows:

"Before the imposing proportions of the American statistics, the fear may arise that amputation and resection at the hip-joint may be banished from the practice of military surgery. But I consider this as giving up all gunshot wounds of the hip-joint, with the rarest exceptions possibly, to certain death. For however great the progress of which conservative surgery is still capable, by far the greater number of gunshot fractures involving the hip-joint will prove fatal if left to themselves."*

METHOD OF OPERATING AND AFTER-TREATMENT.

In order to excise the head or upper portion of the femur, the hip-joint may be exposed by straight or slightly curved incisions, on the outer side of the hip, or by raising a flap. In military practice, the location of the external wound or wounds caused by the projectile influences the choice of the mode of operating. The following plans have been proposed and practised by different surgeons:

LONGITUDINAL INCISION.—This is the method of approaching the joint originally suggested by Charles White (*ante*, p. 8). It was followed by Vermandois and Petit Radel in their experimental researches. It was adopted by Oppenheim in the first excision of the head of the femur for gunshot injury, who remarked that this incision was "the simplest, the least dangerous, and the quickest to heal." It is recommended by Mr. Syme and Mr. Miller, by Malgaigne, by Ravoth, Ried, Hedenus, and B. von Langenbeck. It was the method adopted in at least forty of the eighty-five recorded excisions at the hip for gunshot injury, including six of the successful cases.†

CURVILINEAR INCISION.—Many surgeons are of opinion that the subsequent steps of the operation are more readily performed when the first incision is curved, so as to include the great trochanter in the concavity, which may be directed either backwards or forwards,) (. By making the incision with its convexity backwards, the muscular insertions in the digital fossa are reached more promptly, and the escape of pus is better provided for. With the incision anterior to the trochanter, it is convenient to attack the capsular ligament in front, and disarticulation is easier. M. Chassaignac advises the curved incision back of the trochanter, and Sir William Fergusson recommends the same method. Mr. Costello has found this to afford the readiest mode of exposing the joint in the dead subject. Jæger and Textor preferred a curved incision in front of the femur, with its convexity backwards. In the late war Drs. Gross, Ladd, and Norris adopted this latter modification, while Drs. Hewit, Clements, Ingram, Gilmore, Bontecou, Billings, and Bliss, preferred a slightly curved incision behind the trochanter, with the concavity forwards. This was the method also adopted by Dr. Schönborn in his successful intermediate operation after Sadowa.‡

---

* *Ueber die Schussfracturen der Gelenke und ihre Behandlung*, Berlin, 1868, S. 15.

† *Phil. Trans.* 1769, Vol. LIX, p. 45; *Encyclop. Meth.* T. I. Art. Chirurgie; Oppenheimer, *U. d. Res. d. Hüftgelenkes*, S. 26; Syme's *Principles*, &c. Phila. 1866, p. 695; Miller, *System of Surgery*, 1864, p. 1302; Malgaigne, *Man. de Méd. Op.* 7° ed. p. 248; Ravoth, *Grundriss d. Akiurgie*, Zw. Auf. S. 416; Ried, *D. Resect.* S. 390; Hedenus (cited by Velpeau). Langenbeck, *U. d. Schussfract d. Gelenke*, S. 24.

‡ Chassaignac, *De la Fract du Col. du Fémur*, Paris, 1835; Costello, *Cyclopædia of Practical Surgery*, Vol. IV, p. 46; Fergusson, *System, etc.*, 4th ed. p. 461; Jæger, in *Dieffenbach's Zeitsch, f. d. g. Med.* B. 1, S. 137; Schönborn (cited by B. von Langenbeck.)

CRUCIAL INCISION.—Various complex incisions in +, ⊣, ×, ∟, ⌋, and T, have been recommended by different surgeons as affording facilities for exposing the hip-joint. Montfalcon advocated an ∟-shaped division of the integument, and Dr. Baum (*ante*, p. 13) adopted this plan. Rossi advised an incision in the shape of a reversed ⌋, the horizontal arm carried far backwards, and the triangular flap thus formed reflected. J. F. Heyfelder, Mr. Erichsen, and Mr. T. Holmes, recommended a T-shaped preliminary incision. Roser made a short vertical and long transverse incision, dividing the tensor vaginæ femoris, the rectus, sartorius, and part of the iliacus, and preserving the rotators. Maisonneuve adopted this method. In excisions for gunshot injury, Drs. Crerar, Wagner, McCall, A. A. White, and Allen have employed some form of crucial incisions.*

QUADRILATERAL OR TRIANGULAR FLAP.—Champion relates that he often saw the elder Moreau demonstrate excision at the hip in the cadaver by raising from the outside of the joint a quadrilateral flap, with its base upwards. Percy, though condemning the operation, directed that it should be done by this method, if at all, and P. F. Blandin gave the same advice. Roux adopted this plan on the living subject and recommended it, but he has had few imitators. Rossi advised a V-shaped incision, with the apex of the triangular flap downwards. In some of their operations Jæger and Cajetan Textor made oblique incisions in front and behind the great trochanter and uniting at an acute angle above; in other words, they raised a triangular flap, with the base downwards.† In the operations of the late war the V-shaped incision was employed by Surgeon Reamer (*ante*, p. 31).

SEMI-LUNAR FLAP.—Guthrie, Ballingall, Bégin, Velpeau, Sédillot, Vidal, (de Cassis,) Billroth, Fock, Pancoast, Gross, H. H. Smith, and Chisolm, advise that the joint should be exposed by forming an external semi-lunar or semi-circular flap, some recommending, with Sédillot, that its convexity should be upwards, while others prefer, with M. Velpeau, a dependant flap, its convexity downwards. Sédillot's illustration of the former method is reproduced or imitated by Pitha and Billroth, Bernard and Huette, and in most of the modern manuals of operative surgery.‡

APPRECIATION.—In the operations for excision of the head of the femur in the living subject, which I have witnessed, and in those which I have performed upon the cadaver, I have never observed any difficulty, even in fleshy subjects, in excising the head and upper extremity of the femur through a straight incision, commencing one or two inches above the great trochanter, and carried downwards in the axis of the limb a little behind the prominence of the trochanter, for six or seven inches. With such an incision carried deeply, the subsequent steps of the operation are readily accomplished. The muscular insertions are then divided, and, if the edge of the knife is kept close to the bone, an artery

---

* J. T. Heyfelder, *Ueber Resect. und Amput.* S. 155; Erichsen, *Science and Art of Surgery*, 5th ed. 1869, Vol. II, p. 242; Holmes, *System of Surgery*, Vol. III, p. 815; Roser, *Handbuch der Anatomischen Chirurgie*, S. 630; Maisonneuve, *Gazette des Hôpitaux*, 1847, No. 25; Schillbach, *Beiträge zu den Resect. der Knochen*, Jena, 1851.

† Champion, *Traité de la Résection*; Percy, *Dict. des Sciences Médicales*, T. XLVII, p. 554; Blandin, *Dict. de Méd. et de Chir. Pratiques*, T. XIV, p. 226; Roux, *De la Résection ou du Rétranchment des Os malades*; Rossi, *Méd. Opératoire*, T. II, p. 225; Ried, *Resectionen der Knochen*, S. 301.

‡ Guthrie, *Commentaries*, p. 77; Ballingall, *Mil. Surgery*, p. 397; Bégin, *Nouv. Elém. de Chir.*, Paris, 1833, T. II, p. 221; Velpeau, *Op. cit.*, T. 1, p. 583; Sédillot, *Traité de Méd. Op.*, T. 1, p. 519; Vidal, *Traité de Path. Ext.*, T. V, p. 735; Billroth, *Handbuch*; Fock, *Loc. cit.*; Pancoast, *Treatise on Op. Surg.*, p. 129; Gross, *System of Surgery*, Vol. II, p. 1019; H. H. Smith, *Prin. and Pract. of Surg.*, Vol. II, p. 689; Chisolm, *Military Surgery*, p. 493.

requiring ligature will rarely be wounded. It is best to open the capsular ligament freely on its anterior and inferior aspect. This can readily be accomplished if the external straight incision is long enough. Otherwise, the incision should curve forward slightly at its upper part. Unless the preliminary incisions are faulty, the division of the round ligament and exarticulation present no difficulty. The next best plan is by the semi-lunar flap with its convexity upwards. This affords greater facility in exarticulation and in resecting the upper extremity of the femur; but it involves the risk of hæmorrhage, and leaves a more serious wound. After repeatedly performing the latter operation until the year 1866, Dr. B. von Langenbeck (*Op. cit.*, p. 24) discarded it as "too dangerous," and substituted a longitudinal incision The other forms of flap operation are to be condemned, as well as the complex crucial incisions, although it may be occasionally advisable to vary the direction of the external incision in order to include the entrance or exit wounds of the ball. It is well to have regard to the position of wounds made by the projectile within certain limits; but to make the preliminary incisions through the nates or adductor muscles, as practiced by Drs. Cullen and Dubois, because the entrance wounds were in these localities, is to expose the patient to unnecessary perils.

Many surgeons recommend that the great trochanter should always be removed, whether injured or not, a precept based apparently on theoretical grounds, and not deduced from experience.

The method of operating, though important, is far less essential than the after-treatment. In this, complete immobility of the injured limb is the first requisite. In operations in field hospitals, the surgeon should be assured that he can secure at least temporary rest for his patient, and subsequent safe transportation. The premature removal of patients who have undergone excisions, from permanent hospitals, cannot be too severely reprehended. When the patient is to be undisturbed, many surgeons prefer to simply support the limb in a proper position by cushions and pillows, with addition, perhaps, of slight extension by weights, according to the plan of Dr. Buck. Others prefer the flexed position, and, in the primary operations in the American War,

FIG. 70.—Dressing for excision of the head of the femur. [After Fergusson.]

it was common to suspend the limb in the wire splint devised by Dr. N. R Smith, as the safest mode of providing for the transportation in an ambulance wagon over a rough road, to which the patient would probably be subjected. When patients were treated in stationary hospitals, the advantages of treatment in the straight

FIG. 71.—Bracketted splint for excisions of the head of the femur. [After Erichsen.]

position, with moderate extension, were generally recognized. Sir W. Fergusson's plan of making the counter-extension from the opposite thigh was found to answer well. The apparatus recommended by this distinguished surgeon, which is figured in the accompanying wood-cut, (FIG. 70,) accomplished the three-fold object of keeping the parts steady, securing moderate extension, and straightening the limb. In two of the American operations the bracketted straight splint, applied as indicated in the wood-cut, (FIG. 71,) was found to permit the application of dressings with little annoyance to the patient and little trouble to the dresser..

The external wound is sometimes closed by sutures and adhesive strips, and sometimes left open. Unquestionably it is best to afford most free exit for pus, and no good purpose is subserved by a close approximation of the external wound. With a free incision, left open, drainage tubes and such appliances will be superfluous. Water dressings have commonly been applied, or, when the wounds assumed an unhealthy appearance, water dressings medicated with solutions of chlorinated soda, carbolic acid, and the alkaline permanganates, or Condy's fluid, in some form.*

In the general treatment, a generous diet, easy to be assimilated, has been found to be as preferable to the old low-diet plan as in the after-treatment of the other major operations. In some of the cases stimulants were administered to an immoderate extent, and did harm. The use of these agents and of opiates, will be governed by general principles.

Having stated the facts in regard to many hundred cases of gunshot injury of the hip-joint, and enunciated the practical teachings which I believe may be legitimately derived from them, it only remains for me to respectfully call your attention to the illustrations and tables accompanying this report. The plates are faithfully lithographed by Mr. J. Bien, from photographs made in this Office. The wood-cuts have been copied with minute fidelity from the specimens in the Museum, by Mr. H. H. Nichols, who has been engaged for the last five years upon anatomical work, under my immediate supervision. Mr. G. N. Whittington has computed the ratios, and has rendered me valuable assistance in this, as in former reports, in arranging the statistical material.

To facilitate reference, tables of the cases of excision at the hip for traumatic cause, chronologically arranged, and a bibliographical list of the works on the subject, are subjoined in two appendices.

I have the honor to remain, General,

Very respectfully,

Your obedient servant,

GEORGE A. OTIS,

Assistant Surgeon, U. S. Army.

---

* In the *London Lancet*, Am. ed., November, 1868, p. 678, Mr. P. R. Cresswell reports a case of gunshot fracture of the neck and trochanteric region of the left femur, treated on the expectant plan, with dressings of carbolic acid and boiled linseed oil, with a layer of "antiseptic putty" over this, and asserts that this dressing will revolutionize the treatment of compound fractures. Though the author arrives at a different conclusion, his description plainly indicates that the fracture was extracapsular.

APPENDIX A.

PRIMARY EXCISIONS AT THE HIP FOR GUNSHOT INJURY.

No.	Date.	Operator.	Patient.	Injury.	Operation.	Result.	Authority.
1	May 5, 1829	Oppenheim	A Russian chasseur.	Fracture of neck of left femur and rim of acetabulum by a musket ball.	Straight incision. Ball and head, neck, and great trochanter removed.	Died May 23, 1829, with plague.	S. Oppenheimer. Ueber die Resectionen des Hüftgelenkes. Würzburg, 1840, p. 25.
2	Dec. 5, 1832	Seutin	Private Lisieux, 25th Infantry.	Fracture of neck, trochanters, and upper fourth of femur by a large ball.	Long vertical and short transverse incision. Upper third of femur removed.	Died in nine days with gangrene of the limb.	Paillard, Relation Chirurgicale du Siège de la Citadelle d'Anvers, p. 105.
3	1854	Professor Baum	A subaltern officer.	Comminution of the neck of the femur by a musket ball.	L-shaped incision. Head and neck excised.	Died in twenty-two hours.	Lohmeyer. Die Schusswunden. Göttingen, 1859, S. 199.
4	June, 1855	Mr. G. E. Blenkins.	Private C. Monsterey, 3d Batt. Grenadier Guards.	Fracture of the neck and trochanters of the right femur, by a fragment of shell.	Longitudinal incision of five inches. Upper fifth of femur removed.	Died at the end of the fifth week.	Blenkins. Additions to Cooper's Dictionary. 8th ed., Vol. 1, p. 638.
5	Aug. 6, 1855	Surgeon J. Crerar.	Private W. Smith, 1st Royals.	Fracture of trochanter and neck of left femur by a grenade.	Crucial incision. Head, neck, and trochanters excised.	Died August 21, 1855.	Guthrie. Commentaries, &c. 5th ed., p. 622.
6	Aug. 19, 1855	Mr. T. C. O'Leary.	Private T. McKevens, 68th Regiment.	Fracture of the trochanter and neck of the left femur by a shell fragment.	Vertical incision. Femur sawn below lesser trochanter.	Recovered	Med. and Surg. Hist. of the British Army which served in the Crimea, Vol. II, p. 378.
7	Sept. 8, 1855	Dr. George Hyde	Corporal D. Sheehan, 41st Regiment.	Comminution of trochanter and neck of femur by a grape shot.	Straight incision four inches long. Femur sawn an inch below trochanter minor.	Died September 14, 1855.	Macleod. Notes on the Surgery of the War in the Crimea, p. 344.
8	Aug. 30, 1861	Professor G. C. Blackman.	Private J. McCulloch, recruit.	Upper extremity of femur shattered by a musket ball.	Vertical incision. Head, neck, and trochanters removed.	Died August 30, 1861.	Circular No. 2, S. G. O., 1869, p. 20.
9	Oct. 5, 1861	Asst. Surgeon J. W. S. Gouley, U. S. A.	Private T. Greely, 74th New York Vols.	Fracture of the neck and head of the left femur by a round musket ball.	Longitudinal incision seven inches long. Head and neck removed.	Died October 12, 1861.	New Jersey Medical Reporter, Vol. VIII, p. 76.
10	Aug. 28, 1862	Surg. P. Pineo, U. S. V.	U.S. soldier of Gen. King's Division.	Fracture of the neck and trochanters of the left femur by a conoidal musket ball.	Vertical incision. Upper fourth of femur removed.	Died in a short time. Ball in pelvis.	Circular No. 6, S. G. O., 1865, p. 62.
11	Aug. 30, 1862	Surg. John McNulty, U. S. V.	Private 1st Army Corps.	Fracture of left femur at junction of head and neck by a conoidal musket ball.	Vertical incision six inches long. Head and neck removed.	Died Aug. 31, 1862. Ball in pelvis.	Circular No. 2, S. G. O., 1869, p. 22.

PRIMARY OPERATIONS. 133

PRIMARY EXCISIONS AT THE HIP FOR GUNSHOT INJURY—Continued.

No.	Date.	Operator.	Patient.	Injury.	Operation.	Result.	Authority.
12	Sept. 17, 1862	Surg. John McNulty, U.S. Vols.	Private 1st Army Corps.	Comminution of upper extremity of left femur by a shell fragment.	Vertical incision six inches long. Head and neck removed.	Died in ten hours. Pelvis injured.	Circular No. 2, S. G. O., 1869, p. 22.
13	Sept. 17, 1862	Surg. M. Storrs, 8th Conn. V.	Captain F. M. Barber, 10th Conn. Vols.	Fracture of the neck and trochanters of the right femur.	Straight incision four inches long. Femur sawn at trochanter minor.	Died September 20, 1862.	Circular No. 2, S. G. O., 1869, p. 22.
14	Dec. 13, 1862	Surg. Hunter McGuire, C. S. A.	Confederate private soldier.	Fracture of neck and trochanter of right femur by a fragment of shell.	Shaft divided just below the trochanter.	Died in two or three days.	Circular No. 2, S. G. O., 1869, p. 23.
15	Jan. 11, 1863	Surg. M. T. Carey, 49th Ohio V.	Private John Coon, 60th Indiana Vols.	Comminution of neck and upper part of shaft of right femur by a conoidal musket ball.	Semicircular incision.	Died Jan. 21, 1863.	Circular No. 2, S. G. O., 1869, p. 23.
16	May 3, 1863	Surg. James, 16th South Carolina Regiment.	Soldier of a South Carolina Regiment.	Fracture of the neck of the femur by a musket ball.	....	Died May 6, 1863.	Circular No. 2, S. G. O., 1869, p. 24.
17	May 14, 1863	Surg. H. S. Hewitt, U. S. V.	Sergeant J. M. Tolman, 18th Wisconsin Vols.	Comminution of head and neck of left femur by a conoidal musket ball.	Curvilinear incision.	Died May 19, 1863.	Circular No. 6, S. G. O., 1865, p. 66.
18	July 3, 1863	Surg. J. McNulty, U. S. V.	Captain T. R. Robeson, 2d Mass. Vols.	Fracture of head and neck of right femur by a conoidal musket ball.	Straight incision six inches long.	Died July 14, 1863. Pelvis penetrated.	Circular No. 2, S. G. O., 1869, p. 25.
19	Nov. 16, 1863	Surg. J. S. D. Cullen, P. A. C. S.	Private J. Melear, 8th Michigan Vols.	Fracture of the head and neck of the femur by a conoidal musket ball.	Incision ten inches long thro' the buttock.	Died December 2, 1863.	Circular No. 2, S. G. O., 1869, p. 25.
20	Feb 25, 1864	Surg. N. W. Abbott, 86th Illinois Vols.	Private B. Dempsey, 4th U. S. Artillery.	Crushing of the trochanter and neck of the right femur by a shell fragment.	Longitudinal incision five inches long.	Died February 28, 1864.	Chicago Medical Examiner, October, 1864, p. 612.
21	May 6, 1864	Surg. J. J. Dement, P. A. C. S.	Private Cannon, 49th Georgia Regiment.	Fracture of the neck and trochanters of the left femur by a conoidal musket ball.	Vertical incision six inches long.	Recovered. Died November 23, 1865, of diphtheria.	Circular No. 2, S. G. O., 1869, p. 26.
22	May 5, 1864	Surg. Hunter McGuire, C. S. A.	Private of Ewell's Corps.	Fracture of the neck of the left femur by a conoidal musket ball.	Longitudinal incision.	Died May 22, 1864.	Circular No. 2, S. G. O., 1869, p. 26.
23	May 6, 1864	Surg. J. T. Gilmore, C. S. A.	Private O'Rourke, 18th Mississippi Regiment.	Fracture of the neck of the right femur by a musket ball.	Curvilinear incision four or five inches long.	Died May 9, 1864.	Circular No. 2, S. G. O., 1869, p. 27.
24	May 7, 1864	Surg. C. B. Gibson, C. S. A.	Private J. J. Phillips, 61st Virginia Regiment.	Fracture of upper part of the shaft of the left femur by a conoidal musket ball.	Long vertical incision.	Died May 11, 1864.	Circular No. 2, S. G. O., 1869, p. 27.
25	May 10, 1864	Unknown	Private of Fifth Army Corps.	Fracture of the trochanteric portion of left femur by a musket ball.	....	Died May 13, 1864.	Circular No. 6, S. G. O., 1865, p. 68.

PRIMARY EXCISIONS AT THE HIP FOR GUNSHOT INJURY—Concluded.

No.	Date.	Operator.	Patient.	Injury.	Operation.	Result.	Authority.
26	May 13, 1864	Surg. C. B. Gibson, C. S. A.	Private G. W. Mayo, 25th Batt. Virginia Reserves.	Fracture of neck and shaft of right femur by a conoidal musket ball.	Long straight incision.	Died May 15, 1864.	Circular No. 2, S. G. O., 1869, p. 28.
27	May 12, 1864	Surg. G. W. Snow, 35th Mass. V.	Lieut. J. A. McGuire, 148th Penn. Vols.	Trochanters and neck of right femur smashed by a conoidal musket ball.	Longitudinal incision.	Died May 15, 1864.	Circular No. 2, S. G. O., 1869, p. 28.
28	June 3, 1864	Unknown	Private of Eighteenth Army Corps.	Trochanters and neck of right femur crushed by a shell fragment.		Died June 6, 1864.	Report of Asst. Surg. J. S. Billings, U. S. A., Acting Medical Inspector A. of P.
29	June 18, 1864	Surg. Ladd, 50th North Carolina Regiment.	Private of 50th North Carolina Regiment.	Comminution of neck of right femur by a musket ball.	Curvilinear incision.	Died Aug. 16, 1864. Absence of proper diet.	Transactions American Medical Association, Vol. XVIII, p. 263.
30	June 24, 1864	Surg. J. F. Grant, P. A. C. S.	Private T. J. Hobson, 32d Tennessee Regiment.	Comminution of neck and trochanters of the femur by a conoidal musket ball.	Linear incision ten inches long.	Died June 27, 1864.	Circular No. 2, S. G. O., 1869, p. 30.
31	July 23, 1864	Surgeon W. V. White, 57th Mass. V.	Sergeant E. T. Brown, 21st Mass. Vols.	Comminution of upper extremity of left femur by a fragment of a mortar bomb.	Longitudinal incision.	Died July 27, 1864.	Circular No. 2, S. G. O., 1869, p. 30.
32	July 31, 1864	Surg. G. S. West, C. S. A.	Private J. T. Goode, 6th Virginia Regiment.	Fracture of upper extremity of left femur by a conoidal musket ball.	Linear incision in the axis of the thigh.	Died Aug. 2, 1864.	Circular No. 2, S. G. O., 1869, p. 30.
33	Aug. 20, 1864	Surg. F. C. Reamer, 143d Penn. V.	Private E. A. McDonald, 149th Pa. Vols.	Fracture of head and neck of right femur by a conoidal musket ball.	V-shaped incision.	Died Sept. 4, 1864.	Circular No. 6, S. G. O., 1865, p. 70.
34	Aug. 23, 1864	Surg. A. A. White, 8th Maryland V.	Private Charles Beard, 12th Mississippi Regiment.	Comminution of head and neck of right femur by a conoidal musket ball.	⌐-shaped incision.	Died Aug. 25, 1864. Pelvis injured.	Circular No. 6, S. G. O., 1865, p. 70.
35	Oct. 19, 1864	Surg. A. P. Clark, 8th New York Cavalry.	Sergeant S. Grimshaw, 6th New York Cavalry.	Fracture of head, neck, and shaft of left femur by a shell fragment.	Straight incision seven inches long.	Died Nov. 5, 1864. Pelvis injured.	Circular No. 2, S. G. O., 1869, p. 32.
36	Oct. 26, 1864	Surg. N. Y. Leet, 76th Penn. V.	Private T. G. Pease, 117th New York Vols.	Fracture of trochanters and neck of right femur by a musket ball.	Longitudinal incision.	Died Oct. 29, 1864.	Circular No. 2, S. G. O., 1869, p. 32.
37	Oct. 27, 1864	Surg. N. Y. Leet, 76th Penn. V.	Lieut. D. Beebe, Adjutant 3d New York Vols.	Fracture of neck and trochanters of right femur by a conoidal musket ball.	Straight incision.	Recovered	Circular No. 2, S. G. O., 1869, p. 32.
38	Oct. 27, 1864	Surgeon Clark	Private R. Cole, 29th Connecticut Regiment.	Fracture of upper extremity of right femur by a musket ball.	Longitudinal incision.	Died Oct. 29, 1864.	Circular No. 2, S. G. O., 1869, p. 33.
39	Mar. 29, 1864	Surg. Wm. Fuller, 1st Michigan Volunteers.	Private C. Morrison, 185th New York Vols.	Fracture of trochanter and neck of right femur by a conoidal musket ball.	Longitudinal incision.	Died Apr. 26, 1865.	Circular No. 2, S. G. O., 1869, p. 33.

INTERMEDIATE EXCISIONS AT THE HIP-JOINT FOR GUNSHOT INJURY.

No.	Date.	Operator.	Patient.	Injury.	Operation.	Result.	Authority.
1	May 13, 1849	Dr. H. Schwartz.	O*****. A Danish soldier.	Gunshot fracture of trochanters of the left femur. The tuber ischii was injured.	Longitudinal incision four inches long.	Died May 20, 1849.	ESMARCH. Ueber Resectionen, S. 125.
2	July 5, 1855	Dr. G. H. B. Macleod.	Couch, A soldier of the Rifle Brigade.	Fracture by a rifle ball of the neck of the femur.	The wound was enlarged downwards.	Died July 12, 1855, of cholera.	MACLEOD. Notes, etc., p. 338.
3	1855	Dr. Coombe, R. A.	A British artillery soldier.	Gunshot fracture of the neck of the femur.	Longitudinal incision.	Died in a fortnight.	GUTHRIE'S Commentaries, p. 622.
4	March, 1862	Surg. A. H. Thurston, U. S. V.	Corp. H. F. Smith, 1st Wis. Vols.	Fracture of neck of femur by a musket ball.	Long straight incision.	Died Mar. 15, 1862.	Circular No. 2, S. G. O., 1869, p. 34.
5	April 16, 1862	Surg. G. C. Blackman, U. S. V.	Private D. M. Noe, 46th Ohio Vols.	Fracture of neck of left femur by a conoidal musket ball.	Longitudinal incision.	Died Apr. 24, 1862.	Circular No. 2, S. G. O., 1869, p. 34.
6	May 20, 1862	Asst. Surg. J. S. Billings, U.S.A.	Priv. T. C. Christopher, 18 S.C. Reg.	Fracture of neck and head of left femur by a musket ball.	Curvilinear incision.	Died May 24, 1862.	Circular No. 2, S. G. O., 1865, p. 62.
7	Sept. 4, 1862	Asst. Surg. J. S. Billings, U.S.A.	Private soldier	Fracture of neck of femur by a musket ball.	Straight incision.	Died September 24, 1862.	Circular No. 6, S. G. O., 1865, p. 64.
8	Sept. 27, 1862	Asst. Surg. B. A. Clements, U.S.A.	Private C. E. Marston, 1st Mass. Vols.	Comminution of head and neck of right femur.	Curved incision five inches long.	Died September 30, 1862.	Circular No. 6, S. G. O., 1865, p. 64.
9	Sept. 20, 1862	Asst. Surg. B. A. Clements, U.S.A.	Private F. Macblin, 11th Penn. Vols.	Comminution of neck of right femur by a musket ball.	Straight incision five inches long.	Died September 21, 1862.	Circular No. 6, S. G. O., 1865, p. 64.
10	Sept. 29, 1862	Asst. Surg. J. H. Bill, U. S. A.	Private C. Callaghan, 2d Del. Vols.	Comminution of trochanter of femur by a shell.	Curvilinear incision.	Died October 4, 1862.	Circular No. 6, S. G. O., 1865, p. 64.
11	Jan. 21, 1863	Surgeon Sennet, 94th Ohio Vols.	Sergeant D. W. Hade, 101 Ohio V.	Gunshot fracture of upper portion of femur.	Longitudinal incision.	Died January 31, 1863.	Circular No. 2, S. G. O., 1869, p. 38.
12	July 8, 1863	Surg. G. W. Avery, 9 Conn. Vols.	Private John Miller, 102d N.Y. Vols.	Fracture of neck of left femur by a conoidal musket ball.	Straight incision.	Died September 21, 1863.	Circular No. 2, S. G. O, 1869, p. 39.
13	Oct. 3, 1863	Surg. F. H. Gross, U. S. V.	Private M. Welsh, 10th Ky. Vols.	Fracture of neck of left femur by a conoidal musket ball.	Curvilinear incision.	Died October 25, 1863.	Am. Jour. Med. Sci., Vol. LV, p. 410.
14	Nov. 9, 1863	Surg. J. B. Read, C. S. A.	Lieut. J. M. Jarrett, 15th N. C. Reg't.	Fracture of upper part of shaft of femur by a musket ball.	Straight incision.	Recovered ...	C. S. Med. and Surg. Jour., Vol. 1, p. 5.
15	May 27, 1864	Act. Asst. Surg. G. A. Mursick.	Private Hugh Wright, 87 N.Y.V.	Fracture of neck of femur by a musket ball.	Straight incision.	Recovered ...	N. Y. Med. Journal, Vol. 1, p. 424.
16	May 12, 1864	Surg. C. B. Gibson, C. S. A.	Private M. Smith, 38th Virginia Reg.	Fracture of the upper part of the femur by a musket ball.	....	Died May 13, 1864.	Circular No. 2, S. G. O., 1869, p. 43.
17	May 19, 1864	Asst. Surg. Alex. Ingram, U. S. A.	Private C. C. Cleaver, 2d U. S. Inf.	Fracture of neck and trochanters of right femur by a conoidal musket ball.	Curved incision six inches long.	Died May 23, 1864.	Circular No. 6, S. G. O., 1865, p. 60.
18	May 19, 1864	Act. Asst. Surg. J. F. Thompson.	Private A. Ewing, 140th Penn. Vols.	Comminution of upper part of left femur by a conoidal musket ball.	....	Died May 24, 1864.	Circular No. 6, S. G. O., 1865, p. 68.

EXCISIONS AT THE HIP.

INTERMEDIATE EXCISIONS AT THE HIP-JOINT FOR GUNSHOT INJURY—Concluded.

No.	Date.	Operator.	Patient.	Injury.	Operation.	Result.	Authority.
19	June 3, 1864	Asst. Surg. C. A. McCall, U. S. A.	Captain J. Phelan, 73d N. Y. Vols.	Fracture of the neck of left femur.	Crucial incision..	Died June 21, 1864.	*Circular* No. 6, S. G. O., 1865, p. 68.
20	Aug. 5, 1864	Asst. Surg. W. Thomson, U.S.A.	Priv. Peter Boyle, 59th Mass. Vols.	Fracture of neck of femur by a musket ball.	Straight incision.	Died Aug. 7, 1864.	*Circular* No. 6, S. G. O., 1865, p. 70.
21	1864	Neudörfer	An Austrian soldier	Gunshot fracture of the upper extremity of the femur.		Fatal	HEINE. *Die Schussverletzungen*, S. 659.
22	1864	Neudörfer	An Austrian soldier.	Gunshot fracture of the upper extremity of the femur.		Fatal	DERSELBE.
23	April 3, 1865	Surg. D.W. Bliss, U. S. V.	Lieut. D. N. Patterson, 40th Va. Reg't.	Fracture of head, of left femur by a musket ball.	Curvilinear incision.	Died Apr. 7, 1865.	*Circular* No. 6, S. G. O., 1865, p. 72.
24	April 4, 1865	Asst. Surg. H. Allen, U. S. A.	Corporal H. C. Senet, 122d N.Y. Vols.	Fracture of head of femur by a conoidal musket ball.	T-shaped incision.	Died Apr. 8, 1865.	*Circular* No. 6, S. G. O., 1865, p. 72.
25	April 6, 1865	Asst. Surg. W. F. Norris, U. S. A.	Private H. Phillips, 146th. N Y. Vols.	Fracture of the neck of left femur by a musket ball.	Curved incision.	Died Apr. 21, 1865.	*Circular* No. 6, S. G. O., 1865, p. 72.
26	April 27, 1865	Surg. A. McMahon, U. S. V.	Priv. T. E. Foulke, 2d Alabama Reg't.	Fracture of neck of femur by a musket ball.		Died June 5, 1865.	*Circular* No. 6, S. G. O., 1865, p. 74.
27	April 28, 1865	Surg. A. McMahon, U. S. V.	Priv. G. W. Brantley, 2d Ala. Regt.	Fracture of neck of left femur by a conoidal musket ball.		Died May 2, 1865.	*Circular* No. 6, S. G. O., 1865, p. 74.
28	June 27, 1866	Dr. L. Stromeyer.	A debilitated subject.	Intra-capsular gunshot fracture of the neck of the femur.	Crucial incision.	Died two days after the operation.	STROMEYER. *Erfahrungen über Schusswunden.* Hanover. 1867, S. 52.
29	June 29, 1866	B. Von Langenbeck.	Emil Bauer	Gunshot fracture of the neck of the femur.		Fatal	B. VON LANGENBECK, *ueber die Schussfracturen* Berlin, 1868, S. 151.
30	July 3, 1866	B. Von Langenbeck.	An Austrian soldier.	Gunshot fracture of the neck of femur.	Semicircular flap over the trochanter.	Died July 12, 1866.	LANGENBECK. *Ueber die Schussfracturen.* Berlin, 1868, S. 151.
31	July 3, 1866	Dr. Schönborn	Maxim Glutschak..	Gunshot fracture of the head of the left femur.	Curved incision behind the great trochanter.	Recovered	DERSELBE.
32	July 27, 1806	Bernhard Beck	O. F. S	Gunshot fracture of the neck of femur and trochanter major.	Slightly curved incision on the outside of the thigh.	Died five days after operation.	BERNHARD BECK. *Kriegs-Chirurgische.* Freiburg. 1867, S. 357.
33	July 31, 1867	Glover Perin, Surgeon U.S.A.	Pri. Francis Ahearn,	Gunshot fracture of the upper part of the right femur.	Straight incision..	Died twenty hours after operation.	*Circular* No. 2, S. G. O., 1865, p. 60.

SECONDARY OPERATIONS. 137

SECONDARY EXCISIONS AT THE HIP-JOINT FOR GUNSHOT INJURY.

No.	Date.	Operator.	Patient.	Injury.	Operation.	Result.	Authority.
1	1847	C. Textor	A man of 44 years.	Caries of the head of the femur, consequent upon gunshot fracture.		Died on the tenth day.	O. Heyfelder. Lehrbuch, S. 88.
2	May, 1850	Dr. Ross	Schleswig Holstein soldier.	Caries resulting from a gunshot fracture of the neck of the femur.	Two years after the reception of the injury.	Died shortly after the operation.	Esmarch. Ueber Resectionen, S. 126.
3	Aug. 21, 1862	Asst. Surg. R. Bartholow, U.S.A.	Priv. J. W. Nelling, 1st Mass. Vols.	Comminution of neck of right femur by a musket ball.	Vertical incision four inches long.	Died in four days.	Circular No. 2, S. G. O., 1869, p. 48.
4	Feb. 23, 1863	Asst. Surg. H. A. Dubois, U.S.A.	Private E. Hunt, 71st Penn. Vols.	Fracture of neck of right femur by a conoidal musket ball.	Enlargement of entrance wound.	Died in two days.	Circular No. 2, S. G. O., 1869, p. 49.
5	Mar. 21, 1863	Surg. D. P. Smith, U. S. V.	Private Jos. Brown, 3d Mich. Vols.	Fracture of left femur by a conoidal musket ball.	Longitudinal Incision.	Recovered	Circular No. 2, S. G. O., 1869, p. 50.
6	July, 1863	B. Von Langenbeck.	S. Von Kucharsky, aged 19 years.	Musket ball fractured left trochanter and lodged.	Semicircular incision.	Died in fourteen days.	Langenbeck, Ueber die Schussfract. S. 15.
7	Aug. 12, 1863	Surg. J. D. Read, P. A. C. S.	Private Alfred Toney, 16th N. C. Reg't.	Fracture of left femur by a conoidal musket ball.	Straight incision.	Died in eight days.	Circular No. 2, S. G. O., 1869, p. 51.
8	July 1, 1864	Surg. R. B. Bontecou, U. S. V.	Private H. Woodworth, 4th Vt. Vols.	Grooving of left femur by a conoidal musket ball.	Curved incision.	Died in one day.	Circular No. 2, S. G. O., 1869, p. 52.
9	Aug. 2, 1864	Surgeon J. D. Read, P. A. C. S.	Ensign W. J. Henry, 21st Miss. Reg't.	Comminution of the left femur by a conoidal musket ball.	Straight incision of seven inches.	Died in six days.	Circular No. 2, S. G. O., 1869, p. 52.
10	Sept. 27, 1864	Asst. Surgeon C. Wagner, U.S.A.	Priv. J. Zuborouski, 7th Conn. Vols.	Fracture of neck of right femur by a musket ball.	Crucial incision.	Died in one day.	Circular No. 2, S. G. O., 1869, p. 53.
11	Mar. 9, 1865	Surg. John Varment, U. S. A.	Priv. J. Roth, 189th New York Vols.	Fracture of head of femur by a round musket ball.	Straight incision.	Died June 17, 1865.	Circular No. 2, S. G. O., 1869, p. 53.
12	Mar. 24, 1865	Surg. A. McMahon, U. S. V.	Private H. Train, 31st Mass. Vols.	Fracture of neck of left femur by a conoidal musket ball.	Straight incision.	Died in six days.	Circular No. 2, S. G. O., 1869, p. 54.
13	Aug. 14, 1868	Asst. Surg. J. R. Gibson, U.S.A.	Private Charles F. Read, 37th U.S. Inf.	Fracture of head of left femur by a conoidal musket ball.	T-shaped incision.	Recovered	Circular No. 2, S. G. O.,

SUMMARY.

Excisions at the Hip for Gunshot Injury.	Cases.	Primary.	Intermediate.	Secondary.
Prior to 1861	12	7	3	2
During the American War	63	32	22	9
Later cases	10		8	2
Total	85	39	33	13

THE EIGHTY-FIVE CASES RESULTED AS FOLLOWS:

	Cases.	Died.	Recovered.	Death Rate.
Primary	39	36	3	92.3
Intermediate	33	30	3	90.9
Secondary	13	11	2	84.6
Aggregate	85	77	8	90.6

APPENDIX B.

BIBLIOGRAPHY OF EXCISIONS AT THE HIP.

ABBOTT, N. W. *Chicago Medical Examiner*, October, 1864; p. 612.
ADAMS, Z. B. *Excisions of Joints for Traumatic Cause*, in Boston Medical and Surgical Journal, April, 1867; p. 229.
BALLINGALL, Sir GEORGE. *Outlines of Military Surgery*. Fifth ed. Edinburgh, 1855, p. 397.
BALLANC, D. P. *Beiträge zur Statistik der Hüftgelenkes Resection*. Leipzig, 1868.
BARWELL, RICHARD. *Treatise on Diseases of the Joints*. Am. ed. Philadelphia, 1861, p. 431, and *Trans. Path. Soc.*, London, Vol. XVII, p. 239.
BAUDENS, M. L. *Clinique des Plaies d'Armes a Feu*. Paris, 1836; p. 445.
BECK, BERNHARD. *Zur Statistik der Amputationen und Resectionen*. In *Langenbeck's Archiv*. B. 5, S. 245, 256, and *Kriegs-Chirurgische Erfahrungen während des Feldzuges 1866 in Süddeutschland*. Freiburg, 1867. S. 266, 351, and *Die Schusswunden*. Heidelberg, 1850, S. 332.
BELL, JOSEPH. *A Manual of the Operations of Surgery*. London, 1866, p. 111.
BÉRARD. *Dictionnaire de Médecine, par MM. ADELON, BÉCLARD, BÉRARD, etc*. Paris, 1837. Tome 15, p. 82.
BERNARD & HUETTE. *Illustrated Manual of Operative Surgery*. Am. ed. New York, 1855; p. 105, and *Précis inconographie de méd. op. et d'anat. chir*. Paris, 1850.
BILLROTH. *Ueber den Resectionen* in *Deutsche Klinik*. B. V, S. 229, and *Handbuch*, B. 1, Abth. 2, S. 483.
BLANDIN. *Dictionnaire de Médecine et de Chirurgie Pratiques*. Art. Résection, Paris, 1835; Tome 14, p. 266.
BLENKINS, G. E. *Additions to Cooper's Dictionary of Practical Surgery*. London, 1861. Vol. I, p. 839.
BONINO, EDOUARD. *De la Résection de la Tête du Fémur*, In *Annales de la Chirurgie Française et Etrangère*. Tome X, p. 385.
BOURGERY, J. M. *Iconographie d'Anatomie Chirurgicale*. Paris, 1832. Tome VI, p. 224.
BOWMAN. *Resection of the Hip-Joint*, in *Med. Times and Gazette*. Dec., 1860; p. 210.
BRINTON, J. H. *Consolidated Statement of Gunshot Wounds*. Circular No. 9, S. G. O., Washington, July 1, 1863; p. 12.
BRIOT, M. *Histoire de l'Etat et des Progrès de la Chirurgie Militaire en France pendant les Guerres de la Révolution*. Besançon, 1817; p. 177.
BUSCH, W. *Chirurgische Beobachtungen*. Berlin, 1854. S. 258.
BUSH. *Encyclopædisches Worterbuch*. B. IX, S. 188.
CHAMPION. *Traité de Résection*, Paris, 1817.
CHAUSSIER. *Mém. de la Soc. Méd. d'Emulation*, T. III, p. 399, and *Magazin Encyclopédique*, T. VII, p. 248.
CHEEVER, D. W. *Boston Med. and Surg. Jour.*, Vol. 77, p. 281.
CHELIUS, J. M. *System of Surgery*, by J. F. South. Vol. III, p. 735. Philadelphia, 1847.
CHENU, J. C. *Rapport au Conseil de Santé des Armées sur les Résultats du Service Médico-Chirurgical pendant la Campagne d'Orient en 1854-'55-'56*. Paris, 1865; p. 677.
CHISOLM, J. JULIAN. *A Manual of Military Surgery for the use of Surgeons in the Confederate States Army*. 3d ed. Columbia, S. C., 1864; pp. 390, 495.
COLE, J. J. *Military Surgery in India during the years 1848 and 1849*. London, 1852; p. 137.
COOTE, H. *British Medical Journal*, Jan. 2, 1858.
COSTELLO, WILLIAM B. *The Cyclopædia of Practical Surgery*. London, 1861. Vol. IV, p. 46.
COULSON, WM. *On the Disease of the Hip-Joint*. London, 1867; p. 104.
DECAISNE. *Des Moyens d'éviter les Amputations et les Résections Osseuses*. Bruxelles, 1855.
DEMME, HERMANN. *Specielle Chirurgie der Schusswunden nach eigenen Erfahrungen in den Norditalienischen Lazarethen von 1859*. Würzburg, 1864. S. 355, and *Allgemeine Chirurgie der Kriegswunden*, Würzburg, 1861; and *Allgemeine Chirurgie der Schusswunden*, Würzburg, 1863.
DIEFFENBACH. *Zeitschrift für die gesammte Medicin*, 1836.
DIRCKS, CH. J. M. *Diss. inaug. de resectione capitis femoris*. Wircob, 1846, p. 27.
DRUITT, ROBERT. *The Surgeon's Vade Mecum*, 9th London ed., 1865, pp. 307, 800.
EMMERT, CARL. *Lehrbuch der Chirurgie*. Band 4, S. 256. Stuttgart, 1867.
ERICHSEN, JOHN. *The Science and Art of Surgery*. London, 1869; Vol. II, p. 239, and *London Lancet*, Oct., 1856, Mar., 1857, and *British Med. Jour.*, 1860.
ESMARCH, FRIEDRICH. *Ueber R.sectionen nach Schusswunden*. Kiel, 1851, S. 123.
EULENBERG, ALBERT. *Beiträge zur Statistik und Würdigung der Hüftgelenkes Resection bei Caries*. In *Langenbeck's Archiv*. Band VII, S. 701.
EVE, P. F. *A Contribution to the History of the Hip-Joint Operations Performed during the late Civil War*, in *Transactions Am. Med. Association*, Vol. XVIII, pp. 256, 263.
FERGUSSON, Sir WM. *A System of Practical Surgery*. 4th London ed., 1867, p. 461, and *Medico-Chirurgical Transactions*, Vol. 28, London, 1845, p. 571, and *London Lancet*, April 7th, 1849, p. 359. Ibid, 1858, Vol. 1, p. 75. Ibid, 1860, Vol. 2, p. 142.

FISCHER, H. *Verletzungen durch Kriegswaffen.* In *Pitha & Billroth's Handbuch.* B. I, Abth. 2, S. 483, 496.
FOCK, C. *Bemerkungen und Erfahrungen über die Resection im Hüftgelenk.* In *B. Langenbeck's Archiv für Klinische Chirurgie.* Berlin, 1861. Band I, S. 172, and *Archives Gén. de Méd., Nov. and Dec.,* 1860.
GERDY, J. V. *De la Résection des Extrémités Articulaires des Os.* Paris, 1839; p. 157.
GUERINI, A. *Vade Mecum per le Ferite d'Arma da Fuoco.* Milano, 1866; p. 134.
GOSSELIN. *Résection de la Hanche,* in *Bulletin de l'Acad. de Méd.* Oct. 15th, 1861.
GRAEFE & WALTHER'S *Journal,* B. XXIV, Hft. 4.
GRITTI, ROCCO. *Delle Fratture del Femore per arma da Fuoco.* Milano, 1866, p. 83.
GROSS, SAMUEL D. *A System of Surgery.* Third ed. Philadelphia, 1864. Vol. II, p. 1016.
GROSS, S. W. *Military Surgery,* In *Am. Jour. Med. Sci.* Vol. LIV, 1867; p. 443.
GUERSANT. *Dictionnaire de Médecine,* par MM. ADELON, BÉCLARD, BÉRARD, etc. Paris, 1843. Tome 27, p. 402.
GÜNTHER, G. B. *Die Operationen an den untern Extremitäten mit Einschluss des Hüftgelenkes,* in *Lehre von den blutigen Operationen,* Zweite Abth. Leipzig & Heidelberg, 1857. S. 204.
GUTHRIE, G. J. *Commentaries, etc.* Sixth ed. London, 1855; pp. 76, 145, 620, 645.
GURLT, E. *Amputationen, Exarticulationen, Resectionen,* in *Jahresbericht über die Leistungen und Fortschritte in der Gesammten Medicin.* Berlin, 1868; B. 11, Abth. 2, S. 411, and *Resection im Hüftgelenk* in *Langenbeck's Archiv.* Band VIII, S. 903.
HAMILTON, F. H. *A Treatise on Military Surgery.* New York, 1865; p. 518.
HANCOCK, HENRY. *On Excision of the Hip-Joint.* Lancet, 1857. Vol. II, p. 84.
HARGRAVE, WILLIAM. *A System of Operative Surgery.* Dublin, 1831; pp. 267, 514.
HEDENUS, AUG. GULIELMUS. *Commentatio Chirurgica de Femore in Cavitate Cotyloidea Amputando.* Lipsiae, 1823; pp. XII, 63.
HEINE, C. *Die Schussverletzungen der unteren Extremitäten, nach eigenen Erfahrungen im letzen Schleswig-Holsteinschen Feldzuge.* Berlin, 1866. S. 369.
HENSSER, F. *Cases of Resection,* in *Deutsche Klinik.* Oct. 20, 1859.
HEYFELDER, J. F. *Ueber Resectionen und Amputationen.* Breslau & Bonn, 1854; S. 154.
HEYFELDER, O. *Lehrbuch der Resectionen.* Wien, 1863; S. 73, and *Traité des Résections.* Translated, with notes and additions, by E. Bœckel. Paris, 1863; p. 59.
HODGES, RICHARD M. *The Excision of Joints.* Boston, 1861; p. 90, and *The Excision of Joints for traumatic cause,* in *Military Medical, and Surgical Essays,* Philadelphia, 1864; p. 513.
HOLMES, T. *A System of Surgery.* London, 1861. Vol. II, p. 82; Vol. III, p. 813. Article by Mr. Longmore, and *Report on Surgery,* in *Biennial Retrospect of Medicine and Surgery for the New Sydenham Society for* 1867, p. 287, and *Trans. Path. Soc.,* London, Vol. XVII, p. 229.
HOLT, B. *London Lancet,* 1863. Vol. I, p. 38.
HUETER, C. *Die Resectionen,* in *Langenbeck's Archiv.* B. VIII. S. 94.
HYRTL, J. *Handbuch der Topographischen Anatomie.* Wien, 1865; B. II, S. 534.
JAEGER. *Operatio Resectionis Conspectu Chronologico Adumbrata.* Erlangw, 1832; and *Die Resectionen der Knochen.* Nürnberg, 1860.
KINLOCH. *Charleston Medical Journal and Review,* May, 1857.
KLEINERT. *Algemeines Reportorium,* X Jahrg, VI. Hft., S. 104.
KNOX. *London Medical Times,* June 1, 1851.
KOELER, G. L. *Experimenta circa regenerationem ossium.* Göttingae, 1786.
KRETSCHMAR, G. A. *Ueber Hüftgelenks resection, Inaug. Diss.* Jena.
LANGENBECK, C. J. M. *Nosologie und Therapie der chirurgischen Krankheiten.* Göttingen, 1830; Vierter Band, S. 397, and *Bibliothek für die Chirurgie.* Göttingen, 1807. B. II. S. 734.
LANGENBECK, B. *Ueber die Schussfracturen der Gelenke und ihre Behandlung.* Berlin, 1868. S.
LARGHI, B. *Résection de la Tête et du Col du Fémur,* in *Gazette Méd. de Paris.* Tome Douzième, p. 8, 1857.
LARREY, H. *Histoire Chirurgicale du Siége de la Citadelle d'Anvers.* Paris, 1833, and *Bulletin de l'Académie Impériale de Médecine.* Paris, 1861-'62. T. XXVII; p. 124.
LEE, H. *Excision of Hip-Joint,* in *Br. Med. Jour.* Vol. II, p. 362.
LE FORT, LÉON. *De la Résection de la Hanche pour plaies par Armes à Feu,* mémoire lu a l'Académie Impériale de Médecine. Paris, 1861; p. 567, and *L' Union Médicale,* September 6, 1860.
LEGOUEST, L. *Traité de Chirurgie de Armée.* Paris, 1863; p. 752.
LEPOLD, FELIX *Ueber die Resection des Hüftgelenkes.* Würzburg, 1834. S. 33.
LOHMEYER, C. F. *Die Schusswunden.* Göttingen, 1859. S. 199.
LÜCKE, A. *Beiträge zur Lehre von den Resectionen,* in *Langenbeck's Archiv.* Band III, S. 316.
LYON, IRVING W. *Excision of the Knee and Hip-Joints.* Am. Jour. of the Med. Sci. Vol. XLIX, 1865; p. 49.
MACLEOD, GEORGE H. B. *Notes on the Surgery of the War in the Crimea.* London, 1858; p. 338.
MAISONNEUVE. *Gazette des Hôpitaux* 1847, p. 98, 1849. p. 54, and *Clinique Chirurgicale,* Paris, 1863. T. 1, p. 391.
MALGAIGNE, J. T. *Manuel de Médecine Opératoire.* Sept. ed., Paris, 1861; p. 250.
MATTHEW, in *Medical and Surgical History of the British Army which served in Turkey and the Crimea during the war against Russia in the years* 1854-'55-'56. London, 1858. Vol. II, p. 378.
MILLER, JAMES. *A System of Surgery.* Edinburgh, 1864; p. 1301.
MINER, J. F. *Buffalo Med. and Surg. Jour.* Vol. V, p. 275.
MOTHA, JOHANN. *Beiträge zu den Resectionen der Knochen.* Jena. 1866.

APPENDIX B.

MONTFALCON. *Mémoire sur l'État Actuel de Chirurgie*, Paris, 1816; p. 103.
MOTT, V. *Velpeau's New Elements of Operative Surgery.* New York, 1847. Vol. II. pp. 779, 845.
MURSICK, GEORGE A. *A Successful Case of Excision of the Head and Neck of the Femur for Gunshot Fracture.* In *New York Medical Journal.* Vol. I, 1865, p. 424.
NEUDÖRFER, J. *Handbuch der Kriegschirurgie.* Leipzig. 1864; S. 115, 405.
OCHWADT, A. *Kriegschirurgische Erfahrungen.* Berlin, 1865; S. 53.
OPPENHEIMER, S. *Ueber die Resection des Hüftgelenkes.* Würzburg, 1840. S. 25.
O'LEARY, T. C. *London Lancet*, July 12, 1856; p. 45.
OLLIER, L. *Traité expérimental et clinique de la Régénération des Os.* Paris, 1867; p. 165.
OTIS, G. A. *Surgical Report* in Circular No. 6, S. G. O., 1865, p. 61, and *Amer. Jour. Med. Sci.*, Vol. LVI, p. 129. and *Buffalo Med. & Surg. Journal*, Vol. VIII, p. 21.
PAILLARD. *Relation Chirurgicale du Siége de la Citadelle d'Anvers.* Paris, 1833; p. 105.
PAGENSTECHER. *Zur Resection des Hüftgelenkes*, in *Langenbeck's Archiv.* Band II, S. 312, 315.
PAGET & STANLEY. *Catalogue of the Pathological Specimens contained in the Museum of the Royal College of Surgeons of England.* London, 1847, Vol. II, p. 230.
PANCOAST, J. *A Treatise on Operative Surgery.* Philadelphia, 1846. 2d Ed., p. 129.
PARK, H. *Cases of Excision of Carious Joints.* Glasgow, 1806.
PAUL, HERMAN JULIUS. *Die Conservative Chirurgie der Glieder.* Breslau, 1859; S. 38, 193.
PERCY. *Dictionnaire des Sciences Médicales, par MM. ADELON, ALIBERT, BARBIER, BÉGIN, etc.* Paris, 1820. Tome 47, p. 553.
PETIT—RADEL. *Encyclopédie Méthodique*, T. I, Art. Chirurgie.
PIROGOFF, N. *Grundzüge der allgemeinen Kriegschirurgie.* Leipzig, 1864. S. 812, 815, 1116.
PIRRIE, WILLIAM. *The Principles and Practice of Surgery.* London, 1860; p. 443.
PITHA, F. *Krankheiten der Extremitäten.* Erlangen, 1868; S. 203.
PORTA, L. *Della Disarticolazione del Cotile.* Milano, 1860.
PRICE. *London Lancet*, April 28, 1860.
RAVOTH, FR. *Grundriss der Akiurgie.* Leipzig, 1863. S. 416.
RAVOTH & VOCKE. *Chirurgische Klinik.* Berlin, 1852. S. 795.
READ, J. B. *Resections of the Hip-Joint*, in *Confederate States Medical and Surg. Journal.* Vol. I, p. 5.
RIED, FRANZ. *Die Resectionen der Knochen mit besonderer Berücksichtigung der* von Dr. Michael Jäger ausgeführten derartigen Operationen. Herausgegeben von Dr. Franz Ried. Nürenburg, 1860. S. 385.
ROGERS. *New York Med. Record*, Vol. II, p. 17.
ROSSI, F. *Elémens de Médecine Opératoire.* Turin, 1806. Tome II, p. 224.
ROUX, P. J. *De la Résection ou du Rétranchment de portions d'Os Malades*, Paris, 1862; p. 49, and *Gazette des Hôpitaux*, 1847, No. 28.
RUST. *Handbuch der Chirurgie.* B. V. S. 559.
SALLERON, M. In *Recueil de Mém de Méd. et de Chir. mil.* 2d série, Tome 21, 1858, p. 320.
SALZER. *Ueber Resectionen im Hüftgelenk.* Wochenblatt der K. K. Gesellsch. der Aerzte in Wien. No. 45, S. 336.
SANTESSON. *Om Höftleden och Ledbrosken, etc.* Stockholm, 1849, and *Dublin Journal of Med. Sci.* Vol. XI. p. 432.
SAUNDERS, D. D. *Excision or Resection of the Bones and Joints of the Lower Extremity*, in *Memphis Medical Monthly*, April, 1866; Vol. I, p. 77.
SAYRE, LEWIS, A. M, D. *Report on Morbus Coxarius, or Hip Disease*, in Transactions of the American Medical Association for 1860, Vol. XIII, pp. 534, 558, and *New York Jour. of Med.*, June, 1855.
SCHEDE, M. *De resectione articulationis coxae.* Diss. Inaug. Halis Saxonum. 1866.
SCHILLBACH, LUDWIG. *Beiträge zu den Resectionen der Knochen.* Jena, 1861. S. 3.
SCHWARTZ, HARALD. *Beiträge zur Lehre von den Schusswunden.* Schleswig 1854. S. 142.
SÉDILLOT, CH. *De la Résection Coxo-Fémorale.* In *Gazette Méd. de Paris.* Tome XXI, 1866, p. 691, and *De l'Évidement sous-périosté des Os.* Paris, 1867; p. 165, and *Traité de Médecine Opératoire.* Paris, 1865. Tome I, p. 515, and *Compte Rendu de l'Acad. des Sciences.* Séance du 15 Oct., 1866.
SENFTLEBEN, H. *Beobachtungen und Bemerkungen über die Indikationen, den Heilungsprocess und die Nachbehandlung der Resectionen grösserer Gelenke*, in *Langenbeck's Archiv.* B. 3, S. 79.
SIEBERT, LUCAS. *Statistik der Resectionen.* Jena, 1863.
SIMON, GUSTAV. *Mittheilungen aus der Chirurgischen Klinik.* Prag, 1868. S. 93.
SKEY, F. C. *Operative Surgery.* Am. ed. Philadelphia, 1851; p. 374.
SMITH, STEPHEN. *Handbook of Surgical Operations*, 3d ed. New York, 1862; p. 221.
SMITH, HENRY. *Medical Times and Gazette*, Dec. 4, 1852, and *London Lancet*, April 1st and 15th, 1848, Jan. 2d, 1849.
SMITH, D. P. *Exsection of the Head of the Femur*, in *Am. Med. Times.* Vol. VII, p. 659.
SMITH, G. K. *The Insertion of the Capsular Ligament of the hip-joint, and its relation to intra-capsular fractures of the neck of the femur*, in *Med. and Surg. Reporter*, Vol. VII, pp. 214, 270, 381, 416, 508, 536, 577, 605, and *American Medical Times*, Vol. III, p. 389.
SMITH, H. H. *The Principles and Practice of Surgery.* Philadelphia, 1853; Vol. II, p. 659.
SOLLY, S. *London Lancet*, August 14th, 1852.
STEVENS, GEORGE F. *On Excisions in Cases of Gunshot Wounds*, in Trans Med. Soc. of New York, 1866; p. 132.
STROMEYER, L. *Maximen der Kriegsheilkunst.* Hanover, 1861. S. 501, and *Erfahrungen über Schusswunden im Jahre* 1866, Hannover, 1867.

SYME, JAMES. *The Principles of Surgery.* Edited by D. Maclean. Philadelphia, 1836; p. 694, and *Treatise on the Excision of Diseased Joints.* Edinburgh, 1831, p. 125.
SYME, H. SMITH, and NORMAN. *London Lancet,* 1849, Vol. 1, pp. 21, 61, 266, 299, 447, 490. 567, 597, 625.
SWINBURNE, J. *Exsection of the Hip-Joint, and Conservative Surgery,* in *Med. and Surgical Reporter,* Vol. VII, p. 198, and Vol. IX, pp. 377, 400, 425.
SZYMANOWSKI, JULIUS. *Ueber die Resection des Hüftgelenkes,* in *Langenbeck's Archiv.* Band VI, S. 787, and *Additamenta ad Ossium Resectionem.* Dorpati Livonorum, 1856.
TEXTOR, CAJETAN. *Grundzüge zur Lehre der Chirurgischen Operationen.* Würzburg, 1835. S. 310, 348, and *Ueber die Wiedererzeugung der Knochen nach Resect.* Würzburg, 1842.
TEXTOR, KARL. *Der Zweite Fall von Aussägung des Schenkelkopfes mit Vollkommen Erfolg.* Würzburg, 1858.
TEXTOR. *Ueber die Wiedererzeugung der Knochen nach Resectionen,* Würzburg, 1842.
URE. *Excision of the Head of the Femur,* in *London Lancet,* 1860; p. 443.
VELPEAU, A. L. M. *Nouveaux Eléments de Médecine Opératoire.* Paris, 1832. T. I, p. 583, and *Bulletin de l'Académie de Méd.,* Nov. 19, 1862.
VERMANDOIS. *Ancien Journal de Chirurgie, Médecine, et Pharmacie,* T. LXVI, 1786; pp. 70, 200.
VIDAL, AUG. (De Cassis.) *Traité de Pathologie Externe.* Paris, 1861; p. 735.
VOGEL, A. F. *Observationes quasdam chirurgicas defendit.* Kiliae, 1771, and *Bibl. Chir. du Nord.,* p. 391.
VÖLKERS, C. *Beiträge zur Statistik der Amputationen und Resectionen,* in *Langenbeck's Archiv.* Band IV, S. 574.
WACHTER, G. H. *Diss, Chirurgica de articulis extirpandis, etc.* Groningæ, 1810.
WAGNER, ALBRECHT. *On the process of Repair after Resection and Extirpation of Bones.* Translated by F. Holmes. London, 1850; p. 125, and *Ueber den Heilungsprozess nach Resection und Exstirpation der Knochen.* Berlin, 1853. S. 14.
WAGNER, C. *Report of Interesting Surgical Operations.* Beverly, p. 14.
WALTHER, JÄGER, und RADIUS, *Handwörterbuch der Gesammten Chirurgie und Augenheilkunde,* Leipzig S. 371, 402.
WALTON. *London Medical Times,* April 7, 1847, and *London Lancet,* Jan. 4, 1851.
WARREN, EDWARD. *An Epitome of Practical Surgery.* Richmond, 1863; p. 198.
WILLIAMSON, GEORGE. *Military Surgery.* London, 1863; p. 230.
WINNE, CHARES K. *Statistical Inquiry as to the Expediency of Excision of the head of the Femur,* in *Am. Jour. Med. Sci.,* Vol. XLII, 1861, p. 26.
WHITE, CHARLES, F. R. S. *Cases in Surgery.* In Philosophical Transactions. Giving some Account of the Present Undertakings, Studies, and Labors of the Ingenious in many parts of the World. Vol. LIX, for the year 1769. London: Printed for Lockyer Davis, Printer to the Royal Society, near Gray's Inn Gate near Holbourn, MDCCLXX, p. 45.
WOODHULL, A. A. *Catalogue of the Surgical Section of the U. S. Army Medical Museum.* Washington, 1866; p. 244.
ZANG, C. B. *Darstellung blutiger heilkünstlerischer Operationen.* Wien, 1821, Vierter Theil, S. 287, 300.
ZEIS, EDWARD. *Ueber die Heilung des intracapsulären Schenkelhalsbruches durch Knochencallus.* Dresden, 1804.

CONTENTS.

	PAGE.
Circular No. 2, S. G. O., 1869	3
Preliminary Observations	5
Historical Review	8
Excisions at the Hip in the War of the Rebellion	19
Primary Operations	20
Intermediate Operations	33
Secondary Operations	48
Unclassified or Doubtful Cases	54
Recapitulation	58
Additional Cases of Excision at the Hip	59
Temporization and Amputation at the Hip compared with Excision	62
Review of Cases cited by Demme, Pirogoff, and Gross, in favor of expectant treatment	63
Series of one hundred and twenty-two cases of alleged gunshot fracture involving the hip and treated on the expectant plan	65
Series of thirty-seven cases of gunshot fracture of the femur, with slight injury to the acetabulum	85
Series of five cases of gunshot fractures of the acetabulum without injury to the head of the femur	89
Series of twenty-two cases of gunshot injury of the hip-joint without fracture	90
Series of twelve cases of secondary traumatic arthritis	92
Series of seventeen cases of fracture of the trochanters, with possible or probable injury to the hip-joint	94
Series of twelve cases of fractures of the trochanters, with secondary traumatic arthritis	96
Series of seven alleged examples of gunshot injury of the hip-joint	98
Series of twenty-two cases of supposed gunshot injury to the hip-joint, not well-authenticated	99
Series of eighteen cases treated by the extraction of fragments	102
Coxo-femoral amputations	107
Comments by Dr. W. Thomson	113
Dr. Gibson's case	117
Concluding Observations	121
Letter from Dr. Billings	123
Notes by European Surgeons	127
Method of operating and after-treatment	128
Table of primary excisions	132
Table of intermediate excisions	135
Table of secondary excisions and summary	137
Bibliography	138

Corrigenda.

On page 6, 7th line, for *comprizes*, read *comprises*.
On page 13, 7th line, for March 15, read *May 13*.
On page 13, 19th line, for 23d, read **23**.
On page 54, 26th line, for *incisions*, read *excisions*.

CONTENTS.

	PAGE.
Circular No. 2, S. G. O., 1869	3
Preliminary Observations	5
Historical Review	8
Excisions at the Hip in the War of the Rebellion	19
Primary Operations	20
Intermediate Operations	33
Secondary Operations	48
Unclassified or Doubtful Cases	54
Recapitulation	58
Additional Cases of Excision at the Hip	59
Temporization and Amputation at the Hip compared with Excision	62
Review of Cases cited by Demme, Pirogoff, and Gross, in favor of expectant treatment	63
Series of one hundred and twenty-two cases of alleged gunshot fracture involving the hip and treated on the expectant plan	65
Series of thirty-seven cases of gunshot fracture of the femur, with slight injury to the acetabulum	85
Series of five cases of gunshot fractures of the acetabulum without injury to the head of the femur	89
Series of twenty-two cases of gunshot injury of the hip-joint without fracture	90
Series of twelve cases of secondary traumatic arthritis	92
Series of seventeen cases of fracture of the trochanters, with possible or probable injury to the hip-joint	94
Series of twelve cases of fractures of the trochanters, with secondary traumatic arthritis	96
Series of seven alleged examples of gunshot injury of the hip-joint	98
Series of twenty-two cases of supposed gunshot injury to the hip-joint, not well-authenticated	99
Series of eighteen cases treated by the extraction of fragments	102
Coxo-femoral amputations	107
Comments by Dr. W. Thomson	113
Dr. Gibson's case	117
Concluding Observations	121
Letter from Dr. Billings	123
Notes by European Surgeons	127
Method of operating and after-treatment	128
Table of primary excisions	132
Table of intermediate excisions	135
Table of secondary excisions and summary	137
Bibliography	138

www.ingramcontent.com/pod-product-compliance
Lightning Source LLC
Chambersburg PA
CBHW030338170426
43202CB00010B/1167